Observational astronomy

Observational astronomy

D. Scott Birney
Department of Astronomy, Wellesley College

CAMBRIDGE UNIVERSITY PRESS

Cambridge
New York Port Chester Melbourne Sydney

Published by the Press Syndicate of the University of Cambridge
The Pitt Building, Trumpington Street, Cambridge CB2 1RP
40 West 20th Street, New York NY 10011–4211, USA
10 Stamford Road, Oakleigh, Melbourne 3166, Australia

First published 1991

Printed in Great Britain at the University Press, Cambridge

British Library cataloguing in publication data
Birney, D. Scott
Observational astronomy.
1. Astronomical bodies. Observation
I. Title
522.20

Library of Congress cataloguing in publication data
Birney, D. Scott
Observational astronomy/D. Scott Birney.
p. cm.
ISBN 0 521 38199 1
1. Astronomy – Observations. 2. Astronomy – Technique. I. Title.
QB145.B52 1991
522-dc20 90–31052 CIP

ISBN 0 521 38199 1 hardback
ISBN 0 521 39693 X paperback

Contents

Birney, *Observational Astronomy*

ISBN 0 521 38199 1 hardback
ISBN 0 521 39693 X paperback

ERRATA (1991 printing)

p.35, line 23, *for* sixth *read* eighteenth.

p.35, line 24, *for* six *read* eighteen.

p.176, line 13 up *for* left *read* right.

p.178, Figure 11.6b is incorrectly labelled. The angle between the vertical dashed line and the first arrow on the right should be labelled β_1. The angle between the vertical dashed line and the second arrow on the right should be labelled β_2.

p.250, line 9, *for* P *read* P'.

p.268, line 8, *for* right *read* left.

p.268, line 9, *for* left *read* right.

Preface

Students at the level of advanced undergraduates or beginning graduate students have often found that much information needed in the everyday practice of astronomy is not easily accessible. The necessary details are not to be expected in most textbooks, and one must often refer to early copies of some journals or to a professor's notes. It is my intention that this book should provide students with a ready reference of a practical nature.

For many years a course in astronomical techniques has been taught at Wellesley College, and the students there have been able to apply all of the methods described here. This book is thus based on the notes which I have developed while teaching this course. Over the years I have encountered a number of excellent books which were either to serve as texts for practical courses or as general handbooks for the use of amateur astronomers. My feeling has been that none of these covered the topics which I felt were most necessary at the level which I felt could be most useful.

It is my hope that this book will fill a real need in the reference material available to astronomers at many levels.

1

The celestial sphere and coordinate systems

In the night sky the stars appear as bright points on a dark spherical surface. No such surface really exists, of course, but the concept of a celestial sphere is a useful one that goes back thousands of years. Ptolemy described it and so did Pythagoras and many others. Today we no longer have to worry about the reality of that sphere, and so we eliminate the need for speculation on its composition, radius, thickness and so forth. And we don't have to worry about what is on the other side of it. (Fig. 1.1. shows a drawing of an astronomer with his head sticking

Fig. 1.1. An early concept of the celestial sphere. The careful reader may see that there is a serious flaw in the artist's knowledge of the moon.

through the sphere.) On the other hand, even though the celestial sphere is not a physical entity, we have many practical uses for the concept. The observer is always at the center of it, and the direction from the observer to any star may be considered to be a radius of the celestial sphere.

Coordinate systems

The most fundamental application of the concept of a celestial sphere is in the determination of the coordinates of objects which appear in the sky (or perhaps, on the sky). We may approach this problem of coordinates in a very general way and see first of all just what is involved in specifying the location of a point on the surface of any sphere. Assume, to begin with, that the sphere is rotating. This requires the existence of an axis which must pass through the center of the sphere as in Fig. 1.2. The axis is thus defined by the rotation, and the axis defines, in turn, two points – the poles. Following the convention established for the earth, we designate these as the 'north' pole and the 'south' pole. Now consider a plane passed through the sphere in such a way that it is perpendicular to the axis and includes the center of the sphere. In the case of the earth the plane defined in this way is actually the plane of the equator. In the more general case this plane is referred to as the 'fundamental' plane, and it takes on a more specific name in each of several systems of coordinates. Imagine now that we wish to define in a specific way the location of the point A on the sphere shown in Fig. 1.3(*a*). Let us first pass a plane through the sphere in such a way that the

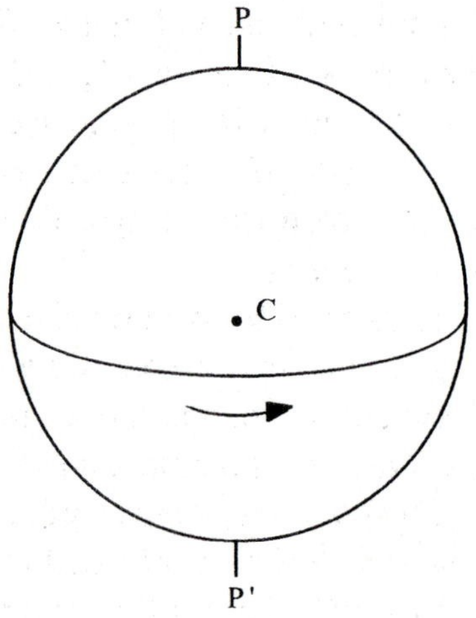

Fig. 1.2. A rotating sphere. The poles and the equator are defined by the rotation.

plane includes both the axis of the sphere and Point A. In Fig. 1.3(*a*) this plane has been indicated by the curve passing through the two poles and Point A. Points B and C have also been indicated. Point C is at the center of the sphere, and point B is the intersection of PA with the equator. It should now be obvious that the angle BCA defines the angular distance of A from the fundamental plane. On the earth an angle similar to BCA is referred to as the 'latitude'. We do not go to the center of the earth in order to determine the latitude of a place, but it is actually this angle that we are talking about when we use the term.

Now let us pass another plane through the sphere. Let this one be parallel to the fundamental plane, and let it pass through A (Fig. 1.3(*b*)). Notice that the radius of this circle is smaller than that of the fundamental circle. At this point we must introduce two definitions. First, a 'great circle' is the intersection of a plane and a sphere in any case in which the plane passes through the center of the sphere. Thus, the fundamental circle is a great circle, and the arc PAP′ is half of a great circle. Second, a 'small circle' is the intersection of a plane and a sphere in any case in which the plane does not pass through the center of the sphere. In the case shown in Fig. 1.3(*b*) the small circle happens to be parallel to the fundamental plane, but a small circle may actually have any orientation.

Returning to Fig. 1.3(*b*) we note now that all points on the small circle are at the same angular distance from the fundamental circle. For the earth we would say that all points on this small circle had the same latitude. In order to be precise about the location of A, therefore, we must specify in some way which of all of the possible great circles through the poles is the one which passes through A. We may do this by means of a second angle measured this time in the fundamental plane. We must first select some arbitrary point as our zero point, and in Fig. 1.3(*c*) such a point has been indicated as D. Now the angle DCB quite specifically defines the great circle through A. Again with reference to the earth the circle PDP′ represents the meridian of Greenwich, England, and the angle, DCB represents the longitude of point A.

The above discussion is intended to show that the location of a point on a sphere can always be specified by means of two angles. One angle is measured at right angles to the fundamental plane, and the second angle is measured in the fundamental plane itself. Astronomers have found it convenient to make use of several coordinate systems, and we shall discuss these below. The individual systems differ most notably in that the positions of the stars are referred to a different fundamental plane in each one.

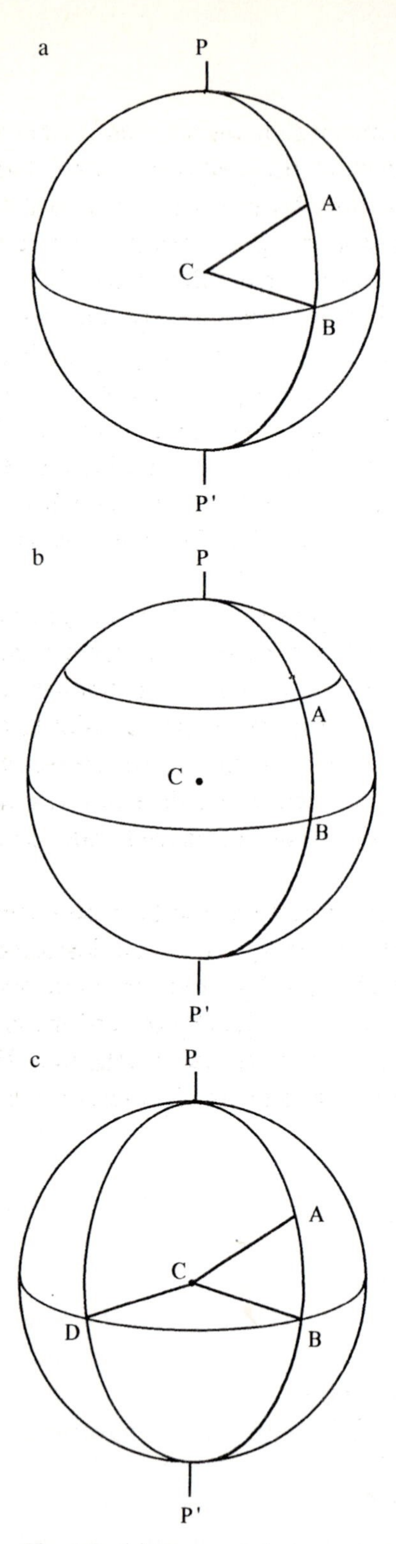

Fig. 1.3. (*a*) The angle between two points on a sphere (*b*) A small circle through point A (*c*) Two angles define the location of point A with respect to the equator and an arbitrarily chosen point D.

The horizon system is perhaps the one which would seem to be the most obvious when we first begin to discuss the positions of stars. Here the fundamental plane is the plane of the observer's horizon, and the pole is the zenith which may be defined as the point directly overhead. The true horizon is by definition ninety degrees from the zenith. The first coordinate is now referred to as the 'altitude'. From our general discussion one should be able to guess that altitude is the angle which the observer would measure from the horizon to an object in the sky and that this angle is measured along a great-circle arc which passes through the zenith and the object. Altitude may vary from zero for an object on the horizon to ninety degrees for an object at the zenith. An object below the horizon would not be visible, but it may be considered to have a negative altitude.

The determination of the second coordinate requires the previous selection of a zero point on the horizon, and here this point has been selected as the north point. The 'azimuth' may now be defined as the angle measured in the plane of the horizon from the north point to the point at which the arc from the zenith through the object crosses the horizon. By convention the azimuth increases in the direction towards the east. Fig. 1.4 illustrates altitude and azimuth measured with respect to the horizon.

This coordinate system is variously referred to as the horizon system, the altitude–azimuth system or as the altaz system. The instrument shown in Fig. 1.5 is a theodolite routinely used by surveyors to measure altitude and azimuth. Since the earth rotates continually toward the east, all celestial objects appear to move from east to west across the sky. This means that in the horizon system the coordinates of all objects are con-

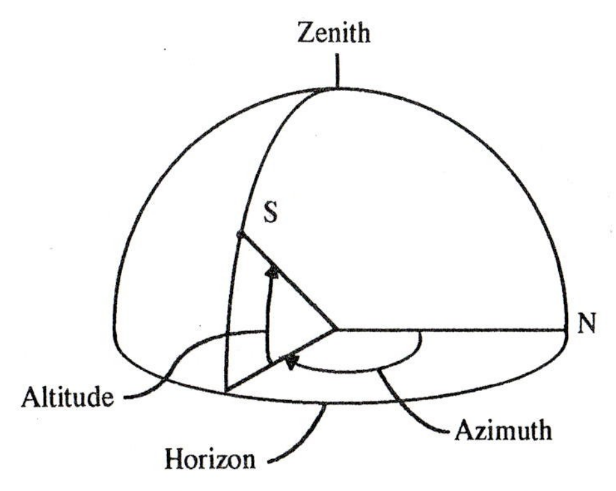

Fig. 1.4. Altitude and azimuth in the horizon system. S is the position of the star.

Fig. 1.5. The theodolite, an instrument for measuring altitude and azimuth (Wellesley College photography).

tinually changing. Such a system does have certain applications as we shall see, but it cannot be used to provide permanent descriptions of the locations of stars on the celestial sphere.

A coordinate system with more useful properties in astronomy is the equatorial system in which the fundamental plane is now the celestial equator. If one imagines a very small spherical earth at the center of a very large spherical sky, then the extension of the earth's axis defines the celestial poles, and the extension of the plane of the earth's equator defines the celestial equator. In the equatorial system, then, coordinates of stars are defined with respect to the celestial equator. The angle measured northward or southward from the equator is called the 'declination', and the angle measured in the plane of the equator is called the 'right ascension'. These two angles are indicated as Dec and RA respectively in Fig. 1.6(*a*). A zero-point for the measurement of angle RA must, of course, be selected, and this has been chosen as the vernal equinox, the point on the sky at which the sun crosses the celestial equator on or near March 21 each spring. This point has been marked V in Fig. 1.6(*a*). There is no obvious way in which one can identify the vernal equinox when looking at the sky, but later in Chapter 4 we shall show that this point can be located quite specifically from very simple observations.

Later in this chapter we shall also describe the methods by which the astronomer determines the declinations and right ascensions of stars.

It is conventional that declination be considered to be positive for objects in the northern hemisphere and negative for objects in the southern. Right ascension increases in the eastward direction on the celestial sphere, and is measured in hours, minutes and seconds.

In Fig. 1.6(*b*) we have shown both the equatorial and the horizon systems in the same sketch. We remind the reader that as the sky appears to rotate around the north celestial pole, the altitude and azimuth will both be changing continuously. The declination and right ascension will, of course, remain fixed. In the diagram the angle ZPS has been marked H, and it is this angle that is commonly referred to as the 'hour angle'. The hour angle decreases progressively when a star is east of the meridian and becomes progressively larger after the star has crossed the meridian into the western sky. It is conventional to regard the hour angle

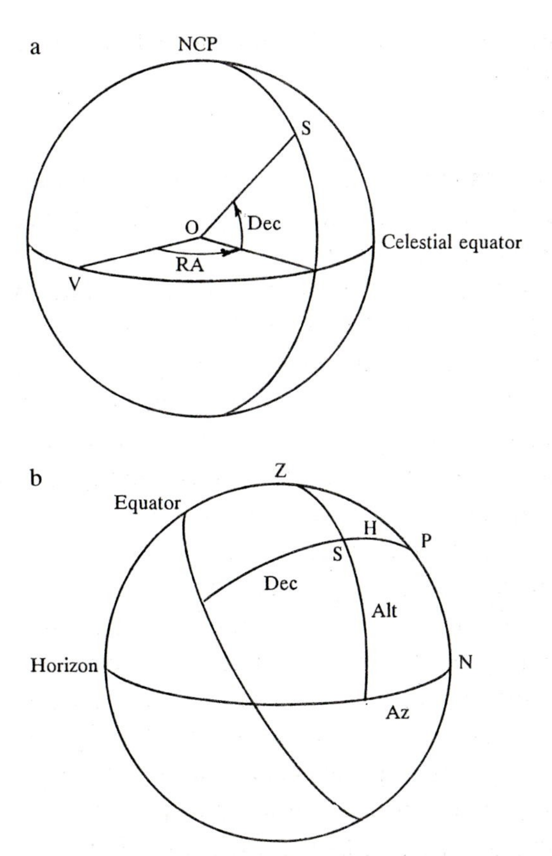

Fig. 1.6. (*a*) The equatorial coordinate system (*b*) The equatorial system superimposed on the horizon system.

as being negative when a star is in the east and positive when a star is in the west. If we are to try to point a telescope toward a particular star, we must first determine the hour angle of the star. The hour angle will depend upon the time at which an observation is being made and the right ascension of the star. We shall return to this subject in Chapter 4 where we shall see how the hour angle is routinely determined.

Another system of coordinates was of considerable use among astronomers several hundred years ago and is used today by those who are engaged in the study of the motions of the planets. In this system the fundamental plane is the ecliptic, the apparent path of the sun around the sky. The zero-point on the ecliptic is again the vernal equinox, and the two coordinates are known as celestial latitude, β, and celestial longitude, λ. Fig. 1.7 illustrates the ecliptic system of coordinates.

Finally, in studies of our galaxy, astronomers make use of a system of coordinates in which the fundamental plane is the galactic equator, a great circle which closely approximates the center line of the Milky Way on the sky. Astronomers have carefully chosen the right ascension and declination of the galactic north pole in such a way that the galactic equator is precisely defined. As illustrated in Fig. 1.8, the two coordinates are known as galactic latitude, b, and galactic longitude, l, and the zero-point on the galactic equator is a point in the constellation Sagittarius, in which we look toward the center of our galaxy. It is worthy of mention here that galactic coordinates cannot be determined by direct observation. They are calculated from the equatorial coordinates of the star, of the galactic pole and of the direction of the center of the galaxy. Details of this computation will be presented in Chapter 4.

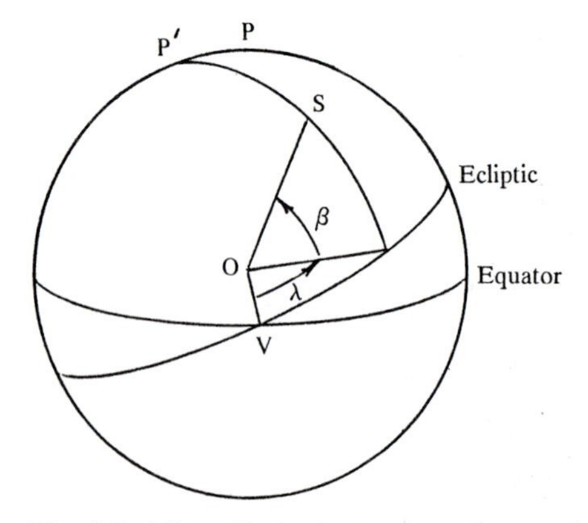

Fig. 1.7. The ecliptic system superimposed on the equatorial system. P′ is the pole of the ecliptic.

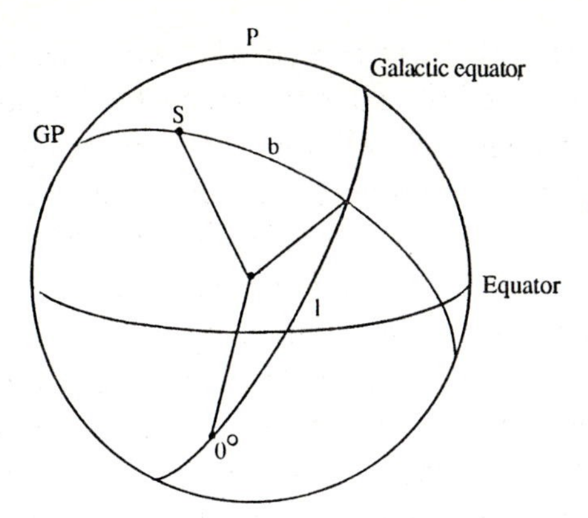

Fig. 1.8. The galactic coordinate system. GP is the galactic pole.

Measuring stellar coordinates

In Fig. 1.9(*a*) the celestial sphere has been drawn as it appears above an observer's horizon. The arc running from the north point on the horizon through the zenith and on to the south point on the horizon is the meridian of the observer at O. The zenith or point directly above O is marked Z. The celestial north pole is marked P. The angle NOP is equal to the latitude of the observer. In Fig. 1.9(*b*) a small circle has been

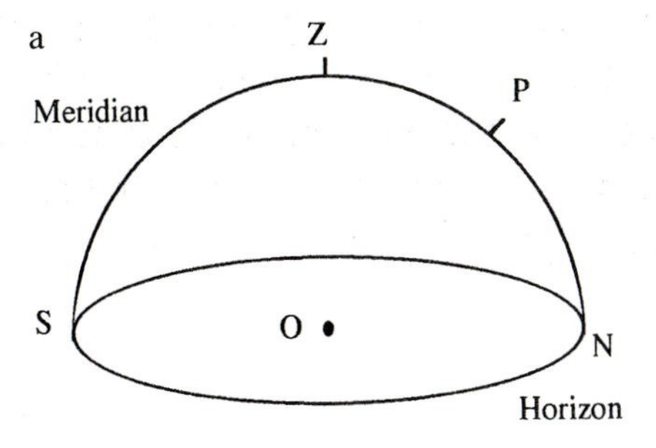

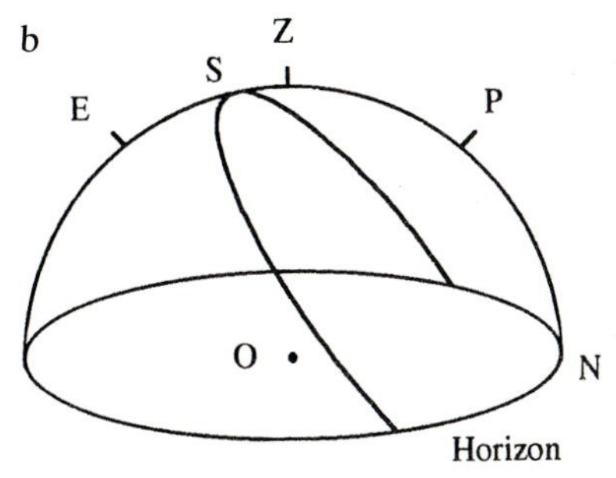

Fig. 1.9. (*a*) Zenith and pole as points on the meridian (*b*) A star crossing the meridian at S.

drawn to represent the path of a star as it rises in the east, crosses the meridian at S and sets in the west. From earlier statements it should be easily seen that angle SOE is equal to the declination of the star. We wish to find a precise, practical way of measuring this angle.

Suppose now that we construct a specialized telescope which is capable of moving only along the meridian. That is, the telescope is to be mounted on an axis oriented in the east–west direction as shown in Fig. 1.10. Such a telescope is known as a meridian telescope or a meridian circle. Small versions are referred to as transit telescopes because the passage of a star across the meridian is called a transit.

A meridian telescope is always equipped with a large and precisely graduated scale or circle so that the angle between successive pointings may be known as accurately as possible. Thus, if such a telescope is pointed first at P and then at S in Fig. 1.9(*b*), the angle POS will be known, and

$$90^{\circ} - < \mathrm{POS} = \text{declination}$$

In principle, then, a star's declination may be found from rather simple measurements.

The graduated circle on the meridian telescope may be adjusted so that it reads 90 degrees when the telescope is pointed toward the north pole and zero degrees when pointed toward the celestial equator. The declination of a star may then be read directly from the scale after a transit. A special reticle is included in the optical path within a meridian telescope (see Fig. 1.11). The observer sees a horizontal line crossed by a series of

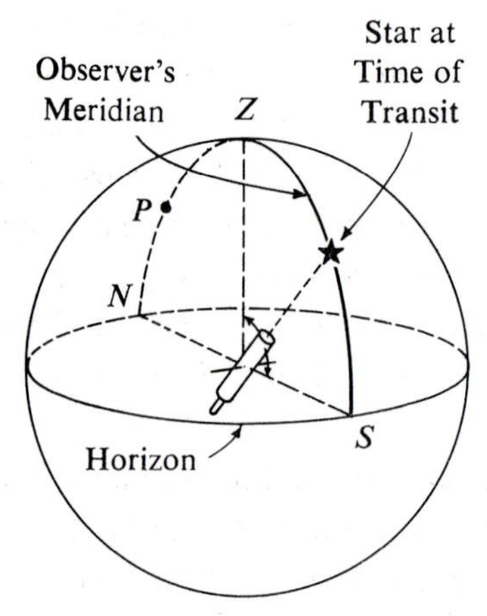

Fig. 1.10. The transit telescope is constrained to move only along the observer's celestial meridian.

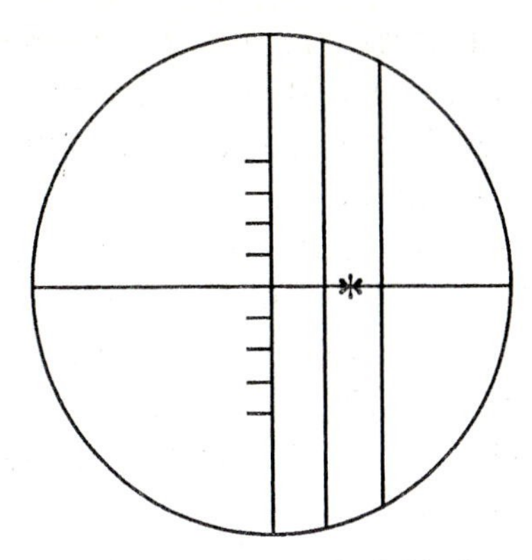

Fig. 1.11. A star in the field of a transit telescope.

vertical lines. The telescope is adjusted to the approximate declination in advance of the predicted time of transit. Then as the observer watches, a star will appear at one side of the field of view, move across the field and disappear on the opposite side. By pressing a key the observer attempts to record the exact moment of transit. Refinements in the declination may be made by noting the appropriate graduations on the vertical scale. Today the transits of stars are recorded photoelectrically and the judgement of the observer has been removed, but the general principles are the same.

The right ascension of a star may be found from the same meridian observation provided that a sidereal clock is available. Sidereal time will be discussed fully in the next chapter, but for the moment we may say that it is based on the rotation of the earth with respect to the stars rather than to the sun. This clock would be adjusted to read zero hours at the moment when the vernal equinox crosses the observer's meridian each day (see p. 19). Then if one records the sidereal time at the moment of the star's transit, that time is the star's right ascension. When the sidereal time is five hours, for example, five hours will have passed since the vernal equinox crossed the meridian and a star with a right ascension of five hours will now be crossing the meridian. This is indicated in Fig. 1.12. Fig. 1.13 is the polar view of the case depicted in the previous figure.

There are in the world only a small number of astronomers who are engaged in the determination of precise coordinates by methods such as those we have described, and these astronomers are only able to work with the brightest stars. The results of these observations are known as fundamental positions and are published in what are known as

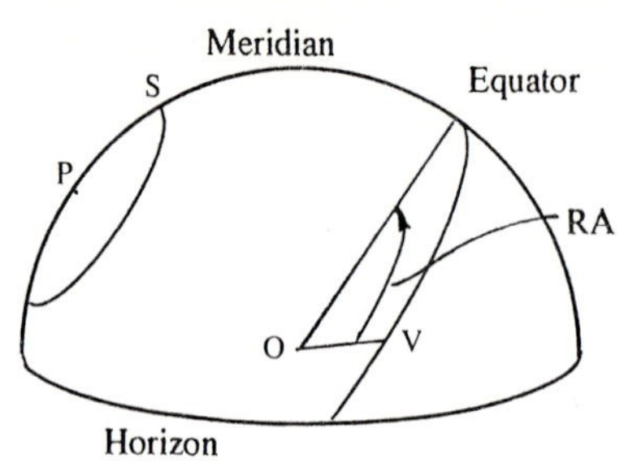

Fig. 1.12. Sidereal time is equal to the right ascension of a star on the meridian.

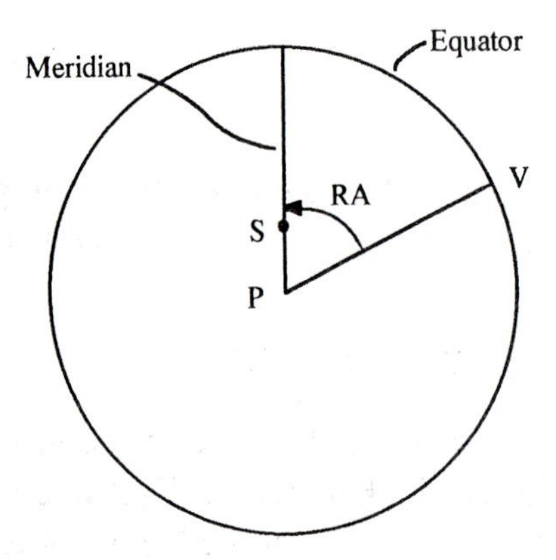

Fig. 1.13. Polar view of the previous figure.

fundamental catalogs (see Chapters 3 and 5). These fundamental positions form the reference system for the determination of the coordinates of fainter stars from the measurement of photographs. The method for doing this will be discussed in Chapter 5.

Finding the celestial pole

As the stars move across the sky each one will cross the meridian at some time in each interval of twenty four sidereal hours. Twelve hours after the initial transit the star will transit again, but this time its altitude at the time of transit will be less. This is seen in Fig. 1.14, and these crossings are referred to respectively as upper transit and lower transit. At the equator all lower transits will be invisible because they will all occur below the horizon. At any other latitude (except that of the North Pole or the South Pole) it is always possible to observe some stars at both upper and lower transits. Such stars are referred to as circumpolar stars.

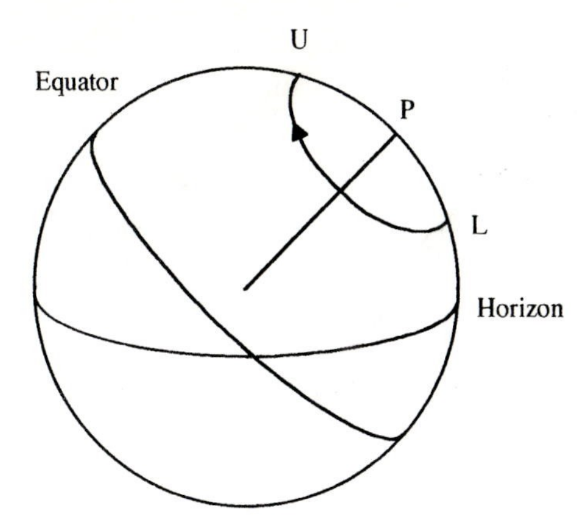

Fig. 1.14. Upper and lower transits of a circumpolar star.

Now if we wish to know the altitude of the pole, we have simply to observe the altitude of a star at both upper transit and at lower transit. Then

$$\text{Alt of pole} = A(\text{L}) + \tfrac{1}{2}(A(\text{U}) - A(\text{L}))$$

In practice the problem is not quite so simple because of the effects of refraction by the atmosphere of the earth. The rays of light from a star are refracted in such a way that the stars appear to be higher in the sky than they actually are. The correction for this effect will be discussed in detail in Chapter 4.

The azimuth of the celestial pole is always 0° regardless of the latitude of the observer. Therefore, as mentioned on p. 10 the meridian telescope's axis of rotation must lie along an east–west line. This alignment may be checked by observing a star six hours before and six hours after its upper transit. As indicated in Fig. 1.15 the altitude of the star should be the same at each of these two times. Furthermore, the angle between

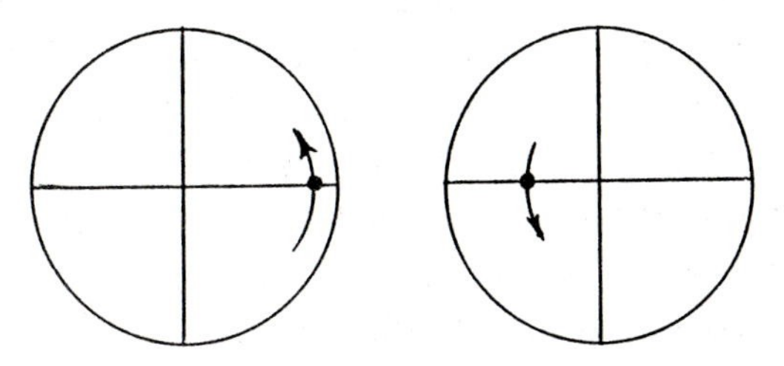

Fig. 1.15. Result of an error in the alignment of the axis of the meridian telescope.

the center of the field and the direction to the star should be the same in both cases. If these angles are not the same, then the mounting must be rotated in azimuth.

As mentioned on p. 6, we also need to know the location of the vernal equinox, our unmarked and invisible reference point in the sky. This point can be located with remarkable precision, and the method for doing so will be described in detail in Chapter 4.

QUESTIONS FOR REVIEW

1. Why is it that two angles are sufficient to define the position of a point on a sphere?
2. Name the two coordinates in each of the four systems described in this chapter.
3. How is the meridian telescope used to determine stellar coordinates in the equatorial system?
4. Why is it that coordinates in the galactic system cannot be measured but must always be calculated?
5. What is a circumpolar star? What effect does the observer's latitude have on the number of circumpolar stars that can be seen?
6. A circumpolar star is observed to have an altitude of 49° 38′ at upper transit and an altitude of 18° 21′ at lower transit. Assume that these are true altitudes after correction for refraction and instrumental errors. What is the latitude of the observer?
7. What is the declination of the star observed in the previous question?

Further reading

Green, R. M. (1985). *Spherical Astronomy.* Cambridge: Cambridge University Press.

McNally, D. (1974). *Positional Astronomy.* Frederick Muller Limited. McNally covers much of the same material as the other authors listed here. This is a useful book for readers who wish to have another perspective.

Pannekoek, A. (1961). *A History of Astronomy.* Interscience Publishers. Material on several early systems of coordinates may be found in the first few chapters of this interesting and authoritative work.

Smart, W. M. (1977). *Textbook on Spherical Astronomy*, 6th edn. Cambridge: Cambridge University Press.

Woolard, E. W. and Clemence, G. M. (1966). *Spherical Astronomy.* Academic Press. The level of detail in this book is such that it can be very useful to those who are working directly in positional astronomy. Chapter 20 is an excellent source for information on the terminology of coordinate systems.

2

Time

Time as we use it in our ordinary lives is based on the rotation of the earth with respect to the sun. As the earth rotates the sun appears to move continually around the heavens, and we define noon as the moment each day when the sun passes an observer's meridian. We could easily use the interval from one noon to the next to define the day, but in order to make all of the daylight hours part of the same calendar day, we start and end our days at midnight. The division of a day into twenty-four hours is strictly arbitrary and goes back to very early times. The Greeks divided the periods of daylight and darkness each into twelve equal parts to make twenty-four divisions in each day. The hours defined in this way were not always the same since the periods of light and dark vary in length with the seasons. The custom of dividing the hour into sixty minutes and the minute into sixty seconds is one of the last vestiges of the sexagesimal system of counting which was in use in the ancient world.

Today we keep track of time with a variety of clocks which range in complexity from sundials of the simplest sort to quartz clocks of the greatest precision. Our choice of one clock over another is made on the basis of our needs in particular situations. We can be on time for most appointments if we know the time to the nearest minute, but astronomers combining data recorded over a long period of time often need to know the time to the nearest thousandth of a second. In the end, however, we always come back to the recognition that the daily motion of the sun is the original basis of our whole system of timekeeping.

Solar Time

Unfortunately, the sun is a bit irregular as a time-keeper. If one were to make daily comparisons between an accurate watch and a sundial, one would soon find that the two were not in close agreement very often. Noon on the sundial would be ahead of noon on the watch during the months of May and November and behind during the months of February and July. The watch, of course, runs at a constant rate, so the sun is apparently moving across the sky at a variable rate.

There are two principal reasons for this difference. The first is that the earth's motion around the sun is not uniform. The earth's orbit is slightly elliptical, so the earth moves faster when it is closest to the sun in January. Fig. 2.1 may help the reader to visualize the reason why the interval from noon to noon is actually longer in the winter than in the summer. At this point we are assuming that the period of the earth's axial rotation is constant, so when the earth's orbital speed is faster, as it is when the earth is closest to the sun, the earth must rotate further before noon will recur.

The second effect arises because of the obliquity of the ecliptic and is illustrated in Fig. 2.2. Imagine that the real sun moves along the ecliptic at a uniform rate and imagine a fictitious or mean sun which moves along the celestial equator at a uniform rate. In Fig. 2.2 the sun is at S and the mean sun is at M. Because we have assumed uniform motion for both S and M, VS=VM. S′ is the projection of the true sun down to the celestial equator. From the time when the real sun crosses the vernal equinox until the time when it reaches the summer solstice in June it is behind the fictitious sun, M. Then it is ahead for three months; then behind and ahead again during the next two quarters. The combined effect of these two factors is shown in Fig. 2.3, and may be computed for any day of any year.

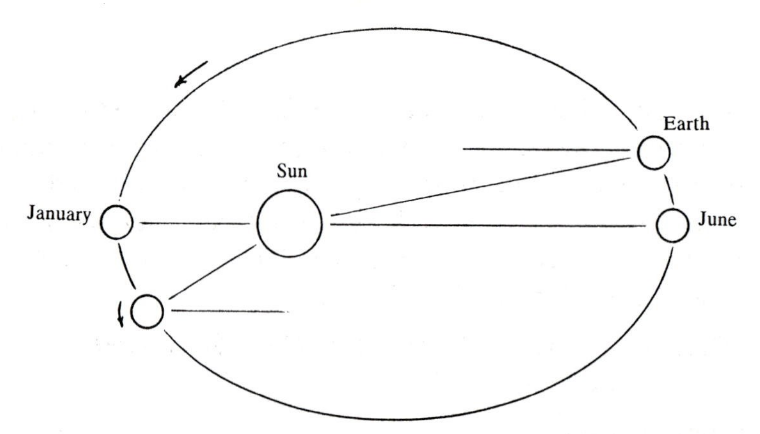

Fig. 2.1. Seasonal variations in the interval from noon to noon. The ellipticity of the earth's orbit is very much exaggerated in this sketch.

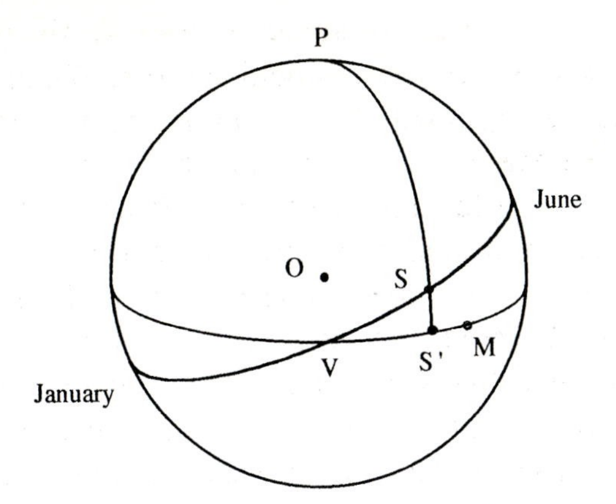

Fig. 2.2. The true sun projected to S′ is lagging behind the mean sun at M. The angles VOS and VOM are equal.

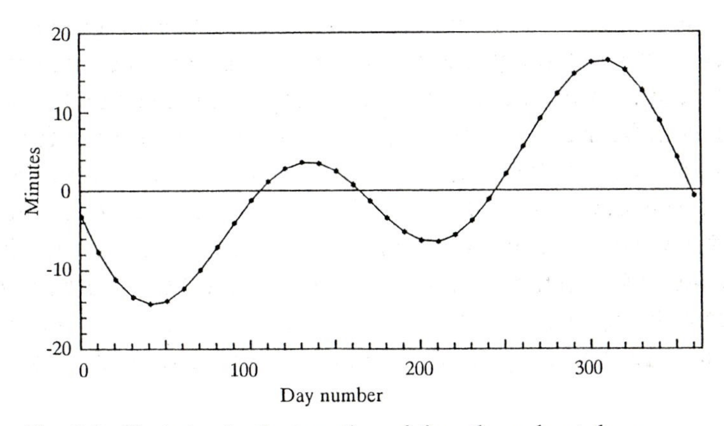

Fig. 2.3. Variation in the equation of time throughout the year.

The non-uniform time recorded by the sundial is referred to as solar time. The watch, however, is designed so that the day which it indicates has the average length of all of the days measured on the sundial in a year. Our watches and clocks are adjusted to measure mean solar time.

The difference between solar time (ST) and mean solar time (MST) is referred to as the equation of time (EqT), or

$$\text{Eq. of Time} = \text{App. Solar Time} - \text{Mean Solar Time}$$
$$\text{EqT} = \text{ST} - \text{MST}$$

This is sometimes written

$$\text{EqT} = \text{RAMS} - \text{RAS}$$

where RAMS is right ascension of the mean sun, and RAS is right ascension of the sun. A value for the equation of time on any particular day of the year may be determined from data published annually in the *Astronomical Almanac* (see bibliography and p. 41). At the equinoxes and at the solstices the equation of time is zero.

Another obvious property of solar time is that it is valid for an observer only at a specific longitude. For an observer at another longitude the solar time will be different (see Fig. 2.4). Up until the middle of the nineteenth century most cities and towns adopted their own local time which was essentially local mean solar time. Travel and communication were slow, and this was good enough for all purposes. With the coming of the railroads and the telegraph a more uniform system was needed and time zones were introduced. For convenience, then, we have arbitrarily marked off twenty-four zones of longitude on the earth, and we maintain the same time for all clocks within each of these zones. The time within each zone is known as standard time. On the eastern coast of North America we use Eastern Standard Time, while on the western coast we use Pacific Standard Time. Each zone is fifteen degrees wide, but for the convenience of the population the boundaries of the time zones are irregular.

In 1884 the meridian of Greenwich, England was chosen as the meridian of zero longitude, and longitude was defined to increase toward

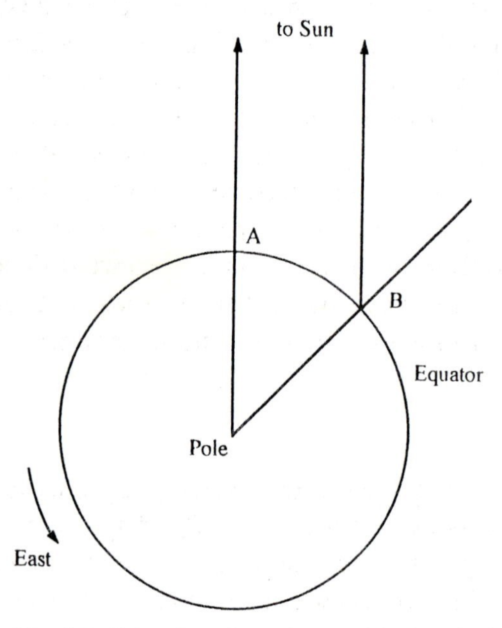

Fig. 2.4. Solar time depends upon the longitude of the observer.

the east and toward the west to a maximum value of 180 degrees. Navigators of ships at sea knew that the difference in local or mean solar time between two places was equivalent to the difference between the longitudes of the two places. Thus, they found it convenient to keep with them always an accurate timepiece set to the mean time in Greenwich. An observation which gave them their local time (i.e., apparent solar time corrected by the equation of time) then gave the longitude.

$$\text{longitude} = \text{Greenwich mean time} - \text{local mean time}$$

Observations of the altitude of the sun, for example, were routinely used to establish the moment of local noon. Today the term 'Greenwich Mean Time' is seldom used. Instead we commonly refer to 'Universal Time', but the meaning is the same.

Ephemeris Time and Dynamical Time

When a particular problem requires that intervals of time be measured with extreme precision, one may use Ephemeris Time. This system is necessary because the actual period of the earth's rotation is increasing in a slow but irregular manner. Therefore, in 1960 a new kind of time which increases in a steady, uniform way was introduced. It was given the name Ephemeris Time. In 1984 the concept of ephemeris time was refined even more, and the name was changed to Dynamical Time. Ephemeris Time and Dynamical Time are determined by adding a correction to the Universal Time in the sense $\mathrm{DT} = \mathrm{UT} + T$. The correction, T, cannot be computed in advance, so the user must always wait until the next value of the correction has been published. At the beginning of 1989 the correction was just over 56 seconds. Instructions for the computation of Dynamical Time are to be found in the *Astronomical Almanac*. There are actually two forms of dynamical time. One, Terrestrial Dynamical Time or TDT, is referred to the earth. The other, Barycentric Dynamical Time or BDT, is referred to the barycenter of the solar system.

Sidereal Time

Because of the earth's orbital motion the solar day does not really represent the true rotation period of the earth. The diagram in Fig. 2.5 may help to illustrate this point. If the earth moves from position 1 to position 2 in the course of a day, then after one complete rotation an observer who was originally facing the sun will now be facing toward V.

Noon will not occur again until the earth has rotated through the additional angle a. Thus the solar day is really longer than the true rotation period of the earth. The stars are the logical frame of reference against which to measure the true rotation period, and as mentioned in Chapter 1, time based on the earth's rotation with respect to the stars is known as 'sidereal time'. It is perhaps obvious that sidereal time is of great importance to astronomers.

In discussing the methods for determining the sidereal time we may note simply that if the earth moves around the sun in 365 days, then in one day the earth will have moved through an angle equal to 360/365 or 0.98 degree. Recall now these correspondences:

$$24 \text{ hours} = 360 \text{ degrees}$$
$$1 \text{ hour} = 15 \text{ degrees}$$
$$4 \text{ minutes} = 1 \text{ degree}$$

From the above it should be clear that the solar day is approximately four minutes longer than the sidereal day. Let us consider some more precise values and see how the correct sidereal time may be calculated. There are several convenient methods by which this may be done.

Finding the Sidereal Time

We mentioned earlier that many of the older observatories were originally equipped with a small transit telescope. Most of these telescopes have since been removed from active use, and their housings have been converted to other purposes. These telescopes were originally used

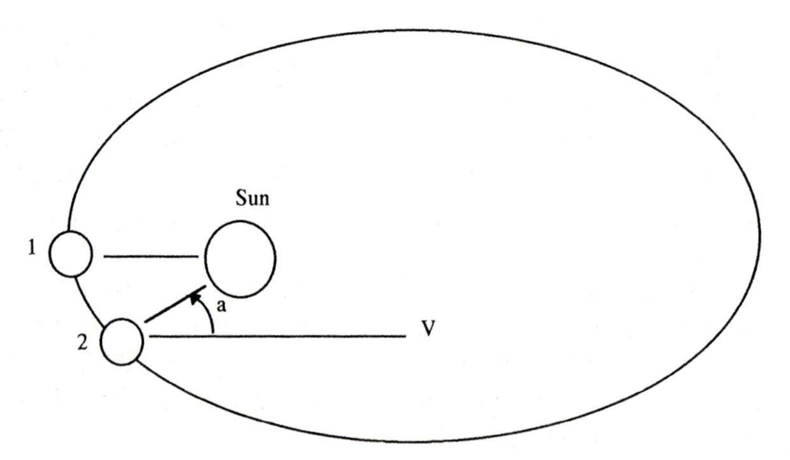

Fig. 2.5. The sidereal day is shorter than the solar day.

to determine the sidereal time directly from observations of stars of known right ascension. The process is just the reverse of that for finding the right ascension. A star is observed during transit, and the local sidereal time at the moment of transit is the same as the previously known right ascension of the star. Local time or standard time may then be calculated.

We could still find the sidereal time from the same sort of telescopic observations, but in the modern world it is much simpler and less time-consuming simply to compute the sidereal time from the known value of standard time. We begin with the standard time which can be quite easily obtained from time signals regularly transmitted *via* radio or telephone by government agencies in a number of countries. A list of the call letters and frequencies and the telephone number on which the signals are transmitted is included in Appendix 2. The time signals consist of a series of tones to mark the seconds. Hours and minutes are identified by frequent vocal statements or by codes. By suitably recording the time signals along with other data, clocks can be set to the nearest thousandth of a second. In many practical situations such precision is not really necessary and the time obtained by dialing an appropriate local telephone number is adequate.

Now as we proceed to methods for the conversion from mean solar time to local sidereal time, let us first note that the *Astronomical Almanac* (see p. 41) contains a table which lists the sidereal time for zero hours UT for every day of the year. These tables are not the same from one year to the next, so one must always use the Almanac for the appropriate year. The sidereal time may be computed by proceeding through the following steps:

Step 1. Convert Standard Time to UT. This is done by adding one hour for each time zone between the observer and Greenwich if the observer is west of Greenwich. For example, an observer in Boston or Quebec would add five hours to convert from Eastern Standard Time to UT. An observer in Chicago would add six hours; in Winnipeg, seven hours, and so forth. Observers east of Greenwich would of course subtract an hour for each time zone.

Step 2. Define the 'solar interval' as the time interval from 0 h UT to the time in question. Convert this solar interval into a sidereal interval. This is done by multiplying by the factor, 1.00273791, the ratio of the number of solar minutes per sidereal day to the number of sidereal minutes per day (1443.9425/1440).

Step 3. Add the sidereal interval to the sidereal time at 0 h UT taken from the Astronomical Almanac. This gives us the Greenwich Sidereal Time.

Step 4. Subtract the longitude of the observer in order to know the Local Sidereal Time.

Example: Find the local sidereal time at 1900 hours on September 29, 1981 for an observer in Wellesley, Mass., USA.

1. 1900+0500=2400 UT=0000 UT on September 30
2. The interval from 0 h UT is 0.
3. The Astronomical Almanac gives 0 h 30 m 40.0511 s as the GST at 0 h UT on September 30, 1981. Thus the Greenwich Sidereal Time is 0 h 30 m 40.0511 s.
4. Subtract the longitude of Wellesley (4 h 45 m 13.3 s) 0 h 30 m 40.0511 s−4 h 45 m 13.3 s=19 h 45 m 26.7 s, the local sidereal time.

This example was obviously simplified because the universal time turned out to be 0 h. Let us consider another example.

Example: Find the local sidereal time at 1100 hours on September 29, 1982 for the observer in the first example.

1. 1100+0500=1600 UT
2. 16 hours=960 min.
   960 min × 1.00273791=962.62839 min or 16 h 2 m 37.7 s. This is the sidereal interval.
3. GST=Sid Int+ST 0 h UT
   16 02 37.7+0 26 43.5042 (ST 0h UT Sept. 29)=16 29 21.2
4. LST=GST−longitude
   16 29 21.2−4 45 13.3=11 44 7.9, the LST.

It is often convenient to convert hours, minutes and seconds into hours and a decimal part of an hour in order to facilitate addition and subtraction of numbers such as these. Most small scientific calculators provide keys for the conversion of degrees, minutes and seconds to degrees, and the same keys work, of course, for the conversion of hours, minutes and seconds into hours.

Those individuals who routinely use a programmable calculator or a computer will find that a second method of computing the sidereal time will be more convenient. In this method there is one constant which must be changed in the program at the beginning of each year. The formula is as follows:

$$\mathrm{GST} = G + 0.0657098232 \times N + 1.0027379093 \times \mathrm{UT}$$

where G is the GST at 0 h UT on the zeroth day of the year. (The zeroth day is obviously the same as the last day of the previous year); N is the number of the day from the beginning of the year (see below); and UT is the universal time in hours. If the computation using the formula gives a number greater than twenty-four hours, then twenty-four hours should be subtracted and the date should be made one day later. The user may find the new value of the constant, G, in the Astronomical Almanac at the beginning of each year and proceed with confidence.

The day number may be found from another formula which lends itself nicely to computer programs:

$$N = \langle 275 \times M/9 \rangle - 2 \times \langle (M+9)/12 \rangle + I - 30$$

where M is the number of the month and I is the day of the month. The brackets $\langle\ \rangle$ indicate that one should use the integer value of the enclosed quantity. In a leap year the coefficient of the second term should be changed from 2 to 1. More information on this method of computing the local sidereal time may be found in *Almanac for Computers* published annually by the U.S. Naval Observatory.

The convenience of calculating the sidereal time means that it is no longer really necessary to maintain an accurate sidereal clock in an observatory. The standard time of an observation may be recorded directly from a clock or a radio, and the sidereal time may be calculated at a later time. However, the sidereal clock is still valuable in providing the observer with a quick means of knowing where to look for an object of a particular right ascension.

Julian Date

There are in astronomy a number of situations in which it is convenient to use a system of consecutively numbered days. Such a sequence is useful whenever the number of days between two events is required. The number of the earlier day may simply be subtracted from the number of the later day. Applications of this type arise, for example, in the study of variable stars, and these will be discussed in Chapter 14. The most widely used scheme for numbering days was introduced in 1582 and is known as the system of Julian Day Numbers. The numbers begin with January 1, 4713 BC, and each day is considered to begin at noon rather than at midnight. Julian Day numbers may be found in the Astronomical Almanac in the tables in which data for the sun are tabulated. See, for example, Fig. 3.9.

In computational work it is often more convenient to calculate the Julian Day number than it is to look it up and enter it from a table. The following formula is quite simple and may be used to find the proper number for any day since the year 1900.

$$\mathrm{JD} = 2\,415\,020 + 365 \times (\mathrm{year} - 1900) + N + L - 0.5$$

Here N is, again, the number of the day counted from the beginning of the year, and L is the number of leap years which have occurred between 1901 and the beginning of the year in question. 2 415 020 is the Julian Day number of January 1, 1900, and the 0.5 accounts for the fact that the Julian Day begins at noon. This calculation gives the Julian Day number at 0 h UT, so one must also add the decimal fraction of a day at the required universal time.

Example: Find the Julian Day Number for April 18, 1983.

First we must calculate N, the number of days which have elapsed from the beginning of the year until April 18.

$$N = \langle 275 \times 4/9 \rangle + 2 \times \langle (4+9)/12 \rangle + 18 - 30$$
$$N = 108$$

The number of leap years from 1901 to 1983 is simply $\langle 82/4 \rangle$ and is 20. Then we may find the Julian Day Number from the formula given above.

$$\mathrm{JD} = 2\,415\,020 + 365 \times (1983\text{–}1900) + 108 + 20 - 0.5$$
$$\mathrm{JD} = 2\;445\,442.5$$

Example: Find the Julian Day Number at 13 h 45 m on April 18, 1984. Note that 1984 was a leap year. Following the procedure used in the previous example:

$$N = \langle 275 \times 4/9 \rangle - 1 \times \langle (4+9)/12 \rangle + 18 - 30$$
$$N = 109$$
$$\mathrm{JD} = 2\,415\,020 + 365 \times 84 + 109 + 20 - 0.5 + 0.5729$$
$$\mathrm{JD} = 2\,445\,809.0729$$

The final term (0.5729) is the fraction of a day equivalent to 13 hours and 45 minutes. Note also that the number of leap years was 20 in both examples even though 1984 was a leap year. The extra day in 1984 was taken care of when the day number, N, was calculated.

This useful method of numbering days was introduced, as we have said, in 1582, and it was devised by Joseph Justus Scaliger, one of the most brilliant men of his time but one who is little known today. He chose the name 'Julian' to honor his father, Julius Caesar Scaliger. There is no connection at all between the Julian Day numbers and the Julian

Calendar which was introduced during the reign of Julius Caesar. Joseph Scaliger defined his Julian Period as 7980 years, the product of the numbers 28, 19 and 15. The first of these is the number of years required for the same day of the week to re-occur on the same day of the year. The second is the Metonic Cycle in which 235 lunar cycles are equal in length to 19 solar years. The third number is the length of the Roman Indiction, a period of years used by the emperor, Constantine, in the fourth century AD. It hardly needs to be said that there is no astronomical significance to Scaliger's Julian Period.

In order to avoid the inconvenience of using very large numbers some people have chosen to count the days only from 1900. This scheme is referred to as the system of Modified Julian Day numbers (MJD). The reader should be able to modify the formula given above to calculate the correct MJD rather than the JD.

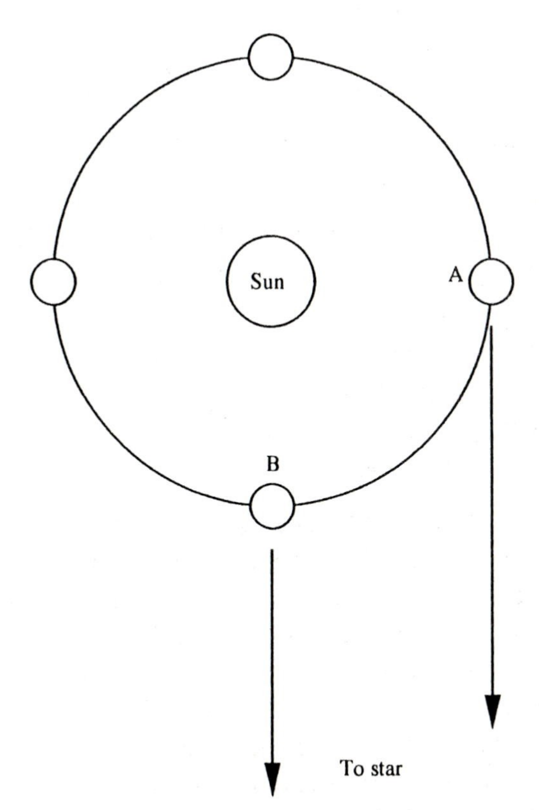

Fig. 2.6. The need for heliocentric time.

Heliocentric Time

When precisely-timed celestial events are observed at several times during the year, we must take into account the time required for light to travel across the earth's orbit. Depending upon the direction of an object in space light can reach the earth as much as eight minutes earlier when the earth is at one position in its orbit than when it is at another. This effect is illustrated in Fig. 2.6. A recurring event such as the eclipse of a binary star would seem to occur earlier when the earth is at position A than it would three months later when the earth is at position B. In order to correct for this, we calculate the heliocentric time, that is, the time at which the event would have been recorded by an observer on the sun.

The method of determining the heliocentric time will be described in detail in Chapter 14 as part of the discussion of variable stars.

QUESTIONS FOR REVIEW

1. What is the basis for our everyday system of timekeeping?
2. Why is the sidereal day not the same length as the solar day? Which is longer?
3. Define solar time and mean solar time. Why is a distinction necessary?
4. Why is the difference (solar time−mean solar time) not constant throughout the year?
5. How can a precise clock be used to determine longitude?
6. Describe the reasons why dynamical time is needed and why it cannot be determined at the time of an observation.
7. Calculate the local sidereal time for an observer in Tucson, Arizona, USA for May 1, 1990 at 21 h 31 m Mountain Standard Time (MST).
8. Calculate the Julian Day Number for May 12, 1990, at 22 h 40 m Eastern Standard Time.
9. Write a computer program to compute local sidereal time.
10. Write a computer program to compute the Julian Day Number.

Further reading

Green, R. M. (1985). *Spherical Astronomy*. Cambridge: Cambridge University Press. Consult Chapter 10 for descriptions of many systems of time.

MacRobert, A. (1989). Time and the amateur astronomer. *Sky and Telescope*, **77**, 378. This is a concise and clear description of the important kinds of time and the need for each.

Pannekoek, A. (1961). *A History of Astronomy.* Interscience Publishers. Good discussions of time-keeping and clocks are scattered throughout this book. Consult the index.

Smart, W. M. (1977). *Textbook on Spherical Astronomy.* Cambridge: Cambridge University Press. Chapter VI is a very complete discussion of the many ways of measuring time.

3

Star charts and catalogs

When we wish to study the characteristics of an individual star, it is crucial that we be able to identify that star with absolute certainty. There are a number of ways in which such identifications may proceed, and the choice of a method depends quite often on the apparent magnitude of the star (see Chapter 7 for a discussion of the magnitude system of measuring brightness). If the star is brighter than about fourth magnitude, the problem is quite simple, for there are not very many stars that are brighter than this. When we have set the telescope to the proper coordinates, there will probably be only one sufficiently bright star in the field of view. On the other hand, when we try to locate a star of eighth or ninth magnitude, there may be six or more in a field thirty minutes of arc in diameter. The numbers of progressively fainter stars increase dramatically. There are so many sources of error involved in setting a telescope that it is not reasonable to expect that the desired star will always be the one nearest the center of the field even if the coordinates are known and the telescope is set with great precision. Therefore, when trying to identify faint stars we must usually rely on maps or charts on which each individual star has been identified in some reliable manner. One can then compare the pattern of stars depicted on the chart with the pattern actually seen in the telescope and make a positive identification. Astronomers today are fortunate that their predecessors went to great lengths to produce several sets of charts and catalogs which greatly simplify our work. Let us review the most commonly used methods for identifying first the brightest stars and then the less bright ones.

Names of stars

In examining star maps or catalogs one should notice that most of the very bright stars have individual names. These names have been in use for many centuries, and many of them go back to the Babylonians in the sixteenth century BC. Some of the names are derived from the location of the star in its constellation, and some of the names are derived from some special characteristics of the star. The prefix 'al' is found in the names of fifteen of the brightest stars. 'Al' is the arabic word for 'the', and it indicates that these stars were some of the many that were named by the Arabic peoples of North Africa and the Middle East. Other names are derived from the location of the star in the figure which the constellation was originally said to represent. For example, the name 'Betelgeuse' is translated as 'in the shoulder of the central one', and it describes the location of this star in the shoulder of the large figure of Orion, the hunter. A very complete study of the origins and meanings of the names of stars is to be found in the book, *Star Names and their Meanings*, by R. H. Allen, listed at the end of this chapter.

Bayer and Flamsteed Catalogs

The transition from naked-eye positions to those determined instrumentally is marked by two other methods of designating or naming individual stars. Looking again at a typical star map one finds that many stars are identified with Greek letters. Betelgeuse, for example, is also specified as 'alpha Orionis', and Rigel is specified as 'beta Orionis'. This use of a Greek letter followed by the possessive form of the latin name of the constellation originated with the astronomer, Johannes Bayer, in 1603. In his catalog Bayer listed for 788 stars coordinates which had been determined earlier by Tycho Brahe using precise, but non-telescopic instruments. Bayer assigned the Greek letters to indicate the relative brightnesses of the stars. Thus, alpha Orionis was the brightest star in Orion, beta Orionis was the next bright, and so forth.

In a second early catalog the stars were designated by a number followed by the possessive form of the name of the constellation. This scheme originated with John Flamsteed, the first Astronomer Royal in England. Flamsteed used an accurate meridian instrument to measure right ascensions and declinations, and he numbered the stars in order of right ascension within each constellation. The Flamsteed catalog contains the positions of 3000 stars and was published in 1725, six years after Flamsteed's death. When we see a designation such as 61 Cygni, we recognize an entry from the Flamsteed catalog. It is the general practice in astronomy today to use the Flamsteed numbers only for stars which were too faint to have been included in the Bayer catalog. Thus, if a star

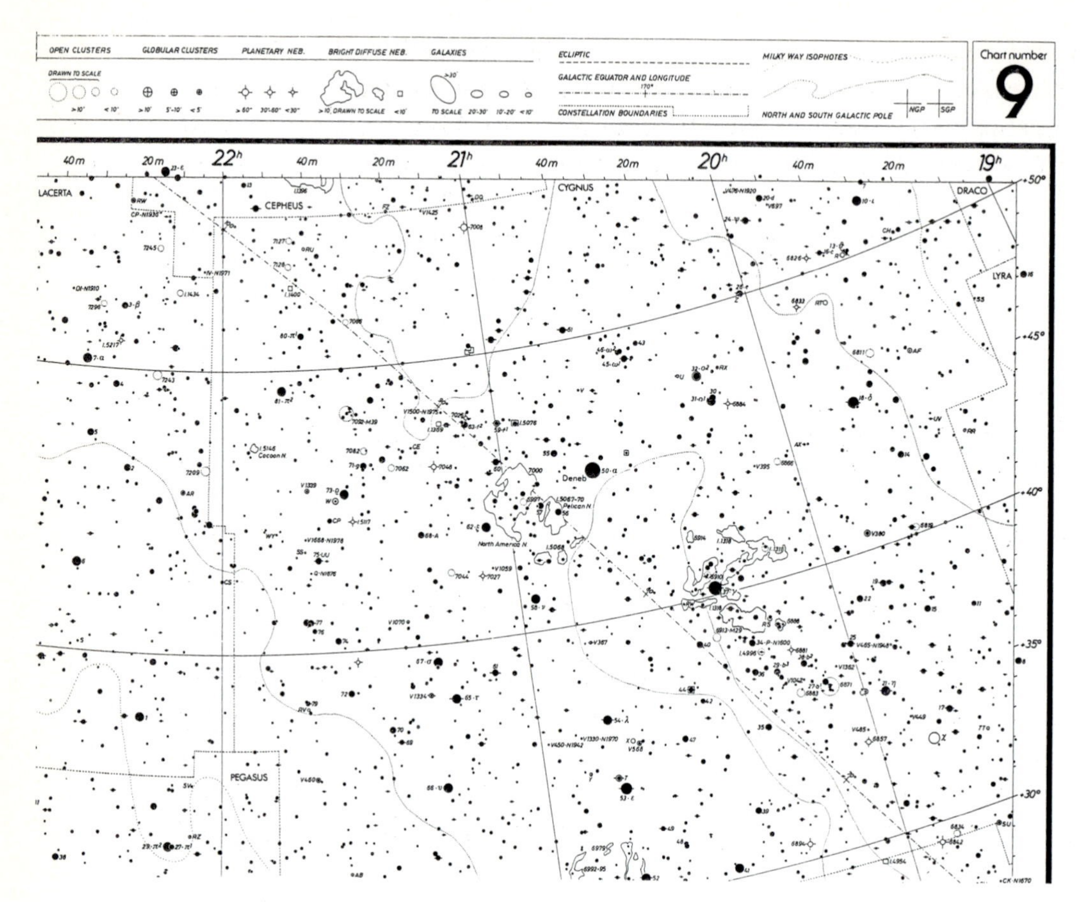

Fig. 3.1. A portion of a typical page from the *Tirion Atlas*. (Copyright 1981 Sky Publishing Corp. Reproduced by permission.)

has been listed by its Flamsteed number, the star is probably not visible to the naked eye. In Fig. 3.1 part of a modern star map has been reproduced, and one may see the use of the Greek letters and the numbers of Bayer and Flamsteed.

Yale Catalog of Bright Stars

The *Yale Catalog of Bright Stars* is perhaps the most complete of the modern standard catalogs in that it contains a great deal more than just the coordinates of the stars. Magnitudes, colors, spectral types, proper motions, galactic coordinates, luminosities and more are listed here for 9091 stars. The limiting magnitude is 6.5. It was first published in 1900, and was revised in 1940, 1964 and 1982. The top portions of two facing pages are reproduced in Fig. 3.2 so that the reader can see how the information is listed. Entries are in order of right ascension around the entire sky. The commonly used abbreviation for this catalog is YBS, and it is general practice to use these letters (or just BS) followed by the catalog number to identify a star. The reader should notice that the entries for the brighter stars in the YBS include the names, the Greek letters of Bayer and the numbers of Flamsteed.

334

HR	NAME	DM °	HD	I	D	VAR	RA (1900) h m s	DEC ° ′ ″	RA (2000) h m s	DEC ° ′ ″	LONG °	LAT °
8251		-4 5489	205423				21 30 4.4	-4 25 44	21 35 17.5	-3 58 59	50.35	-37.86
8252	73 ρ CYG	+44 3865	205435	I		VAR?	21 30 13.1	+45 8 59	21 33 58.8	+45 35 31	90.74	-4.58
8253	8 PSA	-26 15702	205471		W		21 30 23.1	-26 37 3	21 36 10.9	-26 10 18	22.61	-46.64
8254	ν OCT	-77 1510	205478		A		21 30 21.7	-77 50 2	21 41 28.6	-77 23 24	314.29	-35.19
8255	72 CYG	+37 4359	205512	I			21 30 41.4	+38 5 8	21 34 46.5	+38 32 3	86.01	-9.85
8256	7 PSA	-33 15664	205529				21 30 48.5	-33 29 43	21 36 48.8	-33 2 53	12.63	-47.85
8257		+27 4107	205539				21 30 53.1	+27 45 8	21 35 18.9	+28 11 51	78.72	-17.39
8258		+23 4346	205541		15115		21 30 55.5	+24 0 22	21 35 26.9	+24 27 8	75.94	-20.05
8259		+51 3091	205551				21 31 0.6	+51 15 10	21 34 27.4	+51 41 55	94.93	-0.13
8260	39 ε CAP	-20 6251	205637		W	ε CAP	21 31 28.9	-19 54 51	21 37 4.7	-19 27 58	31.94	-44.99
8261		+29 4456	205688		15126		21 31 52.5	+29 36 22	21 36 13.8	+30 3 20	80.23	-16.20
8262		+44 3877	205730	I		W CYG	21 32 14.3	+44 55 36	21 36 2.3	+45 22 29	90.86	-4.98
8263		-1 4180	205765		15142		21 32 25.6	-0 50 20	21 37 33.7	-0 23 25	54.52	-36.38
8264	23 ξ AQR	-8 5701	205767		D		21 32 25.7	-8 18 10	21 37 45.0	-7 51 15	46.45	-40.34
8265	3 PEG	+5 4830	205811		15147A		21 32 44.7	+6 10 8	21 37 43.6	+6 37 6	61.41	-32.28
8266	74 CYG	+39 4612	205835				21 32 56.3	+39 57 51	21 36 56.9	+40 24 49	87.62	-8.76
8267	5 PEG	+18 4827	205852				21 33 4.6	+18 52 7	21 37 45.4	+19 19 7	72.34	-23.99
8268		+34 15163	205872				21 33 5.4	-34 7 42	21 39 6.0	-33 40 45	11.74	-48.38
8269		-52 11911	205877				21 33 10.2	-52 48 38	21 39 59.6	-52 21 32	344.10	-46.53
8270	4 PEG	+5 4834	205924		15157		21 33 31.4	+5 19 13	21 38 31.8	+5 46 18	60.76	-32.96

335

HR	V m	B-V m	U-B m	R-I m	SPECTRAL CLASS	PM(α) ″	PM(δ) ″	PAR ″	RV k/s	vsini k/s	Δm	SEP ″	COMP	N	R
8251	5.77	+1.11	+1.05		gG9	-0.007	+0.005	+.001	- 2V?						
8252	4.02	+0.89	+0.56	+0.50	G8IIICN-0.5Hδ1	-0.026	-0.091	+.002	+ 7	<19:					*
8253	5.73	+0.22			Am	+0.112	-0.020		- 19		8.2	18.4			
8254	3.76	+1.00	+0.89		K0III	+0.057	-0.239	+.053	+ 34SBO			0.1	*		*
8255	4.90	+1.08	+1.02	+0.54	K0.5III	+0.120	+0.101	+.012	- 66	<17					
8256	6.11	+0.22			A7Vn	+0.089	+0.000		- 2						
8257	6.31	+0.35	-0.05		F0IV	+0.127	-0.039		- 42SB2O						*
8258	6.11				A4V	+0.011	-0.007	+.002	- 28SB2	170	0.3	0.3			*
8259	6.15	+0.02	-0.29		B9IIIe	+0.004	+0.003	-.006	- 22V	175					
8260	4.68	-0.17	-0.65	-0.12	B2.5Vpe	+0.010	+0.006		- 24SB?	293	1.3	0.0	O	3	*
8261	6.36R				G8III-IV	-0.064	+0.066		- 20		4.4	2.1			
8262	5.53	+1.58	+1.24	+2.14	M5IIIae	+0.048	+0.010		- 14V						*
8263	6.25	+0.06	+0.05	+0.02	A2V	-0.018	-0.019		+ 17V?	156	3.3	31.5			*
8264	4.69	+0.17	+0.13	+0.10	A7V	+0.113	-0.023	+.012	- 21SBO	154	2.0	0.1	O		*
8265	6.18	+0.02	+0.02		A2V	+0.058	-0.003		+ 3SB	89	1.5	39.2	AB	3	*
8266	5.01	+0.18	+0.10		A5V	-0.006	+0.017	+.020	+ 7V?	171					
8267	5.45	+0.30	+0.14		F1IV	+0.102	+0.015	+.007	- 25V?	134					*
8268	6.28	+0.92			K0	+0.066	-0.048								
8269	6.21	+0.60	+0.32		F7III	-0.015	+0.014								
8270	5.67	+0.25	+0.08		A9IV-Vn	+0.112	+0.028	+.033	- 19V?	195	6.4	27.2			*

Fig. 3.2. A sample of facing pages from the *Yale Catalog of Bright Stars*, fourth edition (courtesy of Dorritt Hoffleit).

The heading for the first column on both the left and right hand pages indicates that the BS number is the same as the HR number. HR refers to the *Harvard Revised Photometry*, an earlier catalog of stellar magnitudes published at the Harvard Observatory.

Bonner Durchmusterung

One of the most useful tools for the identification of stars was produced more than one hundred years ago in Germany. Under the direction of Friedrich Argelander, astronomers at the Bonn Observatory spent seven years (1852–1859) working on a catalog and atlas which listed the positions and magnitudes of roughly 324 000 stars. This phenomenal work includes stars down to approximately magnitude 9.5, and is a tribute to the foresight of Argelander and the diligence of his small staff. The observers used a telescope with an aperture of about three inches, and they worked on just one zone of declination at a time. With the telescope pointed to some point on the meridian, the observer could watch as one star after another passed through the field of view. When he called out that the star was on the meridian (represented by a north–south line crossing the center of the field of view) his assistant wrote down the sidereal time, which was equivalent to the right ascension of the star.

As the star crossed the meridian the observer was also able to note its position above or below a horizontal center line and thus estimate the declination to the nearest tenth of a minute of arc. The observer's estimate of the declination was also recorded by the assistant (see also p. 9).

In this manner the positions of stars from the north celestial pole to a declination of −2 degrees were recorded. All parts of the project were repeated so that there could be a check for completeness and consistency. The positions of stars were then used to produce a set of charts which can be compared to photographs or visual observations when positive identifications are needed. In each zone of declination one degree wide all of the stars are numbered in order of their right ascensions. The identifying number for a star thus consists of two parts: the declination zone and the number within that zone. A typical designation might be, for example, BD +26 4782 or BD −1 148. In Fig. 3.3 we have reproduced a portion of one of the pages from the *Bonner Durchmusterung* (BD) catalog and in Fig. 3.4 the appropriate portion of one of the maps or charts from the BD atlas.

-- 342 --

+39° **21h–22h**

4601—4660	21h		+39°	
m	′	″	′	
8.8	29	43.4	33.0	K
8.1		50.3	28.8	K
8.8		57.0	13.9	K
9.0	30	2.8	54.2	
9.4		3.0	10.9	
8.3		8.4	32.9	K
9.3		14.1	22.8	B
9.5		19.5	13.7	
9.3		29.3	56.3	
9.5		29.5	32.8	B
8.9		58.5	17.6	
*– 5.0	31	8.7	45.8	K
9.5		15.4	1.7	
9.5		29.0	57.5	
9.1		39.4	25.3	
9.5		41.3	28.5	
9.4		47.2	32.2	
9.2	32	5.2	20.4	
9.0		13.1	12.5	
9.5		13.7	7.4	B
9.5		17.2	48.4	B
9.4		42.0	57.7	
9.3	33	21.9	42.9	
9.4		26.2	8.8	
8.7		27.4	40.9	K
9.5		31.3	21.2	
8.0		36.3	59.1	K
9.1		38.2	34.2	
9.4		40.7	50.3	
9.0		42.8	17.2	K
9.3	34	0.6	1.6	
9.5		2.1	16.2	
9.3		3.5	45.1	
9.5		23.8	30.5	
9.5		28.4	40.8	
9.5		47.2	46.0	
9.5	35	13.1	28.0	
9.5		15.9	53.3	
8.7		18.0	17.7	
9.5		42.6	10.2	
9.4		48.4	53.5	B
7.8		58.3	52.2	K
9.5	36	0.8	30.5	
9.1		21.3	36.7	
9.2		21.5	4.1	
9.4		29.9	58.3	
9.3	37	5.4	20.6	
9.5		15.5	55.3	
9.5		16.9	47.5	
8.6		19.9	48.7	K
9.5		33.0	56.1	
9.5		59.3	26.1	
9.4	38	1.6	8.1	
9.5		20.4	4.1	
9.5		57.8	54.7	aB
9.5	39	6.4	2.6	
9.1		8.9	23.4	b
8.7		16.4	48.1	K
9.1		24.3	8.5	
9.1		32.8	7.1	
10 pr.		*24.16*	*2.69*	

4661—4720	21h		+39°	
m	′	″	′	
8.8	39	36.9	45.5	K
9.5		40.1	20.9	
8.8		54.8	49.0	K
9.2	40	0.0	4.5	
9.4		18.3	32.9	
9.4		19.0	28.6	
8.1		29.8	11.0	K
9.2		37.1	43.4	
9.5		45.9	24.0	
9.4		49.7	4.1	B
9.5	41	3.4	14.4	
9.1		5.7	5.5	Kb
9.5		44.3	41.0	
9.4	42	10.1	49.6	
9.0		10.2	52.6	
9.5		19.0	2.2	B
8.3		57.6	56.6	K
9.5	43	9.5	11.7	
9.3		14.9	9.7	B
9.3		29.4	14.1	
9.5		39.8	41.6	
9.5		51.9	20.9	
9.5	44	30.0	4.1	
9.5		42.3	48.6	
9.5		54.3	16.2	
9.5	45	1.7	21.8	
9.3		7.3	11.0	
9.4		8.4	8.2	
9.2		21.8	2.0	B
9.5		22.2	42.0	
8.5		49.6	24.6	
9.5		49.9	40.1	
9.5	46	6.2	30.9	
8.6		13.9	7.9	K
9.5		18.8	45.1	
9.1		24.4	43.3	Bb
9.3		29.7	13.5	B
9.0		32.1	7.9	K
•9.0	47	13.8	52.1	
9.2		40.2	41.3	
8.9		48.2	44.0	K
9.4		58.1	23.6	
8.9	48	10.2	11.1	K
9.0		25.7	4.1	
9.5		36.9	0.9	
8.2		43.7	14.5	K
9.5		57.0	4.7	
9.0	49	5.2	19.7	K
9.3		20.1	39.9	
9.5		38.7	30.5	
9.4	50	1.3	39.9	
9.5		10.6	3.1	
9.5		11.0	24.4	
9.5		15.0	23.3	
9.5		28.4	19.0	
9.1		41.7	59.3	Kb
9.4		44.3	29.8	
9.5		47.2	34.0	
9.3	51	6.8	43.3	
8.8		14.1	13.2	
10 pr.		*24.62*	*2.78*	

4721—4780	21h—22h		+39°	
m	′	″	′	
9.4	51	39.2	4.9	B
9.5		54.1	4.1	
9.4		57.0	12.2	
9.1	52	2.5	51.2	b
9.5		17.6	43.7	
9.1		30.2	58.6	
9.0		49.1	29.8	
9.5	53	4.7	39.6	
9.5		10.2	18.7	
9.5		40.1	51.7	
9.5		49.1	4.3	
9.5		49.4	52.4	
9.4		52.7	24.9	
9.5	55	11.4	20.1	
9.4		17.2	26.3	
9.4		51.2	35.4	b
9.5	56	48.1	31.9	
9.5	57	4.7	53.1	
9.5		44.2	33.0	
8.9		48.7	1.3	K
9.0		51.5	3.3	
9.5		56.1	13.4	
9.5	58	16.6	47.2	
9.5		26.7	20.1	
9.5		42.8	33.0	
9.4		56.8	27.4	B
9.5		53.9	26.4	
9.4	59	21.6	35.4	B
9.5		24.7	25.3	
9.5		33.6	31.6	
9.0	0	0.7	18.4	-
9.5		3.2	54.9	
9.3		5.9	51.0	
9.1		35.7	12.2	
9.0		48.4	57.7	
9.5	1	2.2	54.1	
9.5		16.5	4.7	
9.3		22.7	19.1	
9.4		23.2	29.1	
8.5		37.3	24.6	K
9.1	2	14.5	49.7	
9.5		25.4	0.7	
8.0		28.1	24.6	K
9.1		30.4	45.4	
9.5		38.7	53.5	
9.2		58.3	21.5	B
9.0	3	7.7	57.3	K
9.5		20.0	45.5	B
8.5		48.2	59.4	B
9.3		51.3	4.9	B
9.3		58.2	31.9	
9.0	4	2.3	21.5	K
8.7		5.2	11.1	
8.0		29.9	32.6	K
7.8	5	6.3	28.0	B
9.3		29.5	56.1	
9.3		35.2	34.2	B
8.9		37.5	5.9	K
9.3		57.7	7.7	
9.0	6	16.2	15.2	K
10 pr.		*25.20*	*2.89*	

4781—4840	22h		+39°	
m	′	″	′	
9.3	6	21.9	59.1	
8.6		22.9	51.0	
9.5		26.7	11.8	
9.5		45.3	39.6	
9.5		46.0	5.3	
9.5		49.1	32.0	
9.0	7	46.2	45.8	Kb
9.5		56.8	40.0	
9.3	8	2.4	15.9	
8.0		5.1	46.8	K
9.5		57.2	27.0	
8.0	9	2.1	2.7	L
9.5		41.8	15.5	
9.3		58.5	45.8	
8.8	11	9.2	54.2	
8.7		51.2	3.5	
8.0		53.4	13.6	K
9.5	12	22.3	46.1	
9.5		37.0	16.9	
9.5		59.8	36.7	
9.5	13	8.6	48.9	
9.5		10.7	21.1	
9.5		18.2	17.6	
9.5		19.6	52.4	B
9.1		24.8	59.4	Bb
9.5		44.7	46.5	
9.4		46.6	4.6	B
9.5	14	0.2	29.1	
8.9		10.4	38.5	
9.5		18.3	17.4	
9.4		23.7	28.4	B
9.3	15	19.1	51.3	
8.2		29.0	15.9	K
6.5		37.1	55.8	Lb
9.4	16	2.7	17.0	
9.5		12.1	43.6	
9.2		21.5	4.8	
9.3		37.0	28.7	
9.0		45.5	37.1	
9.5	17	7.0	58.2	
9.5		12.9	34.7	
8.7		15.4	50.7	
9.4		18.2	56.2	
9.5		23.3	14.2	
8.9		39.7	25.0	
9.4		44.8	59.0	
9.5	18	49.6	50.7	
9.5		52.7	34.7	
8.5	19	2.8	25.3	
9.5		41.3	8.3	
9.5		52.2	33.3	
9.1		56.4	54.3	
9.5	20	15.6	44.8	
9.4		16.9	15.6	
9.0		18.7	56.2	
9.5		27.8	58.1	
8.2		28.0	48.3	K
9.5		55.1	35.9	
9.2		59.9	56.2	
9.5	21	5.9	36.8	B
10 pr.		*25.82*	*2.99*	

4841—4900	22h		+39°	
m	′	″	′	
6.0	21	6.2	4.8	K
8.8		6.7	44.0	K
9.4		26.7	39.2	
9.3		27.3	13.2	B
9.2		29.2	47.2	
9.2		44.7	26.1	K
9.5		49.2	42.7	
9.1		57.3	50.7	
9.5	22	4.7	41.3	
8.5		7.2	8.6	K
var		39.7	33.6	K
9.4	23	9.1	10.3	
9.4		12.4	36.0	
9.5		19.2	28.0	
8.8	24	6.0	1.6	
8.0		7.7	27.5	B
8.8		20.7	14.5	
8.8		24.1	21.5	
9.5		24.2	39.6	
7.5		32.0	59.7	K
9.5		43.3	12.1	
9.1		49.5	24.2	K
9.3		57.7	3.1	
9.2	25	14.3	32.5	B
9.0		22.1	32.5	K
9.5		26.4	50.7	
9.1		34.6	53.8	
9.2		41.5	38.6	K
8.5		43.4	3.4	K
8.2		59.3	52.5	G
6.2	26	2.0	2.1	B
9.5		6.3	31.7	
9.4		8.1	33.7	B
9.0		12.9	34.8	K
9.5		34.4	0.2	B
9.3		40.4	6.2	
9.4		40.9	7.1	
9.1		50.7	35.0	K
9.3	27	16.7	0.4	
9.5		24.9	11.1	
9.5		31.4	46.8	
9.4		45.5	31.5	
8.7	28	2.0	53.8	K
9.5		17.2	45.2	
9.4		20.9	30.8	
8.3		26.3	58.1	K
9.5		42.2	10.0	
9.3	29	6.7	8.3	
9.1		7.4	7.1	
9.0		21.5	20.8	
9.5		27.0	58.3	
8.0		50.7	20.8	K
9.2		56.9	27.5	
8.5		57.1	38.5	K
9.5	30	35.0	34.7	
9.4		45.1	8.0	K
9.1		59.9	25.4	K
8.8	31	10.6	24.3	K
9.5		13.4	44.7	
9.5		42.0	55.3	
10 pr.		*26.32*	*3.07*	

4699a 9.5 | 47 16.0 | 53.3 | a

Fig. 3.3. A representative page from the *BD Catalog*.

In 1886 Argelander's successor, E. Schonfeld, published an extension of the BD to declination −23°. At about the same time astronomers at the Cordoba Observatory in Argentina completed the *Cordoba Durchmusterung* (CoD) following the same procedures used at Bonn. This

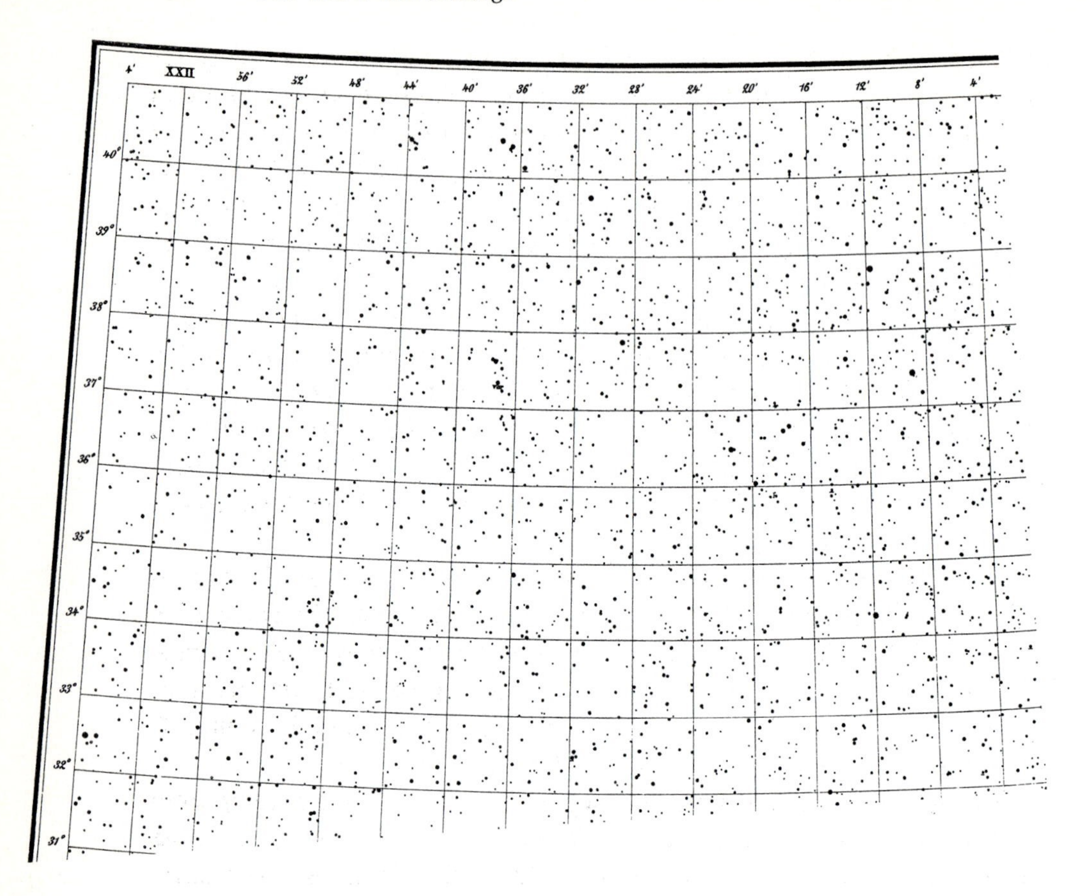

Fig. 3.4. A portion of BD Chart No. 23.

extends the coverage all the way to a declination of −90°, but the magnitude limit is 10.0 rather than 9.5 as in the earlier BD.

One should notice that the BD catalog also contains a column for the magnitudes of the stars. These magnitudes are estimates made by the observer at the time when each star crossed the meridian, and they should not be thought of as having been determined with great precision.

Because of the extensive use which astronomers still find for the *Bonner Durchmusterung*, we shall devote a few paragraphs to the specific ways in which it is used. The astronomer might be given the BD number of a star and have a need to identify it on a photograph, or the astronomer might wish to find the BD number of a star as a means of identifying it. Let us see what steps are involved in the first case for the star BD +39 4612.

Step 1. Find the catalog entry for this star. We locate the section of the BD catalog in which the stars of declination zone +39° are found. This will be Volume 2 from page 326 to page 344. Within these pages we locate the page on which star 4612 is found. Notice from Fig. 3.3 that each page is divided into five columns separated by heavy rulings and that each column contains the entries for sixty stars. The two numbers at the top of each column are the numbers of the first and last entries in that column. BD+39 4612 will be the twelfth entry in the first column on page 342, and it has been marked in Fig. 3.3.

Step 2. Now find the chart on which this star will be located. The numbers at the top right of catalog page 342 tell us that the stars on that page all have right ascensions of 21 hours or 22 hours. (The letter 'u' is the abbreviation for the german word for hour – uhre.) We need to find the BD chart which includes the zone at +39° and right ascensions near 21 hours. This will be Chart No. 23, and a portion of it has been reproduced in Fig. 3.4.

Step 3. To find BD +39 4612 on the chart we note that from the catalog the right ascension is 21 hours, 31 minutes, 8.7 seconds. The declination is +39° 45.8′. Referring to the chart we see that lines have been drawn at intervals of four minutes in RA. BD +39 4612 must be in the area bounded in RA by the lines at 21 hours, 28 minutes and 21 hours, 32 minutes and in Dec by the lines at +39° and +40°.

Step 4. Within the block defined above there are 23 stars running from 4595 to 4617. Since the numbers increase toward the east, 4612 is the sixth star from the eastern or righthand side of the block. We may locate the star by counting the first six stars from east to west (right to left) within this block. Confusion can arise because some of the blocks are quite crowded, so it is a good idea to note the declination and magnitude of each star as one counts. In this particular example the star in question, BD +39 4612, is the brightest star in the block.

The reverse case is that in which one begins with a star and tries to find its BD number. The astronomer must first find the star on the chart and then find its number in the catalog by counting stars in one of the defined blocks.

Even after more than 100 years astronomers still find that the catalogs and charts of the *Bonner Durchmusterung* are extremely useful. Students are advised to practice using these materials until they feel confident and can proceed quickly to an identification.

National Geographic/Palomar Sky Survey

The two-hundred-inch telescope at Mt. Palomar was put into service in the late 1940s, and at about the same time a forty-eight-inch

Schmidt type telescope was installed nearby. The Schmidt had the ability to photograph a very large field in a relatively short time. It had been suggested that objects found on plates made with the Schmidt could be studied in great detail with the two-hundred-inch. Early results with the Schmidt showed sharply defined images of stars right out to the edges of plates which covered an area of 6.6 by 6.6 degrees. This early success prompted astronomers at Mt. Palomar to undertake a complete survey of that portion of the sky accessible from the latitude of the observatory. Financial support was provided by the National Geographic Society, and the extensive program was begun in 1946.

With the completion and distribution of the *Palomar Sky Survey* (PSS) in 1954 astronomers acquired an unexpectedly useful tool to help in the identification of faint stars. In the course of this survey the northern sky from the pole to a declination of −27° was photographed twice: once in blue light and once in red light. Altogether there are 879 separate fields, and stars brighter than magnitude 20 are recorded. In about 1970 more photographs were completed to extend the coverage to declination −42°. A small portion of a pair of prints from the PSS has been included here as Fig. 3.5.

In the problem of star identification the prints of the *Palomar Sky Survey* cannot stand alone due to the fact that they contain no identifying marks. It is, however, quite simple to compare the prints to charts such as those of the BD in order to establish the locations of faint objects. The entire astronomical community was impressed with the usefulness of the massive piece of work, and the survey has become an invaluable tool.

Fig. 3.5. A sample field of stars from the *Palomar Sky Survey*. The photograph made with blue-sensitive materials is on the left. The colors of stars are indicated by the relative brightness in the two photographs. © 1960 National Geographic Society/Palomar Sky Survey. Reproduced by permission of the California Institute of Technology.

Ohio State overlays for the PSS

In 1981 the usefulness of the Palomar survey was greatly enhanced by the publication of a set of transparent overlays which bear the identifications for thousands of astronomical objects. A portion of one of these overlays has been reproduced in Fig. 3.6. The complete set of 1037 overlays was produced by Robert S. Dixon and his associates at the Ohio State University. Stars, galaxies, interstellar clouds, radio sources and more are included on the overlays which have turned out to be remarkably easy to use. This great piece of work was produced by storing all of the needed catalogs in the memory of a computer and then programming the computer to plot each of the individual maps. The transparent overlays were then reproduced from the plots made by the computer.

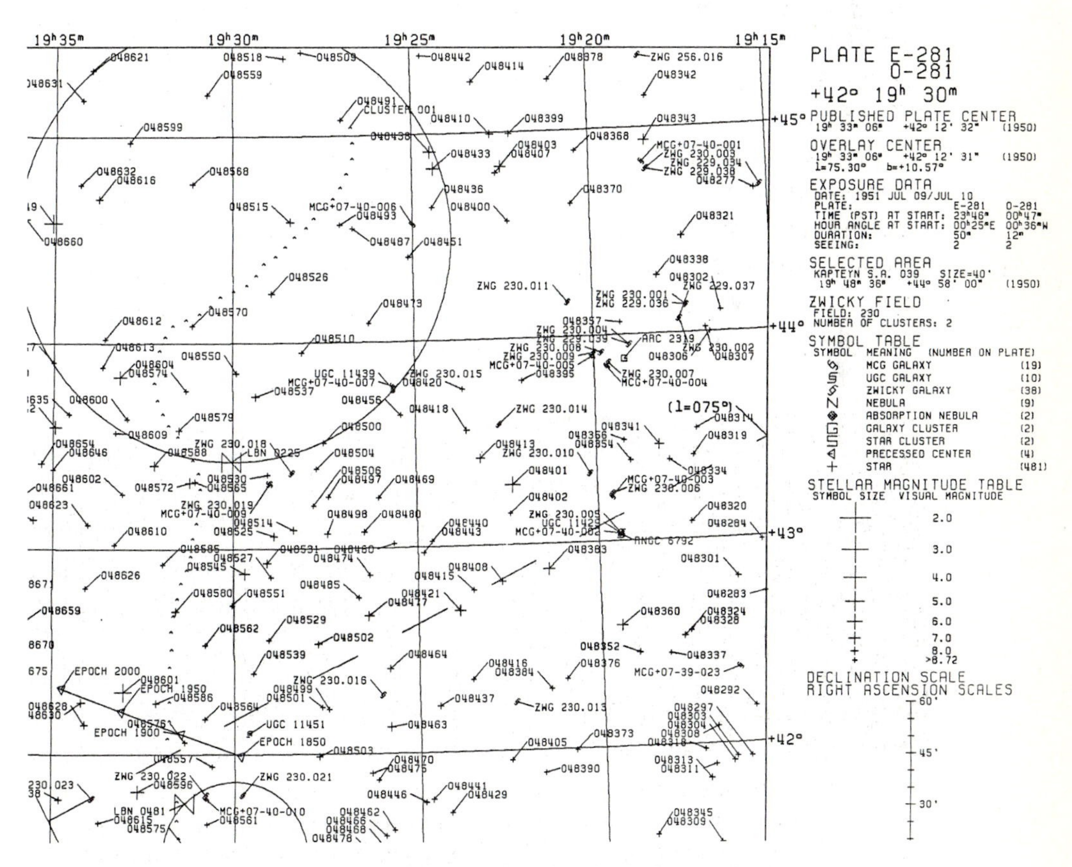

Fig. 3.6. A section of one of the overlays for the *Palomar Sky Survey* (Robert S. Dixon, Ohio State University Radio Observatory).

Stars are represented by the symbol '+', and the size of the symbol indicates the brightness of the star. The numbers next to the + symbols refer to entries in the SAO Catalog to be described below.

SAO Catalog and Atlas

In 1966 an extensive catalog and atlas were compiled and published by astronomers at the Smithsonian Astrophysical Observatory. 258 997 stars are listed in the *SAO Catalog*, and the limiting magnitude is 10 though there are a few stars of magnitude 11 and 12. A sample page from this catalog has been reproduced in Fig. 3.7 so that one may see the large amount of information contained in it. Numbers have been assigned consecutively in zones ten degrees wide in declination. The coordinates are based on the equator of 1950. The cross-listings in the columns on the right of the page are especially valuable for those who are seeking to identify a particular star, due to the fact that the Dixon overlays for the PSS show the locations of all of the stars in the *SAO Catalog*. For example, if one is given the BD number of a star, one may readily find that star in the column headed 'DM Number' in the *SAO Catalog*. This leads to the SAO number for that star. One may then go to the overlays for the PSS to identify the star on the print from the PSS.

Machine-readable catalogs

In a world-wide cooperative effort astronomers have undertaken the transfer of many standard catalogs to magnetic tape or computer memories. This makes the data more readily available in many practical situations. Under the sponsorship of the International Astronomical Union a center for the storage and distribution of catalogs in this new form was established at Strasburg in France. In the United States a similar center was established by the National Aeronautics and Space Administration. The list of commonly used catalogs now available on magnetic tape is quite extensive.

We should point out here that some of the charts and catalogs described in the paragraphs above are not easily accessible to many people who might find them useful. Neither the *Bonner Durchmusterung* nor the *National Geographic/Palomar Sky Survey* can be purchased from stock. Those observatories which own copies handle them with great care, for they are indispensable in many situations. Fortunately, there are microfilm copies of the BD charts, and there are other charts and catalogs

051100			EPOCH 1950							ORIGINAL EPOCH						SP.	SOURCE		+40° 21H
NUMBER	MAGNITUDES m_{pg}	m_v	α_{1950} h m s	μ s	σ_{μ} ".001	δ_{1950} ° ' "	μ' "	$\sigma_{\mu'}$ ".001	σ_{1950} "	α_2 s	σ ".01	ep.	δ_2 "	σ' ".01	ep.	SP.	CAT.	STAR NUMBER	DM NUMBER
1		5.1H	21 34 56.508	-0.0005	01	40 11 17.55	0.017	01	0.04	56.522	02	20.2	17.01	02	17.9	A5	F4	811	39 4612
2	10.7	8.9	34 56.664	0.0035 A	26	40 35 31.22	-0.009	15	0.47	56.591	17	29.7	31.39	17	29.7		AG	2232	40 4582
3	9.4	8.8	35 0.603	-0.0024 A	27	44 44 43.30	-0.009	16	0.49	0.652	16	29.6	43.49	16	29.6		AG	1954	44 3885
4	10.3	8.8	35 6.275	0.0026 A	29	44 58 14.04	-0.012	17	0.51	6.222	17	29.6	14.29	17	29.6		AG	1955	44 3886
5	9.0	8.2	35 8.728	0.0047 A	27	47 13 3.88	-0.024	16	0.49	8.633	16	29.6	4.38	16	29.6	G0	AG	1773	46 3389
6	9.1	8.4	35 26.171	-0.0035 A	24	41 4 6.09	-0.029	15	0.43	26.241	16	29.7	6.69	16	29.7	K0	AG	2083	40 4584
7	10.8	8.4	35 26.796	0.0016 A	24	48 11 50.41	0.006	15	0.45	26.763	17	29.6	50.29	17	29.6		AG	1792	47 3500
8	9.7	9.1	35 29.108	0.0008 A	24	41 53 36.37	0.018	14	0.43	29.091	16	29.7	35.99	16	29.7		AG	2084	41 4212
9		6.1T	35 32.681	-0.0004	07	44 28 16.95	-0.024	05	0.34	32.705	09	96.3	18.28	09	95.5	A3	GC	30278	44 3889
10	8.9	8.9	35 35.196	-0.0007 A	25	43 49 41.37	-0.001	14	0.45	35.211	16	29.6	41.38	16	29.6	A0	AG	2004	43 3989
11	9.5	8.9	35 43.993	0.0025 A	27	45 47 24.05	-0.002	16	0.48	43.942	16	29.6	24.08	16	29.6		AG	1898	45 3609
12	9.7	8.9	35 47.494	-0.0024 A	28	47 53 23.72	0.011	16	0.49	47.543	16	29.6	23.49	16	29.6		AG	1775	47 3501
13	9.2	8.7	35 48.763	0.0050 A	23	40 37 1.97	-0.001	14	0.41	48.661	16	29.7	1.99	16	29.7		AG	2233	40 4587
14	11.3	9.4	35 49.796	0.0017 A	28	43 37 52.43	-0.017	16	0.50	49.761	18	29.6	52.78	18	29.6		AG	2005	43 3990
15	9.2	8.8	35 52.547	0.0023 A	24	42 44 57.30	-0.015	14	0.42	52.501	16	29.7	57.59	16	29.7		AG	2051	42 4162
16	9.8	8.4	35 53.623	-0.0000 A	27	46 30 56.02	-0.003	15	0.47	53.623	16	29.6	56.08	16	29.6	K2	AG	1881	46 3396
17	9.9	8.9	35 56.247	-0.0012 A	26	45 57 35.68	-0.113	15	0.46	56.272	16	29.6	37.98	16	29.6		AG	1899	45 3611
18	8.6	8.7	35 59.478	0.0018 A	26	41 39 17.58	-0.005	15	0.46	59.441	16	29.7	17.69	16	29.7	A0	AG	2085	B* 41 4213
19	10.7	9.0	36 0.462	-0.0020 A	26	47 10 4.45	0.003	15	0.46	0.503	17	29.6	4.38	17	29.6		AG	1777	46 3397
20		7.4T	36 1.400	-0.0010	07	40 51 23.47	-0.052	06	0.35	1.449	17	03.6	26.19	17	97.4	K5	GC	30290	40 4589
21	10.7	9.4	36 5.961	0.0005 A	15	40 14 2.24	-0.002	11	0.32	5.951	17	29.7	2.29	17	29.7		AG	2235	39 4621
22	9.5	8.9	36 6.825	-0.0003 A	24	42 44 9.87	-0.016	14	0.44	6.831	16	29.7	10.19	16	29.7		AG	2052	B* 42 4163
23	7.8	8.3	36 9.535	0.0006 A	26	48 15 40.99	0.010	15	0.47	9.523	17	29.6	40.79	17	29.6	B8	AG	1793	B* 47 3505
24	10.3	9.6	36 13.143	0.0005 A	31	48 14 9.71	0.006	18	0.54	13.133	17	29.6	9.59	17	29.6		AG	1794	47 3506
25	8.4	8.4	36 18.002	0.0020 A	26	40 44 19.01	0.001	15	0.46	17.962	16	29.7	18.99	16	29.7	A2	AG	2236	40 4591
26	9.8	8.8	36 21.452	-0.0005 A	25	41 18 42.02	-0.013	15	0.45	21.462	16	29.7	42.29	16	29.7		AG	2086	40 4592
27	8.1	7.6	36 30.049	-0.0001 A	28	43 3 52.20	0.005	16	0.49	30.051	16	29.6	52.09	16	29.6	K0	AG	2006	B* 42 4164
28	9.4	8.2	36 33.979	-0.0007 A	23	47 31 33.42	0.036	14	0.42	33.993	16	29.7	32.69	16	29.7		AG	1780	47 3509
29	8.2	8.2	36 34.789	0.0013 A	26	46 3 14.49	0.035	15	0.46	34.763	16	29.7	13.78	16	29.7	A2	AG	1882	45 3613
30	11.2	9.1	36 37.718	0.0022 A	26	48 8 37.65	0.017	16	0.47	37.673	18	29.6	37.29	18	29.6		AG	1796	47 3512
31	10.5	8.6	36 37.838	0.0009 A	21	42 44 2.18	-0.006	13	0.40	37.821	17	29.7	2.29	17	29.7		AG	2053	42 4165
32	8.7	8.6	36 37.986	0.0016 A	23	49 12 38.65	0.008	14	0.41	37.953	16	29.6	38.49	16	29.6	A0	AG	1839	48 3452
33	8.9	7.6	36 39.393	-0.0020 A	17	48 29 14.60	-0.014	11	0.34	39.433	16	29.6	14.89	16	29.6	K5	AG	1797	48 3451
34		7.1T	36 40.849	-0.0017	07	49 34 9.63	-0.001	05	0.37	40.933	17	99.2	9.67	15	92.4	G5	GC	30306	49 3579
35	9.4	8.7	36 43.519	0.0003 A	26	46 36 10.62	-0.003	15	0.45	43.513	16	29.7	10.68	16	29.7		AG	1883	46 3399
36	9.0	8.6	36 52.596	0.0002 A	27	43 12 28.88	-0.006	16	0.47	52.591	16	29.7	28.99	16	29.7		AG	2007	42 4169
37	8.6	8.4	36 58.561	-0.0010 A	27	43 21 16.10	-0.009	16	0.47	58.581	16	29.7	16.29	16	29.7		AG	2008	42 4170
38	10.9	8.9	36 58.990	0.0024 A	25	43 31 38.02	-0.003	16	0.46	58.941	18	29.7	38.09	18	29.7		AG	2009	43 3993
39	10.6	9.1	37 0.010	0.0029 A	27	41 37 5.43	-0.013	16	0.48	-59.951	17	29.7	5.69	17	29.7		AG	2087	41 4220
40	8.9	8.9	37 0.737	0.0003 A	24	43 39 48.55	0.038	15	0.43	0.731	16	29.7	47.78	16	29.7		AG	2010	43 3994
41	9.8	9.0	37 5.562	0.0044 A	26	45 45 20.15	0.023	15	0.46	5.473	16	29.6	19.68	16	29.6	A	AG	1900	45 3615
42	9.5	8.7	37 6.094	0.0001 A	26	40 35 49.14	0.003	15	0.45	6.091	16	29.7	49.09	16	29.7		AG	2237	40 4598
43	9.3	8.4	37 9.759	-0.0011 A	26	44 47 15.73	-0.013	15	0.45	9.782	16	29.7	15.99	16	29.7		AG	1958	44 3892
44	9.8	8.9	37 11.373	-0.0015 A	29	49 26 21.53	0.017	17	0.51	11.403	16	29.6	21.19	16	29.6		AG	1842	49 3583
45	9.2	8.7	37 13.487	-0.0013 A	26	49 45 22.71	-0.019	16	0.46	13.514	16	29.6	23.09	16	29.6		AG	1843	49 3584
46	8.8	8.7	37 15.928	0.0004 A	21	43 57 28.15	0.003	13	0.39	15.921	16	29.7	28.09	16	29.7	A5	AG	2012	B* 43 3996
47	9.0	8.4	37 16.878	0.0082 A	14	40 7 6.67	0.009	11	0.30	16.711	16	29.7	6.49	16	29.7		AG	2238	39 4625
48	8.8	8.5	37 18.811	-0.0000 A	26	43 48 43.48	-0.000	15	0.45	18.811	16	29.7	43.48	16	29.7		AG	2013	43 3997
49	8.8	8.4	37 19.232	-0.0039 A	29	41 47 51.36	0.009	17	0.50	19.311	16	29.7	51.19	16	29.7	F0	AG	2088	41 4222
50	7.7	8.2	37 21.262	-0.0000 A	26	45 56 54.93	0.003	15	0.46	21.262	17	29.7	54.88	17	29.7	B9	AG	1901	45 3618
51	7.7	8.1	21 37 24.284	0.0021 A	19	40 24 37.11	0.011	11	0.36	24.241	17	29.7	36.89	17	29.7	A0	AG	2239	39 4627
52	9.6	9.1	37 29.176	0.0013 A	23	42 19 30.85	-0.066	17	0.45	29.151	16	29.7	32.19	16	29.7		AG	2054	41 4223
53	8.5	8.0	37 29.238	0.0013 A	22	41 29 58.42	-0.028	17	0.43	29.212	16	29.7	58.99	16	29.7	K0	AG	2089	B* 41 4224
54	8.7	8.9	37 30.538	-0.0036 A	26	43 3 4.65	0.018	15	0.46	30.611	16	29.6	4.29	16	29.6	A2	AG	2014	B* 42 4172
55	9.1	8.7	37 31.028	-0.0006 A	21	41 30 19.69	0.000	12	0.38	31.041	16	29.7	19.69	16	29.7	K0	AG	2090	B* 41 4225
56	10.1	8.7	37 37.726	0.0017 A	26	44 57 22.16	-0.021	15	0.46	37.692	16	29.7	22.59	16	29.7	M8	AG	1959	44 3895
57	8.4	8.4	37 42.635	0.0066 A	25	44 12 55.02	0.051	15	0.45	42.501	16	29.7	53.98	16	29.7	F5	AG	1961	43 3999
58	8.3	7.9	37 45.591	-0.0001 A	25	46 58 24.06	0.019	15	0.45	45.593	16	29.7	23.68	16	29.7	K0	AG	1887	B* 46 3403
59	8.7	8.8	37 48.649	0.0004 A	24	41 21 52.44	0.017	14	0.42	48.642	16	29.7	52.09	16	29.7	A2	AG	2091	40 4601
60	10.3	9.2	37 56.298	0.0012 A	27	48 18 46.10	0.005	15	0.47	56.273	17	29.6	45.99	17	29.6		AG	1798	B* 47 3518
61		6.7T	37 57.869	0.0019	07	44 12 17.24	-0.004	05	0.31	57.785	12	05.1	17.46	12	99.9	B9	GC	30334	43 4002
62	9.2	8.7	38 6.136	0.0002 A	27	49 1 17.87	-0.015	16	0.47	6.133	16	29.6	18.19	16	29.6		AG	1844	48 3458
63		7.3T	38 6.188	0.0015	36	48 54 19.46	0.004	22	1.12	6.135	22	14.0	19.30	22	11.9	K2	GC	30335K	B* 48 3457
64	9.9	9.2	38 8.637	0.0016 A	27	48 13 9.46	0.018	15	0.47	8.803	16	29.6	9.09	16	29.6		AG	1801	47 3520
65	8.8	8.8	38 9.303	0.0014 A	29	49 59 16.70	-0.010	17	0.51	9.274	16	29.6	16.89	16	29.6	A2	AG	1845	49 3589
66	8.7	8.8	38 10.637	0.0007 A	21	46 2 39.71	0.016	13	0.39	10.623	16	29.7	39.38	16	29.7	A0	AG	1888	45 3625
67		5.4T	38 13.144	0.0049	04	43 2 46.23	0.022	02	0.20	12.857	07	91.9	45.00	06	92.5	K5	GC	30338K	B* 42 4177
68		7.2T	38 22.873	-0.0010	07	49 27 20.52	0.021	06	0.37	22.930	14	95.3	19.27	12	91.1	B9	GC	30343	49 3590
69	10.7	8.8	38 22.901	-0.0025 A	22	43 43 7.05	-0.006	14	0.41	22.951	17	29.7	7.18	17	29.7		AG	2017	43 4004
70	10.4	9.1	38 26.602	0.0011 A	22	43 39 23.80	0.020	14	0.41	26.581	17	29.7	23.38	17	29.7		AG	2018	43 4005
71	8.9	8.9	38 35.747	0.0018 A	20	40 53 45.35	0.008	12	0.37	35.711	16	29.7	45.19	16	29.7		AG	2240	40 4605
72	9.5	9.1	38 41.132	-0.0000 A	27	49 14 33.80	0.010	16	0.47	41.133	16	29.6	33.59	16	29.6		AG	1847	48 3461
73	9.5	9.0	38 46.443	0.0015 A	28	45 5 0.75	-0.021	15	0.45	46.412	16	29.7	1.18	16	29.7		AG	1903	44 3901
74	9.4	8.9	38 47.136	0.0037 A	20	43 59 4.82	0.036	13	0.38	47.061	16	29.7	4.08	16	29.7		AG	2019	43 4006
75	9.1	9.0	38 56.700	-0.0030 A	26	42 46 44.64	-0.008	15	0.45	56.761	16	29.7	44.79	16	29.7		AG	2055	42 4178
76	9.8	8.9	39 5.655	-0.0008 A	24	41 53 6.63	-0.038	14	0.43	5.671	16	29.7	7.39	16	29.7		AG	2092	41 4231
77	8.9	8.5	39 6.900	0.0058 A	27	48 58 14.77	-0.016	15	0.47	6.783	16	29.6	15.09	16	29.6		AG	1802	48 3464
78	8.0	8.4	39 11.686	0.0012 A	21	41 3 7.39	0.025	12	0.38	11.662	16	29.7	6.89	16	29.7	A0	AG	2093	40 4608
79	8.5	8.0	39 13.575	0.0001 A	26	47 36 2.24	-0.022	15	0.47	13.573	16	29.7	2.69	16	29.7	K0	AG	1784	47 3522
80	9.7	9.1	39 15.468	0.0014 A	24	42 25 21.58	0.019	14	0.43	15.441	16	29.7	21.19	16	29.7		AG	2057	41 4232
81	9.7	9.0	39 15.653	-0.0033 A	18	42 44 20.19	-0.079	13	0.36	15.721	16	29.7	21.79	16	29.7		AG	2058	42 4179
82		7.4T	39 17.923	0.0012	22	47 19 0.86	-0.010	18	0.80	17.878	19	13.0	1.24	21	11.1	K2	GC	30369	46 3407
83	9.7	9.3	39 20.048	0.0018 A	36	44 47 42.30	-0.014	21	0.63	20.012	16	29.7	42.59	16	29.7		AG	1963	44 3903
84	9.5	7.9	39 25.300	-0.0001 A	23	48 6 18.81	0.070	14	0.42	25.303	16	29.7	17.39	16	29.7		AG	1803	B* 47 3523
85	9.8	9.0	39 25.979	0.0004 A	26	41 31 54.48	-0.005	15	0.45	25.971	16	29.7	54.59	16	29.7		AG	2094	41 4234
86	10.1	9.0	39 27.112	0.0015 A	26	44 14 43.10	0.020	15	0.46	27.081	16	29.7	42.68	16	29.7		AG	1964	43 4010
87	9.3	9.2	39 29.207	0.0012 A	24	40 34 1.78	0.014	14	0.43	29.182	16	29.7	1.49	16	29.7		AG	2242	B* 40 4610
88	8.7	8.8	39 29.728	-0.0011 A	21	44 16 6.80	0.020	13	0.39	29.751	16	29.7	6.38	16	29.7	A5	AG	1965	43 4011
89		6.0T	39 33.408	-0.0013	03	40 34 39.12	-0.042	02	0.18	33.471	09	01.3	41.53	07	92.1	A0	GC	30376	B* 40 4611
90	10.1	8.8	39 34.507	0.0007 A	27	49 7 24.02	0.026	16	0.47	34.493	16	29.6	23.49	16	29.6		AG	1848	B* 48 3467
91	9.6	8.7	39 37.146	-0.0027 A	24	44 48 53.77	-0.011	15	0.43	37.202	16	29.7	53.99	16	29.7		AG	1966	44 3904
92	10.5	9.4	39 37.813	-0.0009 A	18	40 19 21.17	-0.006	10	0.28	37.831	17	29.7	21.29	17	29.7		AG	2244	39 4641
93	9.6	8.9	39 38.001	-0.0025 A	26	43 1 42.45	0.023	15	0.45	38.051	16	29.7	41.99	16	29.7		AG	2020	42 4181
94	8.7	8.7	39 47.098	0.0013 A	26	43 43 32.39	-0.005	15	0.45	47.071	16	29.7	32.48	16	29.7		AG	2021	43 4012
95	9.7	8.8	39 47.497	0.0003 A	25	44 3 13.37	0.009	15	0.45	47.491	16	29.7	13.18	16	29.7		AG	1967	43 4013
96	8.1	7.6	39 47.501	0.0034 A	26	40 17 49.47	-0.011	15	0.46	47.431	16	29.7	49.69	16	29.7	A0	AG	2245	39 4642
97	8.9	8.8	39 47.599	0.0028 A	23	44 7 20.19	0.010	14	0.42	47.541	16	29.7	19.98	16	29.7		AG	1968	43 4014
98	10.4	8.5	39 53.639	0.0003 A	27	48 58 13.25	-0.012	15	0.47	53.633	17	29.6	13.49	17	29.6		AG	1805	48 3470
99	9.5	8.7	39 53.951	0.0009 A	26	45 19 4.43	-0.017	15	0.45	53.932	16	29.7	4.78	16	29.7		AG	1904	44 3905
00	8.8	8.0	39 56.507	-0.0057 A	27	49 4 18.66	-0.105	16	0.47	56.623	16	29.6	20.79	16	29.6	K0	AG	1849	48 3471

Fig. 3.7. A sample page from the *SAO Catalog* (Smithsonian Astrophysical Observatory).

which are readily available and are much less costly. We mention below only three more.

Atlas Coeli Skalnate Pleso

Both readily available and quite useful is the *Atlas Coeli* published by Antonín Bečváăr of Czechoslovakia in 1948 and revised in 1958. All of the bright stars are identified and so are large numbers of variable stars, gaseous nebulae and galaxies. The limiting magnitude is 7.75. The atlas is accompanied by a catalog which contains data on all of the stars and other objects shown on the charts.

Tirion Atlas and Sky Catalog 2000.0

Fig. 3.1 was reproduced from the *Tirion Atlas* published in 1982. This atlas is similar to the *Atlas Coeli* described above, but it has been drawn with all objects precessed to the year 2000. The overall size of the charts is smaller than in the case of the *Atlas Coeli*, and this makes it more convenient to use at the telescope than its predecessor.

A catalog to accompany the *Tirion Atlas* was published separately in 1982 but was designed to be fully compatible with the atlas. Known as *Sky Catalog 2000.0*, this volume contains 45 269 entries, and a sample page has been displayed here as Fig. 3.8. This atlas and catalog are readily available at reasonable cost.

Uranometria 2000.0

The most recent major atlas is *Uranometria 2000.0* which was published in 1987. Part 1 consists of two hundred and fifty-nine charts which display all stars and many other objects brighter than magnitude 9.5 and north of declination −6°. Part 2 continues the coverage to the south celestial pole. The opening pages of Part 1 give a detailed and interesting history of the making of celestial maps and globes. The charts are accurate and easy to use. They contain most of the stars in the catalogs which were mentioned earlier in this chapter.

A three-volume catalog to accompany the two atlas volumes is in preparation at the time of this writing, and it will give cross-references to the BD and to other catalogs.

As an exercise in the use of the materials described here, the reader should be able to begin with the star BD +39 4612 and find it on all of the

HD	SAO	Star Name	α 2000	δ 2000	μ(α)	μ(δ)	V	B−V	M_V	Spec	RV	d(pc)	ADS	Notes
			h m s	° ′ ″	s	″								
205555	126917		21 35 58.8	+6 04 26	+0.001	−0.01	7.8	0.4	2.6	F0 V		110 s		
205730	51079		21 36 02.2	+45 22 29	+0.004	+0.01	5.53	1.58		Mc	−14			W Cyg, v
205603	107262		21 36 04.9	+15 04 56	−0.006	−0.07	6.7	0.9	3.2	G5 IV	0	50 s		
205732	51082		21 36 08.3	+41 55 56	−0.001	−0.01	7.7	0.9	3.2	G5 IV		80 s		
205697	71504		21 36 08.6	+37 52 35	0.000	+0.01	7.6	0.1	0.6	A0 V		250 s		
205947	19525		21 36 09.0	+63 42 03	+0.002	0.00	7.9	0.1	0.6	A0 V		260 s		
205795	33567		21 36 09.5	+50 30 08	+0.002	+0.03	7.17	0.00	0.4	B9.5 V		210 s	15135	m
205471	190478	8 PsA	21 36 10.8	−26 10 17	+0.008	−0.02	5.73	0.22		A m				m
205716	71507		21 36 12.7	+38 46 30	+0.005	−0.01	8.0	1.1	−0.1	K2 III		270 mx		
205808	33568		21 36 13.1	+50 40 46	0.000	+0.01	7.26	−0.10	0.4	B9.5 V		240 s		
205688	89834		21 36 13.7	+30 03 20	−0.005	+0.07	6.50	0.9	1.8	G8 III-IV	−20	80 mx	15126	m
205584	126918		21 36 13.9	+6 08 14	+0.001	−0.01	7.72	1.26		K2				
205619	126920		21 36 16.7	+9 47 05	−0.002	−0.02	8.0	1.1	0.2	K0 III		360 s		
205462	213128		21 36 20.0	−35 10 50	+0.001	−0.05	7.7	2.1	−0.3	K5 III		170 s		
205527	164508		21 36 22.0	−18 23 31	−0.002	+0.01	7.7	0.4	3.0	F2 V		87 s		
205448	230748		21 36 23.0	−40 19 49	−0.001	−0.01	7.1	1.9	−0.3	K5 III		190 s		
205605	145520		21 36 32.4	−5 03 09	0.000	−0.01	7.0	1.1	0.2	K0 III		230 s		
205744	71514		21 36 32.8	+33 14 04	−0.002	−0.01	7.3	1.1	0.2	K0 III		260 s		
205733	71513		21 36 33.6	+32 06 09	+0.002	+0.02	7.40	1.6	−0.5	M4 III e	−7	360 s		AB Cyg, v
205895	33572		21 36 34.5	+54 49 03	+0.001	0.00	7.90	0.2	2.1	A5 V		140 s	15140	m
205745	71515		21 36 35.3	+32 25 42	−0.001	−0.01	7.7	1.1	0.2	K0 III		300 s		
205606	145521		21 36 36.4	−9 02 48	0.000	−0.02	8.0	1.1	0.2	K0 III		360 s		
205391	247108		21 36 38.9	−51 24 01	+0.001	+0.01	7.7	1.8	−0.1	K2 III		150 s		
205390	247109		21 36 41.1	−50 50 46	+0.045	−0.22	7.15	0.88	6.3	K2 V		15 ts		
205690	107275		21 36 43.1	+11 52 11	0.000	+0.01	7.9	1.1	−0.1	K2 III		390 s		
205577	190487		21 36 43.6	−21 30 08	+0.001	−0.01	7.8	1.0	0.2	K0 III		330 s		
205780	71522		21 36 47.6	+32 42 13	+0.003	−0.01	8.0	0.5	3.4	F5 V		81 s		
205529	213136	7 PsA	21 36 48.7	−33 02 53	+0.007	0.00	6.11	0.22	2.4	A7 V n		55 s		
205702	126926		21 36 53.3	+5 48 53	−0.005	−0.11	7.7	0.5	4.0	F8 V		55 s		
205634	164519		21 36 55.0	−10 10 27	+0.004	+0.02	7.4	0.4	2.6	F0 V		90 s		
205835	51101	74 Cyg	21 36 56.8	+40 24 49	0.000	+0.02	5.01	0.18		A5	+7	22 mn		
205949	33576		21 36 57.2	+54 38 17	−0.001	+0.01	7.9	0.1	0.6	A0 V		260 s		
205530	213139		21 36 57.9	−35 53 05	+0.008	−0.03	7.4	0.6	3.0	F2 V		53 s		
205624	164516		21 36 58.6	−18 44 40	+0.006	−0.07	7.6	0.6	−0.1	K2 III		160 mx		
205489	230754		21 36 59.6	−45 43 35	+0.002	−0.02	8.0	1.3	0.2	K0 III		250 s		
205588	190489		21 37 00.0	−28 36 35	0.000	−0.03	7.5	0.9	0.2	K0 III		300 s		
205762	107280		21 37 03.0	+19 46 57	0.000	−0.06	7.3	1.1	0.2	K0 III		260 s		
205637	164520	39 ε Cap	21 37 04.7	−19 27 58	+0.001	+0.01	4.68	−0.17	−2.3	B3 IV p	−24	250 s		m
205984	33579		21 37 05.9	+55 07 04	+0.002	+0.02	8.0	0.1	0.6	A0 V		280 s		
205746	107279		21 37 06.3	+11 43 09	+0.001	+0.01	7.2	0.1	0.6	A0 V	+12	210 s		
205950	33578		21 37 06.4	+53 22 36	−0.001	−0.02	7.6	0.1	0.6	A0 V		230 s		
206078	19528		21 37 10.2	+62 18 16	+0.005	+0.13	7.13	0.97	0.3	G8 III	−75	230 s		
205734	126930		21 37 14.1	+5 51 05	+0.001	−0.02	8.0	0.4	2.6	F0 V		120 s		
205965	33583		21 37 15.3	+51 21 49	−0.001	0.00	7.68	1.58	−0.3	K5 III		350 s		
205966	33586		21 37 17.3	+51 03 53	−0.003	+0.01	7.18	1.87	−0.4	M0 III	−23	240 s		
205693	164525		21 37 19.8	−11 27 38	0.000	+0.01	7.5	1.6	−0.5	M2 III		400 s		
205674	164524		21 37 20.7	−18 26 28	+0.002	−0.10	7.1	0.5	3.4	F5 V		55 s		
205880	71531		21 37 21.4	+35 19 38	−0.001	−0.05	7.2	1.1	0.2	K0 III		250 s		
205881	71532		21 37 23.7	+34 36 11	+0.005	+0.04	7.9	0.7	4.4	G0 V		49 s		
205939	51109		21 37 27.7	+44 41 47	0.000	−0.02	6.1	0.1	1.7	A3 V	+4	76 s		CP Cyg, q

Fig. 3.8. Part of a page from Sky Catalog 2000.0. (Copyright 1982 Sky Publishing Corp. Reproduced by permission.)

charts and catalogs which have been included. A number of different names or designations have been used, but the reader may make all of the necessary cross-references by going from one catalog to another. It is interesting to note the effect of precession over the years by comparing the coordinates in the *BD Catalog* to those in *Sky Catalog 2000.0*.

Astronomical Almanac

Another essential book for the observational astronomer is the *Astronomical Almanac* published annually by the U.S. Naval

Observatory in the United States and the Royal Greenwich Observatory in the United Kingdom, and a copy should be available in every observatory. Formerly known as the *American Ephemeris and Nautical Almanac*, this volume contains tables of the positions of the sun, moon and planets for every day of the year. We have already pointed out that it also contains the data from which the sidereal time may be calculated. Other tables list the positions of the satellites of the planets and the positions of a selected list of the brightest stars. Portions of two pages containing data on the sun are included here as Fig. 3.9.

The discussion in this chapter has been limited to charts and catalogs which are useful in the process of the identification of celestial objects.

C4

SUN, 1990

FOR 0^h DYNAMICAL TIME

Date		Julian Date	Ecliptic Long. (for Mean Equinox of Date)	Ecliptic Lat. (for Mean Equinox of Date)	Apparent Right Ascension	Apparent Declination	True Geocentric Distance
		244	° ′ ″	″	h m s	° ′ ″	
Jan.	0	7891.5	279 17 12.54	−0.12	18 40 24.73	−23 07 02.4	0.983 3553
	1	7892.5	280 18 22.80	+0.02	18 44 49.99	23 02 32.3	.983 3361
	2	7893.5	281 19 32.93	+0.16	18 49 14.92	22 57 34.6	.983 3210
	3	7894.5	282 20 42.85	+0.28	18 53 39.49	22 52 09.5	.983 3103
	4	7895.5	283 21 52.49	+0.39	18 58 03.68	22 46 17.1	.983 3042
	5	7896.5	284 23 01.79	+0.47	19 02 27.45	−22 39 57.6	0.983 3030
	6	7897.5	285 24 10.71	+0.51	19 06 50.78	22 33 11.1	.983 3072
	7	7898.5	286 25 19.22	+0.53	19 11 13.64	22 25 58.1	.983 3169
	8	7899.5	287 26 27.33	+0.51	19 15 36.01	22 18 18.5	.983 3327
	9	7900.5	288 27 35.04	+0.45	19 19 57.86	22 10 12.8	.983 3546
	10	7901.5	289 28 42.38	+0.37	19 24 19.16	−22 01 41.1	0.983 3829
	11	7902.5	290 29 49.37	+0.27	19 28 39.91	21 52 43.6	.983 4178
	12	7903.5	291 30 56.07	+0.14	19 33 00.07	21 43 20.6	.983 4592
	13	7904.5	292 32 02.49	+0.01	19 37 19.64	21 33 32.4	.983 5071
	14	7905.5	293 33 08.67	−0.12	19 41 38.58	21 23 19.2	.983 5613
	15	7906.5	294 34 14.62	−0.24	19 45 56.89	−21 12 41.2	0.983 6216
	16	7907.5	295 35 20.33	−0.36	19 50 14.55	21 01 38.7	.983 6879
	17	7908.5	296 36 25.81	−0.45	19 54 31.55	20 50 12.1	.983 7598
	18	7909.5	297 37 31.03	−0.52	19 58 47.86	20 38 21.6	.983 8370
	19	7910.5	298 38 35.96	−0.57	20 03 03.49	20 26 07.5	.983 9195
	20	7911.5	299 39 40.57	−0.59	20 07 18.40	−20 13 30.2	0.984 0067
	21	7912.5	300 40 44.82	−0.58	20 11 32.59	20 00 30.0	.984 0986
	22	7913.5	301 41 48.65	−0.54	20 15 46.05	19 47 07.3	.984 1949
	23	7914.5	302 42 52.02	−0.48	20 19 58.76	19 33 22.4	.984 2953
	24	7915.5	303 43 54.85	−0.39	20 24 10.70	19 19 15.7	.984 3996
	25	7916.5	304 44 57.06	−0.28	20 28 21.87	−19 04 47.6	0.984 5076
	26	7917.5	305 45 58.59	−0.15	20 32 32.26	18 49 58.5	.984 6190
	27	7918.5	306 46 59.33	−0.01	20 36 41.84	18 34 48.7	.984 7337
	28	7919.5	307 47 59.20	+0.13	20 40 50.61	18 19 18.7	.984 8517
	29	7920.5	308 48 58.08	+0.27	20 44 58.56	18 03 28.9	.984 9730
	30	7921.5	309 49 55.88	+0.40	20 49 05.69	−17 47 19.7	0.985 0974
	31	7922.5	310 50 52.51	+0.51	20 53 11.98	17 30 51.5	.985 2253
Feb.	1	7923.5	311 51 47.87	+0.59	20 57 17.44	17 14 04.7	.985 3568
	2	7924.5	312 52 41.91	+0.65	21 01 22.07	16 56 59.8	.985 4921
	3	7925.5	313 53 34.56	+0.66	21 05 25.86	16 39 37.1	.985 6316
	4	7926.5	314 54 25.80	+0.65	21 09 28.82	−16 21 57.1	0.985 7755
	5	7927.5	315 55 15.60	+0.60	21 13 30.95	16 04 00.2	.985 9240
	6	7928.5	316 56 03.95	+0.52	21 17 32.26	15 45 46.8	.986 0776
	7	7929.5	317 56 50.89	+0.42	21 21 32.74	15 27 17.3	.986 2364
	8	7930.5	318 57 36.42	+0.29	21 25 32.41	15 08 32.2	.986 4005
	9	7931.5	319 58 20.59	+0.16	21 29 31.28	−14 49 31.7	0.986 5700
	10	7932.5	320 59 03.42	+0.02	21 33 29.35	14 30 16.4	.986 7450
	11	7933.5	321 59 44.95	−0.11	21 37 26.65	14 10 46.6	.986 9253
	12	7934.5	323 00 25.22	−0.24	21 41 23.17	13 51 02.6	.987 1108
	13	7935.5	324 01 04.25	−0.34	21 45 18.94	13 31 04.8	.987 3014
	14	7936.5	325 01 42.04	−0.43	21 49 13.97	−13 10 53.7	0.987 4968
	15	7937.5	326 02 18.62	−0.49	21 53 08.27	−12 50 29.6	0.987 6968

Fig. 3.9. Portions of facing pages from the Astronomical Almanac for 1990.

Astronomers make use of a great many more catalogs in their regular work, and we shall describe many of these in later chapters. There are, for example, catalogs of magnitudes, spectral types, proper motions and radial velocities, and there are catalogs of special classes of objects such as binary stars, variables, pulsars, novae and radio sources. Some are short lists, and some fill many volumes. Most are available only at well-equipped observatories.

We must also point out that many faint stars have no specific designation in any catalog. It is customary for astronomers working with stars such as this to publish a small map or 'finding chart' showing the star of interest and the other stars which an observer would be likely to see when

SUN, 1990 C5

FOR 0^h DYNAMICAL TIME

Date		Position Angle of Axis P	Heliographic		H. P.	Semi-Diameter	Ephemeris Transit
			Latitude B_0	Longitude L_0			
		°	°	°	″	′ ″	h m s
Jan.	0	+ 2.63	−2.89	343.21	8.94	16 15.89	12 03 02.95
	1	2.14	3.01	330.04	8.94	16 15.91	12 03 31.50
	2	1.66	3.13	316.87	8.94	16 15.92	12 03 59.71
	3	1.17	3.24	303.70	8.94	16 15.93	12 04 27.55
	4	0.69	3.36	290.53	8.94	16 15.94	12 04 54.99
	5	+ 0.20	−3.47	277.36	8.94	16 15.94	12 05 21.99
	6	− 0.28	3.58	264.19	8.94	16 15.94	12 05 48.54
	7	0.76	3.69	251.02	8.94	16 15.93	12 06 14.60
	8	1.25	3.80	237.85	8.94	16 15.91	12 06 40.16
	9	1.73	3.91	224.68	8.94	16 15.89	12 07 05.17
	10	− 2.21	−4.02	211.52	8.94	16 15.86	12 07 29.64
	11	2.69	4.13	198.35	8.94	16 15.83	12 07 53.54
	12	3.16	4.23	185.18	8.94	16 15.79	12 08 16.84
	13	3.64	4.33	172.01	8.94	16 15.74	12 08 39.54
	14	4.11	4.44	158.84	8.94	16 15.68	12 09 01.61
	15	− 4.58	−4.54	145.67	8.94	16 15.62	12 09 23.05
	16	5.05	4.64	132.51	8.94	16 15.56	12 09 43.83
	17	5.51	4.73	119.34	8.94	16 15.49	12 10 03.94
	18	5.98	4.83	106.17	8.94	16 15.41	12 10 23.35
	19	6.44	4.93	93.00	8.94	16 15.33	12 10 42.07
	20	− 6.89	−5.02	79.84	8.94	16 15.24	12 11 00.07
	21	7.35	5.11	66.67	8.94	16 15.15	12 11 17.33
	22	7.80	5.20	53.50	8.94	16 15.06	12 11 33.85
	23	8.25	5.29	40.34	8.93	16 14.96	12 11 49.62
	24	8.69	5.38	27.17	8.93	16 14.85	12 12 04.61
	25	− 9.13	−5.46	14.00	8.93	16 14.75	12 12 18.82
	26	9.57	5.55	0.84	8.93	16 14.64	12 12 32.24
	27	10.00	5.63	347.67	8.93	16 14.52	12 12 44.85
	28	10.43	5.71	334.51	8.93	16 14.41	12 12 56.65
	29	10.85	5.78	321.34	8.93	16 14.29	12 13 07.64
	30	−11.27	−5.86	308.17	8.93	16 14.16	12 13 17.79
	31	11.69	5.94	295.01	8.93	16 14.04	12 13 27.11
Feb.	1	12.10	6.01	281.84	8.92	16 13.91	12 13 35.60
	2	12.51	6.08	268.68	8.92	16 13.77	12 13 43.24
	3	12.91	6.15	255.51	8.92	16 13.63	12 13 50.06
	4	−13.31	−6.21	242.34	8.92	16 13.49	12 13 56.03
	5	13.70	6.28	229.18	8.92	16 13.35	12 14 01.18
	6	14.09	6.34	216.01	8.92	16 13.19	12 14 05.50
	7	14.47	6.40	202.84	8.92	16 13.04	12 14 09.01
	8	14.85	6.46	189.68	8.92	16 12.88	12 14 11.70
	9	−15.22	−6.52	176.51	8.91	16 12.71	12 14 13.61
	10	15.59	6.57	163.34	8.91	16 12.54	12 14 14.73
	11	15.95	6.62	150.17	8.91	16 12.36	12 14 15.07
	12	16.30	6.67	137.01	8.91	16 12.18	12 14 14.66
	13	16.66	6.72	123.84	8.91	16 11.99	12 14 13.50
	14	−17.00	−6.77	110.67	8.91	16 11.80	12 14 11.61
	15	−17.34	−6.81	97.50	8.90	16 11.60	12 14 09.00

trying to make an identification. Examples of finding charts will be found in Chapter 14, Variable stars.

Summary of stellar nomenclature

Let us list now the methods described so far for naming stars, and let us also extend the list to include some other naming schemes which will be described in later sections of this book. We give typical numbers and identify the catalog in which each is used.

Aldebaran	traditional Arabic name
alpha Tauri	*Bayer Catalog*
87 Tauri	*Flamsteed Catalog*
BS 1457	*Yale Catalog of Bright Stars*
HR 1457	*Harvard Revised Photometry*
BD +16 629	*Bonner Durchmusterung Catalog*
DM +16 629	*Bonner Durchmusterung Catalog*
SAO 94027	*Smithsonian Astrophysical Observatory Catalog*
HD 29139	*Henry Draper Catalog* of spectral types
ADS 3321	*Aitken's Catalog of Double Stars*
(The above designations all refer to the same star.)	
CD Cygni	*General Catalog of Variable Stars*
V591 Cygni	*General Catalog of Variable Stars*
HDE 083246	*Henry Draper Extension*

QUESTIONS FOR REVIEW

1. The charts reproduced in Figs 3.1 and 3.4 have some areas in common. Compare these two charts and try to match stars on one chart to stars on the other. Hint: Look for patterns. The effects of precession have caused differences in coordinates on the two charts.
2. Find 74 Cygni in Fig. 3.1 and then find the BD number for that star.
3. Explain the method used to construct the *BD Catalog*.
4. In what case might a star be identified without recourse to a star atlas? When would charts be absolutely necessary?
5. Why does the problem of positive identification of stars become more difficult for faint stars?

Further reading

Allen, R. H. (1963). *Star Names, Their Lore and Meaning.* Dover Publications, Inc. Allen's book is the classic reference for information on the names of the stars and the origins of those names. The original edition was published in 1899.

Hoffleit, D. (1982). *Yale Catalog of Bright Stars*, 4th edn., Yale University Press. This excellent catalog is also available in machine-readable form.

Pannekoek, A. (1961). *A History of Astronomy*. Interscience Publishers. Consult the index for references to all of the early catalogs beginning with that of Hipparchus.

Tirion, W., Rappaport, B. and Lovi, G. (1987). *Uranometria 2000.0*, Willman–Bell, Inc. This up-to-date atlas contains an extensive summary of the history of star charts, atlases and catalogs. The charts are of such a size that they are convenient to use at the telescope.

4

Applications of the spherical triangle

On the surface of a sphere a three-sided figure will be formed by the intersections of three great circles. Such a figure is referred to as a spherical triangle, and these figures have some interesting properties. For example, the sum of the angles in a spherical triangle will usually be greater than 180 degrees whereas in a plane triangle this sum is exactly 180 degrees. Consider the spherical triangle formed by the equator and the hour circles of the vernal equinox and a star as in Fig. 4.1. The intersections of the three great circles have been labelled A, B and C. Circles through the poles cross the equator at angles of 90 degrees. Therefore, angles B and C are each equal to 90 degrees, and the sum of the three angles will be greater than 180 degrees.

One should note especially that the angles in a spherical triangle are angles at which two planes intersect each other. Each plane must, of course, pass through the center of the sphere. Thus, angle A in Fig. 4.1 is the angle between the planes specified by the hour circles through point B and through point S. Angle A is also equal to angle BOC.

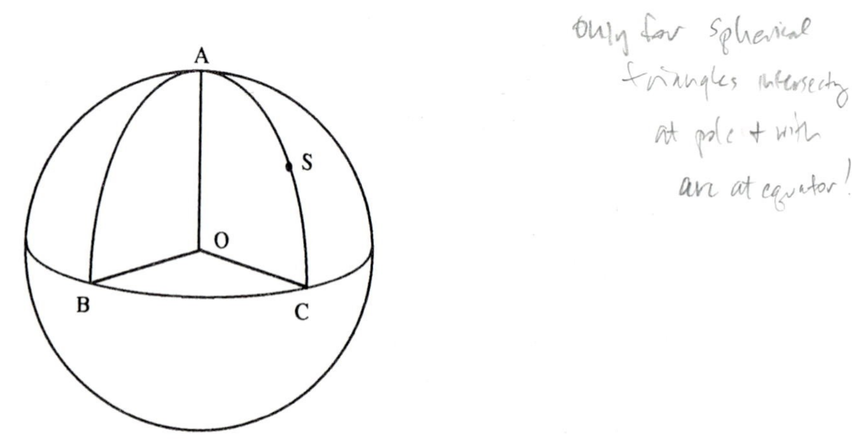

Fig. 4.1. A spherical triangle, ABC, in which two angles are each 90°.

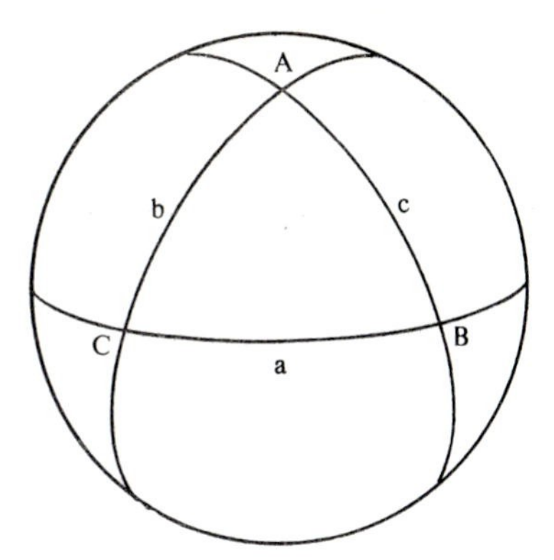

Fig. 4.2. Sides and angles in a spherical triangle.

It is customary, just as in plane trigonometry, to label the sides of a triangle with lower case letters, a, b and c indicating the sides opposite the angles A, B and C. Note here that the length of a side is expressed in terms of an angle measured at the center of the sphere. This has been shown in Fig. 4.2.

To see a good example of a spherical triangle as it might be used in an astronomical problem, consider Fig. 4.3. Here the horizon has been drawn and a star is shown at point S. The equator, the meridian and the star's hour circle have been drawn also. A spherical triangle has been formed by the pole, the zenith and the star. Recalling our earlier definitions of altitude and azimuth, notice that the angle PZS is the star's azimuth, and that the angle SOD is the star's altitude. Angle ZPS is the star's hour angle, and angle POS is 90 degrees minus the declination. An understanding of the relationships between the sides and angles in spheri-

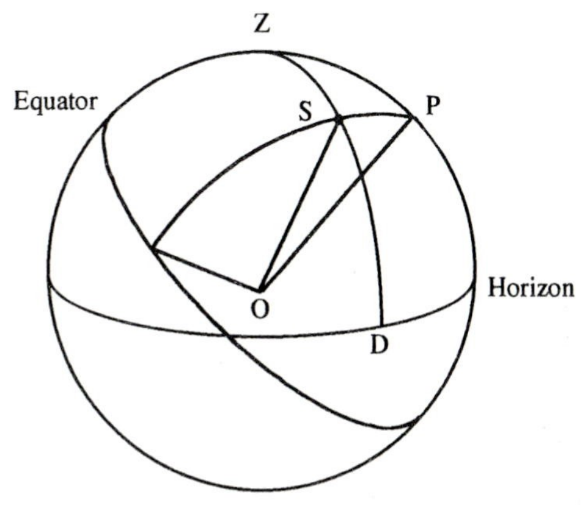

Fig. 4.3. The spherical triangle, ZPS, formed by the zenith, pole and star.

cal triangles leads to methods by which coordinates in one system may be transformed into coordinates in another system.

Basic formulae of spherical trigonometry

Using only the well-known relationships of plane trigonometry, it is possible to derive several formulae which relate the sides and angles of spherical triangles. As mentioned above, it is formulae such as these that make it possible for us to transform coordinates from one system to another. For example, if we know the declination and hour angle of a star, we may quite easily calculate the altitude and azimuth of the star.

Consider first the spherical triangle shown in Fig. 4.4(*a*). The center of the sphere is at O, and the vertices are ABC. The sides are, of course,

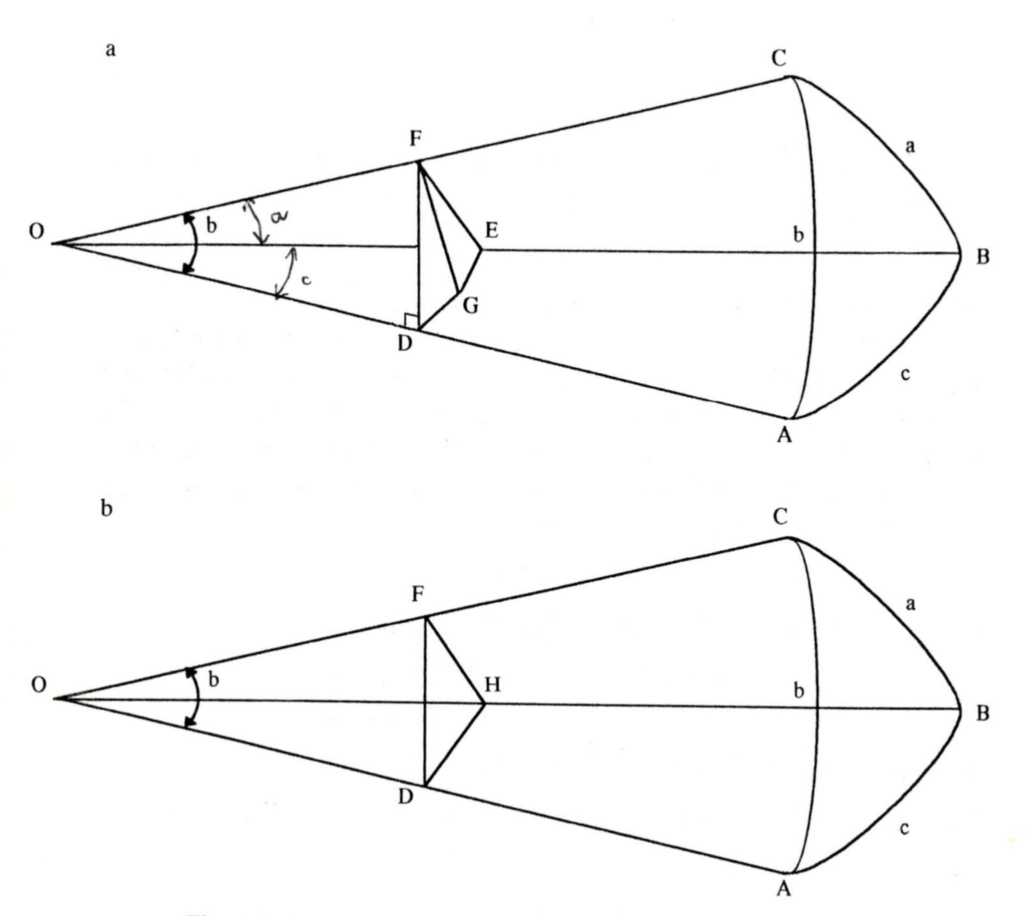

Fig. 4.4. Geometry for derivation of (*a*) the sine formula (*b*) the cosine formula.

abc. The triangle FDG has been drawn in such a way that FD is perpendicular to OA and FG is perpendicular to plane AOB. In a similar way the triangle FEG has been drawn. One may now note the following relationships:

$$\sin A = \sin FDG = FG/DF$$
$$\sin B = \sin FEG = FG/EF$$
$$\sin A/\sin B = EF/DF$$

One may also note:

$$\sin b = DF/OF$$
$$\sin a = EF/OF$$
$$\sin a/\sin b = EF/DF$$

Thus we may write

$$\sin A/\sin B = \sin a/\sin b$$

or

$$\sin A/\sin a = \sin B/\sin b$$

The same type of analysis may be applied to derive relationships involving angle C and side c, so we could also write

$$\sin A/\sin a = \sin B/\sin b = \sin C/\sin c.$$

This is the sine formula of spherical trigonometry.

Another useful formula may be derived if we consider some relationships in Fig. 4.4(*b*). We begin again with a spherical triangle ABC, and the center of the sphere is at O. Points D, F and H are the vertices of a triangle oriented in such a way that sides DF and DH are perpendicular to line OA. Note also that angle FDH is equal to angle A. From plane trigonometry recall the law of cosines:

$$a^2 = b^2 + c^2 - 2bc \cos A$$

Applying this in triangles OHF and DHF we have

$$HF^2 = OH^2 + OF^2 - 2\ OH\ OF \cos a$$
$$HF^2 = DH^2 + DF^2 - 2\ DF\ DH \cos A$$

Subtracting one of these from the other gives

$$0 = OH^2 + OF^2 - 2\ OF\ OH \cos a - DH^2 - DF^2 + 2\ DF\ DH \cos A \quad (1)$$

It is also clear that

$$OF^2 = DF^2 + OD^2$$
$$OH^2 = DH^2 + OD^2$$

Substituting now in (1) gives

$$0=2\ OD^2-2\ OF\ OH\cos a+2\ DF\ DH\cos A$$

This expression may be solved for cos a and rewritten as

$$\cos a=(OD\ OD)/(OF\ OH)+(DF\ DH)/(OF\ OH)\cos A$$

One may now substitute sines and cosines of the angles b and c in the triangles ODF and ODH and write

$$\cos a=\cos b\cos c+\sin b\sin c\cos A$$

This is the cosine formula of spherical trigonometry. It hardly needs to be said that expressions for cos b and cos c may be derived in the same manner as above. Almost all problems involving spherical triangles may be solved by using the cosine or the sine formula or by using both of them successively.

It is possible to derive two other formulae which can often simplify the solution of particular problems. These formulae are stated below without elaboration. Derivations may be found in Smart, *Spherical Astronomy*, Chapter 1.

Formula C:

$$\sin a\cos B=\cos b\sin c-\sin b\cos c\cos A$$

Four-parts formula:

$$\cos a\cos C=\sin a\cot b-\sin C\cot B$$

In the application of any of these formulae to practical problems the student should first try to draw a reasonably precise sketch on which the appropriate spherical triangle may be identified. The known sides and angles may then be recognized and the proper formula may be applied. An alternative is to try to remember specific formulae for specific applications. The latter procedure is not of much help when the problem to be solved is new or different and does not fit into one of the established categories. We repeat, draw the figure, and see what sides and angles are known. Then apply a formula which can be solved for the desired quantity.

Hour angle

In Chapter 1 we introduced the notion of hour angle and we showed this in a simple way in Fig. 1.6(*b*). Let us now be more specific about the determination of hour angle so that we can proceed in the next section to methods for conversions between horizon and equatorial coordinates. With the background of the material presented previously, we may now define hour angle as the angle between the observer's meridian and the hour circle through the star. In the plane of the celestial equator as shown in Fig. 4.5(*a*) the hour angle is angle MOT. The star is at point S, and the figure as drawn here shows a star in the western half of the sky. Fig. 4.5(*b*) is similar but shows a star in the eastern half of the sky. Examination of these two sketches should show the reader how the hour angle of a star is to be found. The sidereal time is represented by angle MOV, and the right ascension is angle TOV. Since the hour angle is angle MOT, we can say that the hour angle is simply the difference between the sidereal time and the right ascension. This convention would indicate that if the hour angle is positive, the star is west of the meridian. If the hour angle is negative, the star is east of the meridian.

This concept of hour angle may be made more clear by examination of Fig. 4.5(*c*) which shows the celestial sphere as it would be seen by an observer looking at the sphere from a point above the north celestial pole. The equator is in the plane of the paper and the north celestial pole is at the center of the circle. The meridian and the star's hour circle appear as radial lines, and so does the hour circle of the vernal equinox. The hour angle is clearly the angle between the plane of the meridian and the plane of the star's hour circle.

Astronomers usually try to make their observations when the hour angle is small, because then the star being observed is close to the meridian. If the hour angle is 12 h, the star will either be below the horizon or, at best, between the pole and the horizon. A quick look at a sidereal clock can tell an observer which stars can be most conveniently observed at that particular time. Stars with right ascensions which are within an hour of the sidereal time will be close to the meridian.

The observer should also take note of the declination of a star in considering the ease with which a star might be observed. One should understand that even when a star is on the meridian, it may be very low in the sky or below the horizon. The declination of the star is the important factor here, for the meridian (or maximum) altitude can be given by

$$\text{alt(max)} = (90° - \text{lat}) + \delta$$

Conversion of coordinates

Let us look now at the application of the formulae of spherical trigonometry to some of the practical problems suggested earlier. Cona

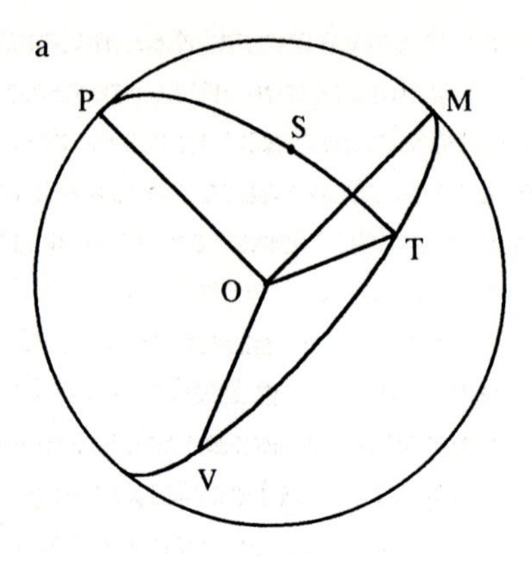

b

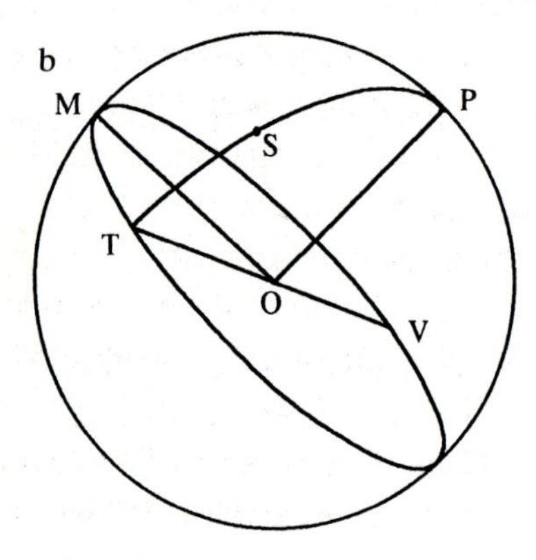

c

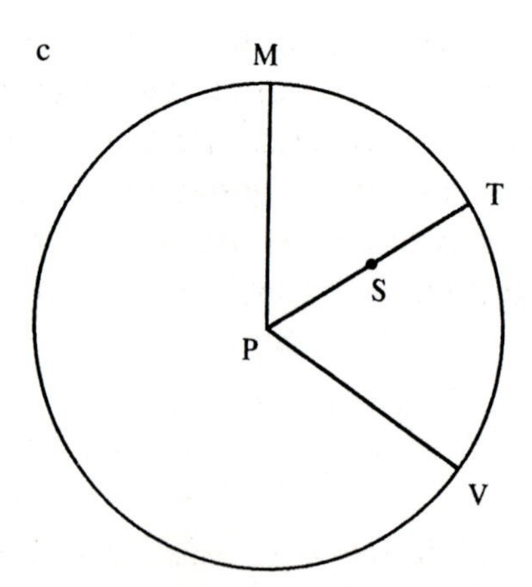

Fig. 4.5. Hour angle (*a*) for a star in the west (*b*) for a star in the east (*c*) polar view.

sider first the determination of the altitude and azimuth of a star at some specified time. We must, of course, know the coordinates of the star and the local sidereal time to begin with, and we assume that the latitude of the observer is also known. We find the hour angle so that we can know whether the star is in the eastern or the western portion of the sky. Then we may draw the appropriate sketch. Assume for example, that the declination is +30 degrees and that the hour angle is one hour and thirty minutes east. The sketch should appear as in Fig. 4.6. Note the triangle formed by the star, the zenith and the north celestial pole. Side PZ is equal to (90°−lat); side ZS is equal to (90°−alt)[1]; and side PS is equal to (90°−dec). The angle at P is the hour angle, and the angle at Z is the azimuth. In order to find the altitude we may use the cosine formula as follows:

$$\cos(90°-\text{alt})=\cos(90°-\text{lat})\ \cos(90°-\text{dec}) + \sin(90°-\text{lat})\ \sin(90°-\text{dec})\ \cos \text{HA}$$

where HA is the hour angle. Once this has been done one may apply the sine formula to find the azimuth:

$$\sin(90°-\text{alt})/\sin \text{HA}=\sin(90°-\text{dec})/\sin \text{az}$$

If one is using a computer or a programmable calculator, the program can be written so that the user needs only to enter the latitude, the declination and the hour angle. If a simple scientific calculator is being used, the computation may be simplified by first applying some trigonometric identities such as cos (90°−a)=sin a. Then we could write instead

$$\sin \text{alt}=\sin \text{lat}\ \sin \text{dec}+\cos \text{lat}\ \cos \text{dec}\ \cos \text{HA}$$
$$\cos \text{alt}/\sin \text{HA}=\cos \text{dec}/\sin \text{az}$$

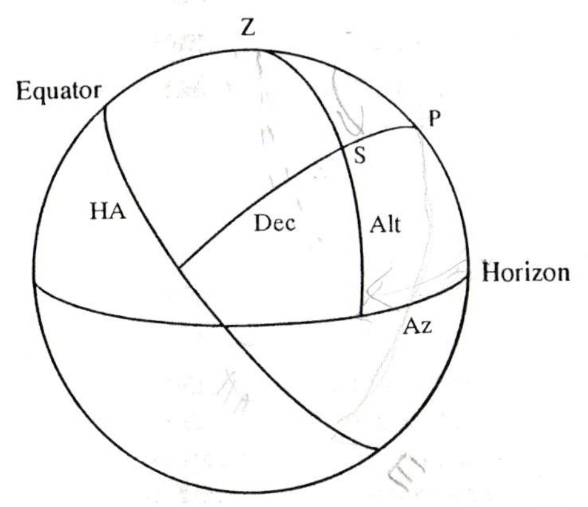

Fig. 4.6. Relation between equatorial and horizon coordinates. HA=hour angle, Dec=declination, Alt=altitude, Az=azimuth.

[1] The term 'zenith distance' is often used in place of (90°–alt).

If in Fig. 4.6 the altitude, azimuth and declination had been known, then the hour angle could have been determined.

Ecliptic coordinates

In order to find the coordinates with respect to the ecliptic when the right ascension and declination are known, one may proceed in a manner similar to that described above. First, one should draw a figure which shows with reasonable accuracy the geometry of the problem. This has been done in Fig. 4.7. Here P is the pole of the celestial equator, and P′ is the pole of the ecliptic. V is the vernal equinox, and a star is shown at S. The points P, P′, and S are the vertices of a spherical triangle. Point V is the pole of a great circle which passes through points P and P′, so the angles PP′V and P′PV are each equal to 90 degrees. Angle VPS (not shown) is the right ascension, and angle P′PS is (90°+RA). Angle VP′S is the celestial longitude, λ, and angle PP′S is $(90° - \lambda)$. Side PS is (90°−dec), and side P′S is $(90-\beta)$ where β is the celestial latitude. The side PP′ is 23.5 degrees, the obliquity of the ecliptic. Clearly there are enough known quantities here for one to solve for side SP′ and for angle PP′S, and thus determine the ecliptic coordinates.

Fig. 4.8 has been drawn to show a similar situation in which the right ascension is greater than six hours. A is the autumnal equinox, so angle APS is (180−RA). Angle P′PS is (90+(180−RA)) or (270−RA). Likewise,

$$<PP'S = (90 - (180 - \lambda)) = \lambda - 90$$

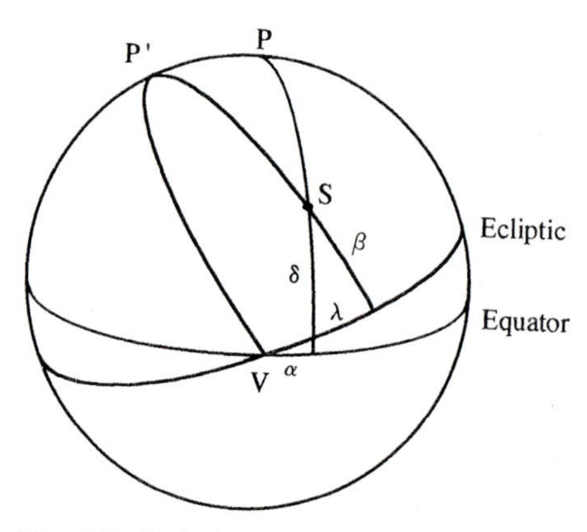

Fig. 4.7. Relation between equatorial and ecliptic coordinates.

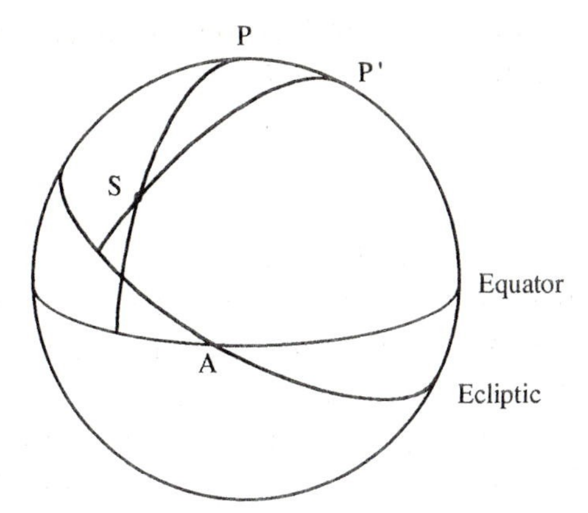

Fig. 4.8. Ecliptic coordinates near the autumnal equinox.

Galactic coordinates

In Chapter 1 we mentioned the system of galactic coordinates in which the approximate central line of the Milky Way is chosen as the fundamental plane or equator. This system is a necessity in studies of the structure of our galaxy and of the sun's position in it, for a mere mention of the galactic coordinates conveys a great deal about the location of a star in the vast, rotating galaxy of which we are a part. The two coordinates are galactic latitude, b, and galactic longitude, l. The scheme is illustrated in Fig. 4.9. The points GP and GC are respectively the galactic pole and the direction toward the center of the galaxy. It should be clear to the reader that the locations of these points give us the actual definition of the system, for the first defines the fundamental plane, and the second defines a zero point on that plane. Astronomers from many

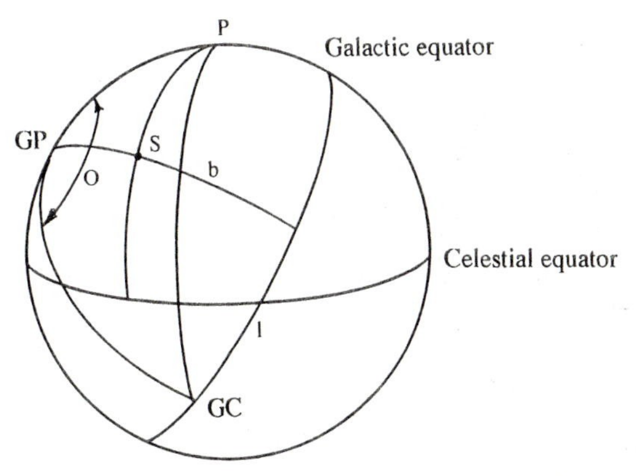

Fig. 4.9. Relation between equatorial and galactic coordinates.

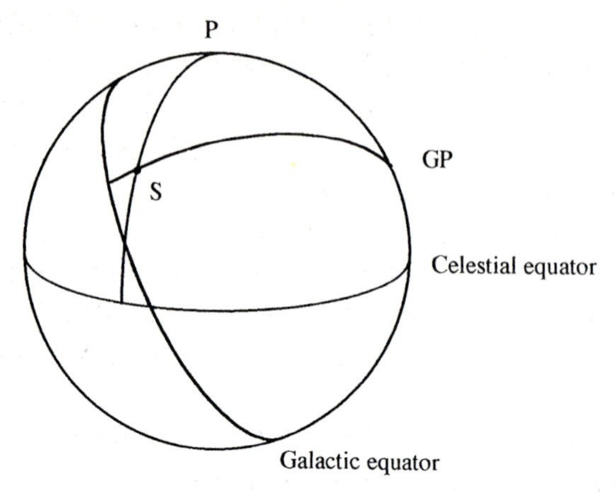

Fig. 4.10. Galactic cordinates on the side of the celestial sphere opposite the galactic center.

countries worked together for a long time to find the best positions, and they announced their conclusions at a meeting of the International Astronomical Union in 1960. The adopted results are

Coordinates of the galactic pole

$$\alpha = 12\text{ h } 49\text{ m} = 192°\ 15' \qquad \delta = +27°\ 24'$$

Coordinates of the galactic center

$$\alpha = 17\text{ h } 42\text{ m} = 265°\ 36' \qquad \delta = -28°\ 55'$$

Referring again to Fig. 4.9 we note the triangle P GP GC. Within this triangle there is enough information for us to calculate the galactic longitude, O, of the north celestial pole, and as we shall see, this angle is useful in other calculations. Using the cosine formula we find that O=123°.

At this point the student should be able to look at the triangle P GP S and see how the cosine and sine formulae should be applied to find the galactic latitude and the galactic longitude.

In Fig. 4.10 we have drawn a new diagram in which right ascensions from 0 h 49 m to 12 h 49 m are on the side facing the observer. It is hoped that this will help to make it easier for the student to recognize the correct triangle and the known sides and angles for other examples.

Location of the vernal equinox

As we mentioned in Chapter 1 there is no obvious mark in the sky at the location of the vernal equinox, and yet it is crucial that we know the location of this point. We can actually find it from careful observations of transits of the sun. Consider Fig. 4.11 which shows the celestial sphere with the equator and the ecliptic. Imagine that the sun is at point S, and the vernal equinox is at V. P and P′ are respectively the equatorial and ecliptic poles. Side VA is the right ascension of the sun, and side SA is its declination. In the triangle SVA angle SVA is the obliquity of the ecliptic, *e*. The value of *e* may be found over a long period of time by noting the maximum value of the declination of the sun. This will be seen at the time of the summer solstice, and an acceptable value of this is 23.5 degrees.

Earlier in this chapter we referred to the four-parts formula as one which had occasional application, and we stated it as

$$\cos a \cos C = \sin a \cot b - \sin C \cot B$$

We apply this in triangle SVA as follows

$$\cos \mathrm{VA} \cos 90° = \sin \mathrm{VA} \cot \mathrm{SA} - \sin 90° \cot e$$
$$\sin \mathrm{VA} \cot \mathrm{SA} = \sin 90° \cot e$$
$$\sin \mathrm{VA} = \tan \mathrm{SA} \cot e$$

Now if we measure the declination of the sun at the time of any meridian passage, we can solve this expression for VA, the sun's right ascension. The sun's hour angle is, of course, zero, so the sun's right ascension is equal to the local sidereal time, and we have found the vernal equinox.

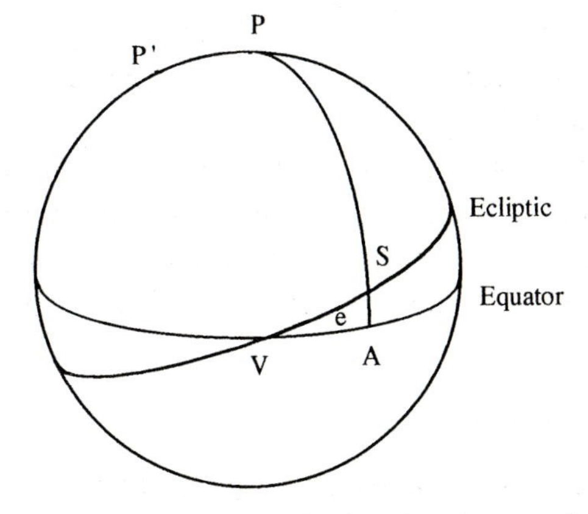

Fig. 4.11. Geometry for locating the vernal equinox.

Precession

The coordinates of all stars are continually changing due to the effect which we know as the precession of the equinoxes, and astronomers need to know how to make corrections for these changes. Those students who have completed an introductory course in astronomy will recall that the rotational axis of the earth precesses due to the gravitational attractions of the moon and the sun on the earth's equatorial bulge. The effect is the same as that noticed when a spinning top begins to slow down. The axis of the top begins to trace out a cone. Similarly, as the axis of the earth precesses, it too traces a conical path. As a result, the north celestial pole is not a fixed point on the celestial sphere. It moves continually in a circular path around the pole of the ecliptic. The period of this motion is 26 000 years, a period so long that one is tempted to think that the effect of precession would not be significant in a lifetime. Only a little experience at the telescope is enough to prove that this is not the case. The correction for precession is vital. We are concerned here only with the astronomical consequences of precession, and we shall leave the dynamical analysis of its causes to works on dynamics or celestial mechanics.

The celestial poles move because the earth's axis moves, and in like manner, the celestial equator moves because the plane of the earth's equator is moving. Since we refer the coordinates of stars to the celestial equator, then it is clear that the coordinates of all stars must be changing continually. If we use a meridian telescope to measure right ascensions and declinations of stars today, we shall find that those coordinates will not be accurate twenty-five years from now. And if we try to set a telescope using coordinates determined fifty years ago, we shall find that the desired stars are not centered in the field of view. One soon finds that the use of the telescope is expedited when the coordinates of stars are corrected for the effects of precession. The corrections are not difficult to compute, and observers should take the time to program their computers for this problem. The derivation of the correction formulae is another good exercise in the application of the methods of spherical trigonometry.

Let us look first at Fig. 4.12. Here the equator and the ecliptic have been shown, and the poles of these two circles are P and P′ respectively. The small circle through P represents the path along which the pole will move due to precession. A future position of the pole is at P″. The vernal equinox is at V, and a future position of the vernal equinox is at V′. ε

(epsilon) is the angle between the ecliptic and equatorial poles and is approximately equal to 23.5 degrees. A star is at S, and its coordinates in the equatorial system are α and δ. Its coordinates in the ecliptic system are λ and β. Measured with respect to the future pole these coordinates are α_1, δ_1, λ_1 and β_1. It may be noted that $\beta=\beta_1$. The celestial latitude is not changed by precession.

Analysis of the effect of precession can lead to the following expressions for the annual changes in right ascension and declination.

$$\delta_1-\delta=\theta \sin \varepsilon \cos \alpha$$
$$\alpha_1-\alpha=\theta\{ \cos \varepsilon+\sin \varepsilon \sin \alpha \tan \delta\}$$

Here θ is the amount of annual precession, that is, the angle through which the vernal equinox has moved in one year. This angle is very close to 50″. ε is the obliquity of the ecliptic which is 23° 27′ 8.26″.

More precise corrections for precession are computed from the expressions:

$$\alpha_1-\alpha=m+n \sin \alpha \tan \delta$$
$$\delta_1-\delta=n \cos \alpha$$

where $m=46''.092+0''.0002797\times t$,
$n=20''.051-0''.0000834\times t$,
and t is the number of years which have elapsed since 1900. We repeat that these formulae give the annual correction for precession. The computer should be programmed to calculate a correction for the first year and add it to the original coordinate. This new coordinate is used in the correction to be applied for the second year and so forth until the coordinate for the desired year has been obtained. In other words, one should

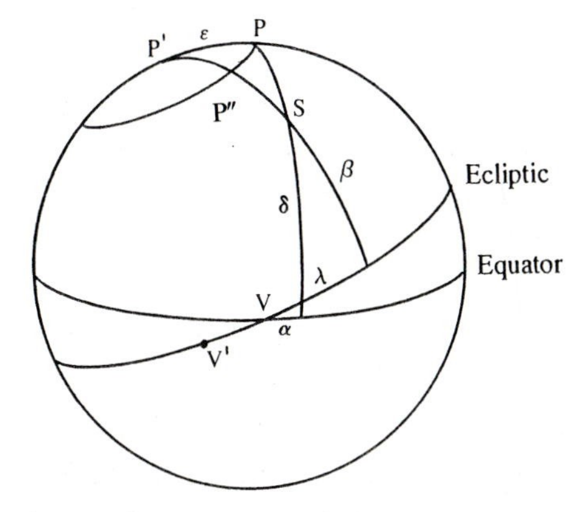

Fig. 4.12. Precession of the equinoxes.

compute a new coordinate for each year from 1900 to the year for which corrected coordinates are needed.

Nutation

We have described precession as an effect caused by the gravitational attraction of the moon and the sun on the earth's equatorial bulge. If we recall that the orbits of the earth and the moon are elliptical rather than circular, then we see that the forces causing precession are not constant. In addition, just as the plane of the earth's equator precesses, so also the plane of the moon's orbit precesses. This causes the nodes N and N′ of the lunar orbit to move progressively around the ecliptic. In Fig. 4.13(*a*) and Fig. 4.13(*b*) these changes in the orientation of the orbit have been shown. In a period of 18.67 years the nodes move all of the way around the ecliptic. The moon's orbit is inclined at an angle of approximately five degrees to the plane of the earth's orbit, and this means that the maximum declination of the moon can vary through a range of ten degrees. When the moon is north of the ecliptic, its declination can be as large as 28.5 degrees. When it is south of the ecliptic, its

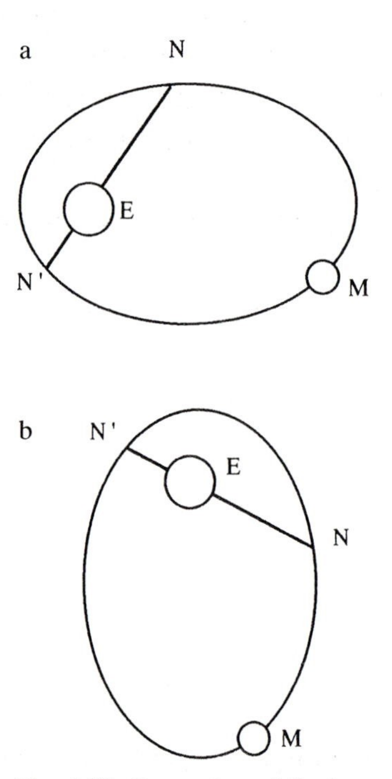

Fig. 4.13. Precession of the lunar orbit.

maximum declination is only 18.5 degrees. Thus, as the moon moves with respect to the celestial equator, there are variations in the direction of the attractive force between the moon and the equatorial bulge.

Altogether, then, the precessional force undergoes changes in magnitude as the earth–sun and earth–moon distances vary, and the force undergoes changes in direction as the declination of the moon varies. The net result is a wobble with an amplitude of 9″ superimposed on the general precession. It is this wobble that is referred to as nutation. If it is to be assumed that in the practical problem of star identification, the use of a finding chart is common practice, then we need not worry much about a correction that is never larger than 9″.

There are, however, situations in which the desired precision requires the application of corrections for nutation. For such cases the user should make use of the procedures and constants found in the *Astronomical Almanac* or in books by Smart or Taff listed in the bibliography.

Proper motion

When we compare old coordinates of stars with those determined more recently, we often find that the difference in position is greater than can be accounted for by the effects of precession alone. Many of the stars seem to show continuous changes in position indicating a certain angular velocity. Such angular velocities are known as 'proper motions', and an example is depicted in Fig. 4.14. The proper motion of an individual star may be as large as several seconds of arc per year.

Two factors combine to give rise to the observed proper motion of a

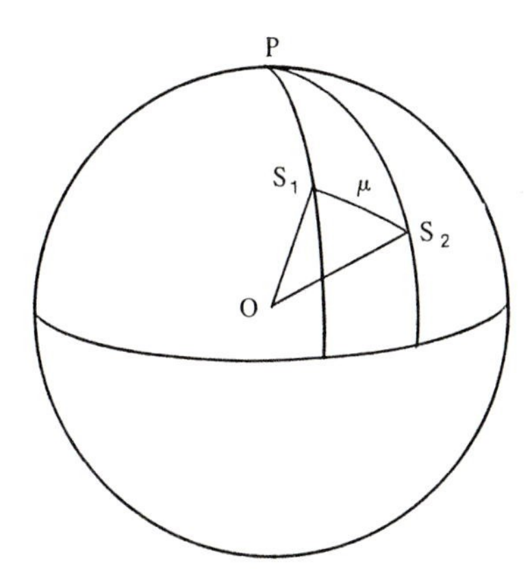

Fig. 4.14. Proper motion indicated by the angle, μ. S_1 and S_2 are the two positions of the star.

star. First, each star may be expected to have its own independent motion through space. One component of this motion will be detected by us as proper motion. The other component is, of course, the radial velocity, and we shall deal with that in Chapter 13. Second, part of a star's observed angular velocity results from the motion of the sun (and with it the earth) with respect to the stars. The point on the sky toward which the sun moves is defined as the apex of the sun's motion, and stars in the general direction of the apex seem to have systematic angular velocities away from that point (see Fig. 4.15). On the opposite side of the sky the stellar motions are directed toward the antapex, the point away from which the sun seems to be moving. This is, of course, a perspective effect as students of drawing will be aware.

We may illustrate these two effects with examples of familiar situations. First, if we watch an airplane flying at a high altitude, we notice that the plane has a certain angular velocity which depends on the speed and course of the plane and its distance from us. This angular velocity may be likened to the proper motion which results from the individual motions of the stars. Second, the driver of an automobile could measure the angle between two points on opposite sides of the highway and quite far down the road. As the driver approaches the two points, the angle between them will increase. As the driver passes between the two points, the angle will be 180 degrees, and thereafter the angle will decrease. In the same way, the motion of the sun through space causes the stars to have systematic proper motions away from the solar apex and toward the antapex. The observed proper motions of stars are the combination of these two effects: the motion of the stars and the motion of the sun.

The reader should notice that in the SAO or the YBS catalogs it is the components in RA and Dec that are actually listed rather than the total

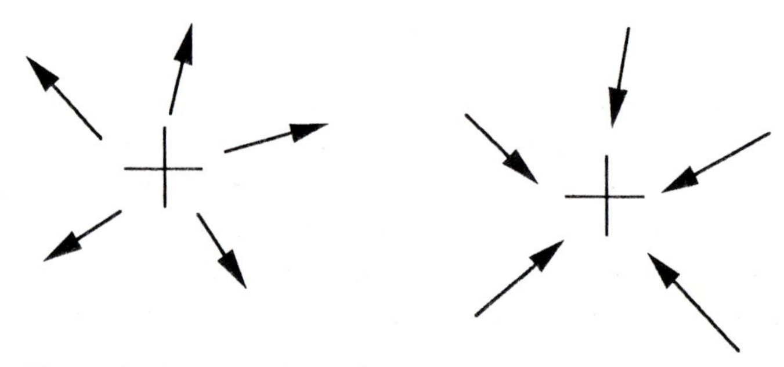

Fig. 4.15. Proper motions of stars near the apex (left) and near the antapex (right).

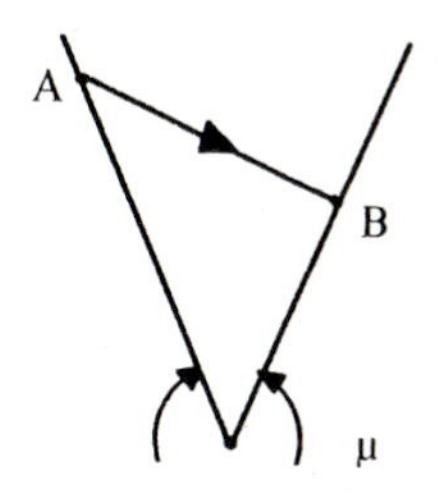

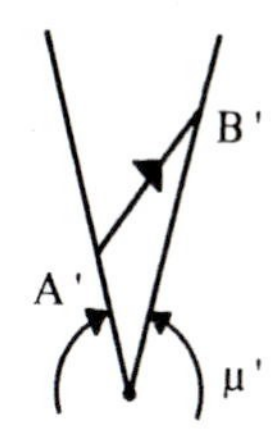

Fig. 4.16. Dependence of proper motion on the direction of motion of a star.

proper motion μ. These components are to be used in applying corrections when precise positions for a particular year are desired.

In many studies the intrinsic motions of individual stars are desired, and the contribution of solar motion must be removed. The solar motion itself is determined through an analysis of the proper motions of a large number of stars.

It should be obvious that proper motion should be largest for stars which are relatively close to us and are moving with high velocity at right angles to the line of sight. A star might be close to us and traveling at a high velocity, but if the motion is along or close to the line of sight, the observed proper motion will be zero or very small (see Fig. 4.16). Proper motion should be essentially unmeasurable for faint, very distant galaxies even though their actual velocities through space might be very large.

Total proper motion

When the two components, μ_α and μ_δ, are known, then the total proper motion, μ, may be calculated. But in order to give the complete description of the proper motion, its direction must also be specified. This is indicated as ϕ in Fig. 4.17 and is known as the position angle. The two quantities, μ and ϕ, can be calculated from the tabulated proper motions in right ascension and declination by means of the following formulae:

$$\mu^2=(\mu_\delta \cos \delta)^2+\mu_\alpha^2$$

$$\cos \phi=\mu_\delta/\mu \quad \text{or} \quad \sin \phi=(\mu_\alpha \cos \delta)/\mu$$

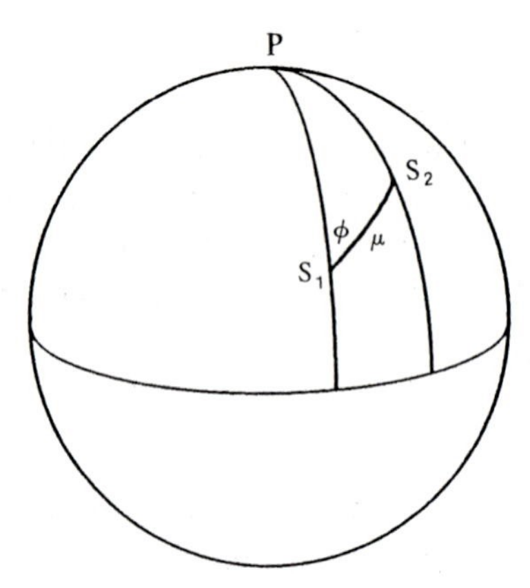

Fig. 4.17. Position angle, ϕ, of proper motion.

The above formulae may be derived in a variety of ways, and the reader is referred to Smart, McNally and van de Kamp for more details. Consideration of Fig. 4.18, however, will indicate some simple relationships which lead to the same results given above. S_1 and S_2 are the two positions of a star with a proper motion μ. The two values of declination are indicated, as is μ_α, the proper motion in right ascension. Now imagine a small circle drawn from S2 parallel to the equator. This circle will cross the arc PS_1 at point Q forming a small triangle S_1S_2Q. The length of side S_2Q is $\mu_\alpha \cos \delta$; S_1Q is μ_δ; and angle S_2S_1Q is the position angle ϕ. Recognizing that the sides of this figure are very small we can see that the triangle approximates a plane triangle within which the relationships stated above are easily seen.

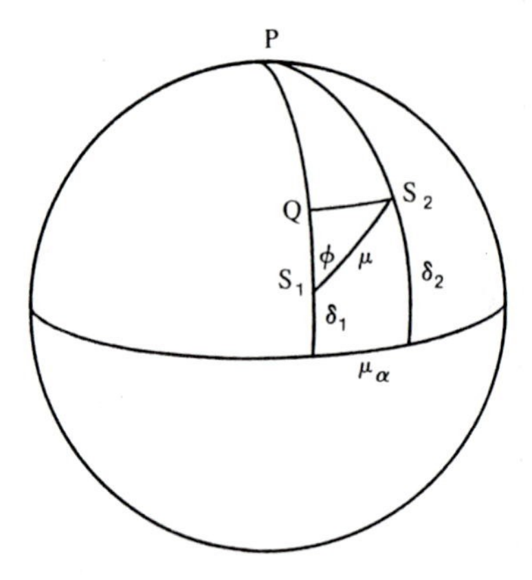

Fig. 4.18. Geometry for the computation of total proper motion from components in right ascension and declination.

Refraction

We recall from elementary optics that when a ray of light passes from a less dense medium into a more dense one, the direction of the ray can be changed by refraction. This effect is quite familiar and is pictured in Fig. 4.19. We can predict the effects of refraction by applying Snell's Law which may be stated as follows:

$$m_1 \sin I = m_2 \sin R$$

Here m_1 and m_2 are the indexes of refraction of the two media (air and water, for example), I is the angle of incidence and R is the angle of refraction.

When a ray of starlight enters the earth's atmosphere, then, we can expect the direction of the ray to be altered by refraction since the earth's atmosphere is more dense than interstellar space. Snell's Law requires that the rays will be bent toward the normal, so we should expect stars to appear higher in the sky than they otherwise would. For an object at an

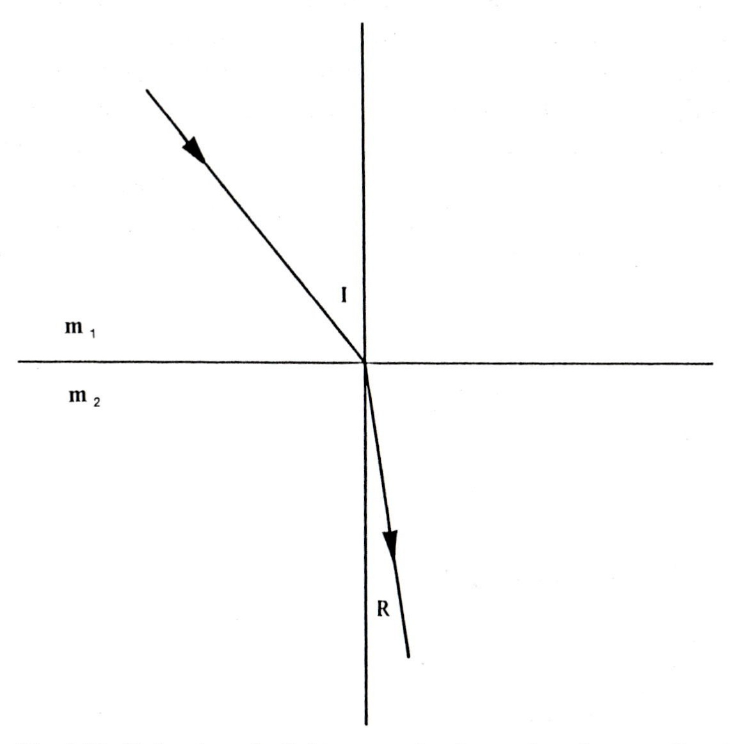

Fig. 4.19. Refraction of a light ray passing from a less dense medium into a more dense one.

altitude greater than forty five degrees the change in direction is less than one minute of arc and is not easily noticed, but for an object close to the horizon the effect can be thirty minutes of arc or more. Such a large amount must be considered when precise positions are required. If, for example, one wishes to make a precise determination of the latitude by measuring the altitude of the celestial pole, then one should correct the measured altitude for refraction.

The analysis of atmospheric refraction is somewhat complex because of the fact that the density of the atmosphere is not uniform. At mean sea level the average atmospheric pressure is 760 mm of mercury, but this pressure decreases rapidly with increasing height. At heights of about eleven kilometers the pressure is only one quarter of its sea level value, and this is the height at which commercial airliners often fly. At heights of fifty kilometers the pressure is less than 0.0001 of the sea level pressure. Since we know how the index of refraction of air changes with pressure, we can predict that the actual path of a ray passing through the atmosphere should be a curve as shown in Fig. 4.20. Notice here that the apparent direction to the star is higher than the true direction. As we consider objects which are progressively lower in the sky the effects of refraction become progressively greater. Low objects appear to be lifted more than higher ones, and the difference increases rapidly as one approaches the horizon. This is seen in quite a striking way when we look at the setting sun. As the sun appears to be just about on the horizon we begin to notice that the sun seems to be flattened. Refraction has raised the lower limb more than it has raised the upper limb to produce the

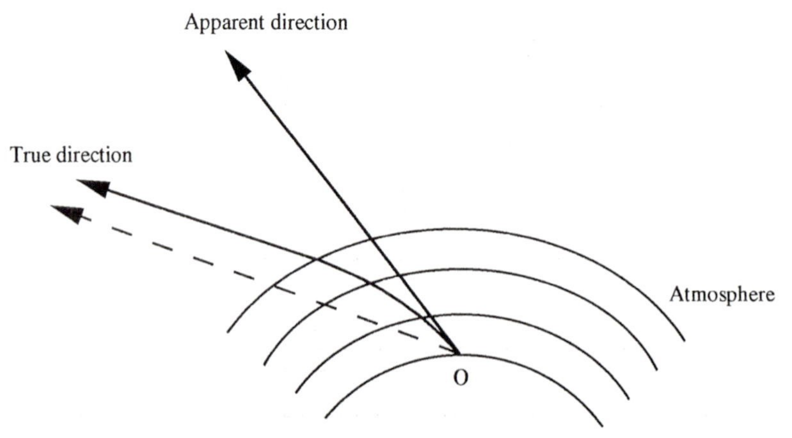

Fig. 4.20. An increase in depth within the atmosphere means an increase in density of the air and an increase in refraction.

familiar, non-circular image. In fact we continue to see the sun when it is about half a degree below our geometrical horizon.

It is desirable that we have a simple usable formula which will make it possible for us to calculate the refraction in a practical way, and this problem may be approached in either a theoretical or an empirical manner. The theoretical analysis usually begins with the assumption that the atmosphere is composed of discrete layers increasing in density downward as indicated in Fig. 4.21. Applying Snell's Law at successive boundaries between layers it is quickly shown that the effect of many layers of increasing density is the same as the effect would be if the lowest layer was the only layer. Going further one may derive this expression

$$r = C \tan z'$$

Here r is the angle of refraction, C is a constant and z' is the observed or apparent zenith distance. For zenith distances less than 45 degrees, C may be given a value of $60''$.

At larger values of the zenith distance the curvature of the earth begins to become significant, and another term must be added. Now we find

$$r = A \tan z' + B \tan^3 z'$$

$A = (\mu - 1) + B$ and B is approximately equal to -0.07. μ is the index of refraction of air at sea level. The constants A and B actually depend upon the temperature and pressure of the air and are usually determined from observations.

The problem of refraction may also be approached experimentally. By measuring the effect of refraction for many stars under a variety of

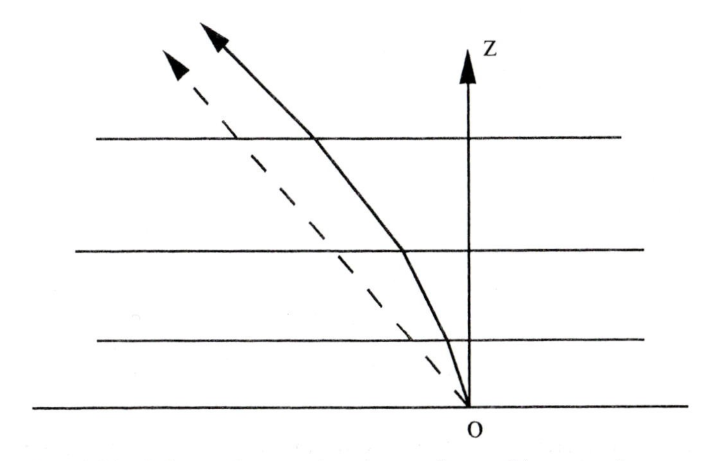

Fig. 4.21. A layered approximation to the earth's atmosphere.

atmospheric conditions empirical formulae have been determined. One such is Comstock's formula:

$$r = 60.4''(b/760)/(1+t/273) \tan z'$$

where b is the barometric pressure in mm of mercury and t is the temperature in degrees Celsius. r is expressed in seconds of arc. Values of the refraction effect computed from Comstock's Formula are shown in Table 4.1.

If we define sunset as the time when the upper limb of the sun disappears below the horizon, then we can see from the above data that the true zenith distance of the sun's upper limb is 90° 35′ at the moment of sunset. The day is therefore lengthened by the time required for the sun to move through this extra angle. As indicated in Fig. 4.22 the sun crosses the horizon at an angle that depends on the observer's latitude, so the day may be lengthened by several minutes.

Since the effect of refraction is to lift all objects toward the zenith, we should understand that this will affect the right ascension and declination of an object under study. The result can be seen in Fig. 4.23 in which a star is at S and its refracted position is at S′. A point, D, is the intersection on the arc, PA, of a small circle drawn parallel to the equator and passing through S′. The arc AB is the change in RA, and the arc SD is the change in declination. The effect of refraction calculated as described above gives the length of the side, S′S, in the small triangle S′SD. If the effects on RA and Dec are to be found, we need to first find the lengths

Table 4.1. *Atmospheric refraction*

Altitude, degrees	Refraction	Altitude, degrees	Refraction	Altitude, degrees	Refraction
0	34′50″	11	4′47″.7	30	1′39″.5
1	24 22	12	4 24 .5	35	1 22 .1
2	18 06	13	4 04 .4	40	1 08 .6
3	14 13	14	3 47 .0	45	57 .6
4	11 37	16	3 18 .2	50	48 .3
5	9 45	18	2 55 .5	55	40 .3
6	8 23	20	2 37 .0	60	33 .2
7	7 19	22	2 21 .6	65	26 .8
8	6 29	24	2 08 .6	70	20 .9
9	5 26	26	1 57 .6	80	10 .2
10	5 16	28	1 48 .0	90	0 .0

Corresponding to temperature of 10°C, and to a barometric pressure of 760 mm Hg.

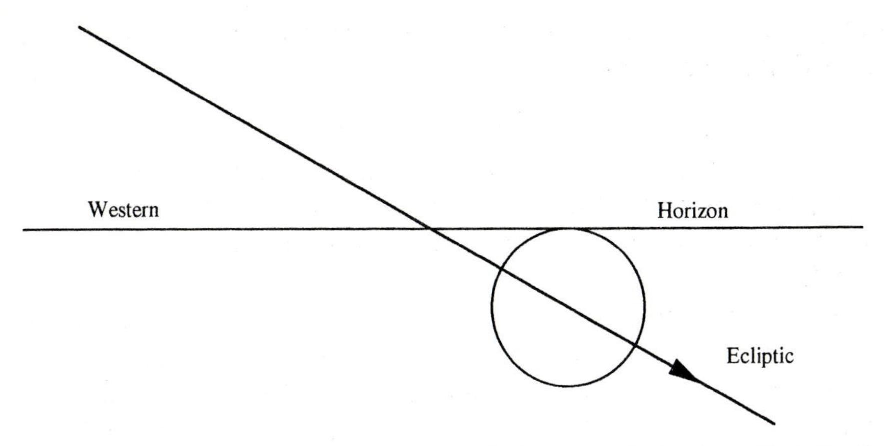

Fig. 4.22. The angle at which the ecliptic crosses the horizon affects the times of sunrise and sunset.

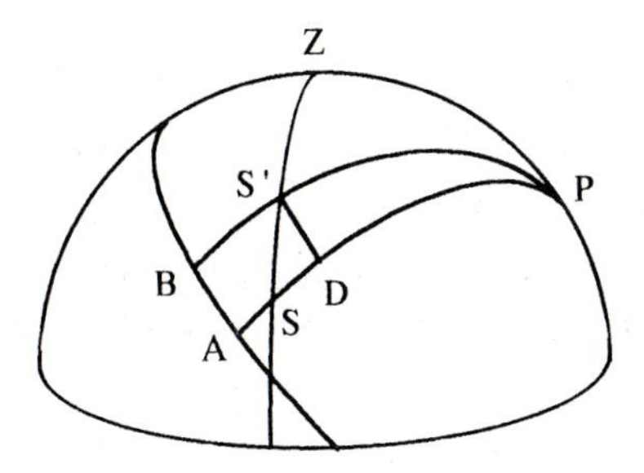

Fig. 4.23. The effect of refraction on the coordinates of a star.

of the two sides, SD and S′D, and we must also know the size of the angle DSS′.

Recognizing that angle PSZ is nearly the same as angle PS′Z, we may apply the cosine rule and solve for angle PS′Z. The lengths of all three sides in this spherical triangle are known.

Triangle DSS′ is always small, so we may safely consider it to be a plane triangle rather than a spherical one. Having solved for the angle DSS′ (i.e. PSZ) we may readily find sides DS and S′D. DS is the correction in declination, and S′D gives us the correction in right ascension, AB, from

$$S'D = AB \cos \delta$$

QUESTIONS FOR REVIEW

1. Draw a realistic sketch of the celestial sphere for an observer at 30°N latitude. Indicate the position of a star with a declination of 45° and an hour angle of −1 h. Be sure to show the star's hour circle, the horizon, the visible celestial pole and the zenith.
2. Calculate the altitude and azimuth of the star in the previous question.
3. Repeat the two questions above for the case when the star's hour angle is +2 h.
4. Repeat Question 1 for a star with a declination of −15°.
5. An observer at a latitude of 52° 13′ measures the meridian altitude of the sun to be 45° 47′. What was the local sidereal time at the moment of transit? Hint: You will need to refer to the *Astronomical Almanac*.
6. Calculate the galactic coordinates of Vega.
7. Draw a realistic sketch of the celestial sphere for an observer at a latitude of −41° 17′. Indicate the position of a star with a declination of −20° and an hour angle of 2 h.
8. Draw an appropriate sketch and determine the ecliptic coordinates of Rigel.
9. Look up the coordinates of Aldebaran for the equinox of 2000.0. Calculate the annual precession and determine the coordinates of Aldebaran for the year 2001.0.
10. An observer at latitude 46° 18′ measures the altitude and azimuth of a star to be 26° 12′ and 258° 51′. The sidereal time is 9 h 20 m. Find the equatorial coordinates of this star.
11. Find the great-circle distance between the radio observatories at Owens Valley, California and Green Bank, West Virginia.

Further reading

Green, R. M. (1985). *Spherical Astronomy.* Cambridge: Cambridge University Press. See Chapter 1 for derivation of all the pertinent formulae.

McNally, D. (1974). *Positional Astronomy.* Frederick Muller Limited. See Chapters III and IV for discussion of some of the material covered in this chapter.

Schaefer, B. (1989). Refraction by earth's atmosphere. *Sky And Telescope*, **77**, 3, 311. This interesting article contains a short computer program for determining the effects of atmospheric refraction.

Schutte, K. (1975). Fundamentals of Spherical Astronomy. In *Astronomy: A Handbook*, ed. G. Roth, p. 164, Sky Publishing Corporation. Applications of the formulae to specific problems are included in this article which is Chapter 6 in Roth's book.

Smart, W. M. (1977). *Textbook on Spherical Astronomy*, 6th edn. Cambridge: Cambridge University Press. Detailed derivations of the formulae relating to spherical triangles may be found in Chapter I. Applications of the formulae may be found in most of the other chapters.

Taff, L. G. (1981) *Computational Spherical Astronomy.* John Wiley & Sons. This book contains still another treatment of coordinate systems and transformations from one system to another.

van de Kamp, P. (1967). *Principles of Astrometry*. W. H. Freeman and Company. This book includes derivations of the principal formulae of spherical trigonometry as well as material on astrometry.

5

Photographic astrometry

The general term 'astrometry' is used to describe methods by which the positions of stars may be determined. We discussed in Chapter 1 the use of the meridian telescope in determining the right ascensions and declinations of stars, and we wish to repeat here that this work is of fundamental importance to all astronomers. However, when it is necessary to determine the coordinates of faint stars, comets, asteroids or other objects, the meridian telescopes usually cannot be employed. In the cases of comets and asteroids the objects are moving too fast for a precise setting to be made with a meridian telescope, and in the cases of faint stars and galaxies, the objects may be too faint to be seen at the eyepiece. It has been possible, though, for astronomers to develop some interesting and accurate methods of computing coordinates from measurements made on photographs, and these methods are described by the term 'photographic astrometry'.

The positions of stars on a photograph are directly related to their actual positions in the sky, so it might seem that the analysis should be simple and straightforward. This is not really the case. The geometry in Fig. 5.1 shows the basic relationships. The center of the lens of a telescope is at C and the focal plane is at FF′. A well-made lens should produce an image of plane GG′ in the plane FF′. The plane GG′ may be thought of as being tangent to the celestial sphere at point A. The sky appears as a spherical dome on which the stars appear as points. Thus, in a photograph a star at S is projected to T on the tangent plane, and an image of T is formed at T′. Measurement of a distance such as T′O on the plate gives a direct indication of the size of angle ACT, so if we know the equatorial coordinates, A and D, of point O, and if we know the direction of north on the plate, the problem of finding the coordinates of T′ should be a simple one. In actual fact, however, two other factors must be considered. First, the plate center is not usually known with precision, and second, we do not have an exact north–south reference line on the plate. The best that we can do is to measure the location of T′ with respect to an arbitrary x–y system with its origin at some arbitrary point located on or off of the plate.

Standard coordinates

We can, however, construct an idealized system in which the center of the plate coincides with the center of the field on the sky and the axes on the plate are parallel to a set of axes on the tangent plane. Rectangular coordinates in such an idealized system are referred to as 'standard coordinates'. Standard coordinates provide a useful system by which coordinates in the sky can be determined from measured positions on a photograph.

Standard coordinates may be said to represent the position of an object with respect to the point of tangency at A in Fig. 5.1, and their geometrical meaning may be understood from examination of Fig. 5.2. Here the

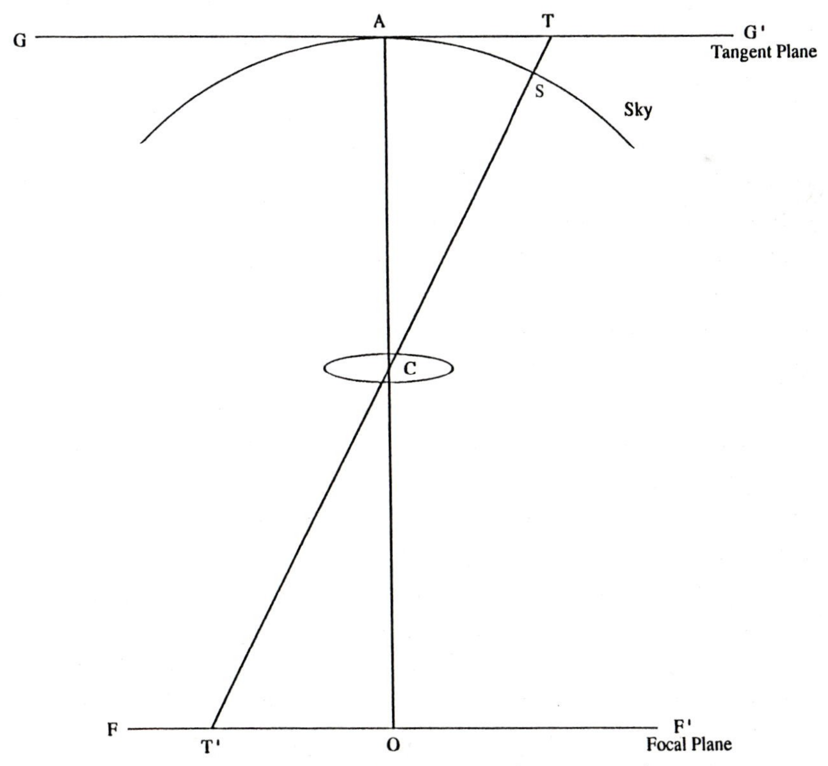

Fig. 5.1. Projection of the celestial sphere onto the tangent plane.

celestial pole is at P, and it is projected to Q in the tangent plane. A star is at S and is projected to T. In the tangent plane, then, T has the coordinates ξ' and η'. In the focal plane of the telescope these two coordinates are ξ and η, and these are the quantities which are referred to as the standard coordinates. ξ and η are expressed in terms of the focal length of the telescope as a unit.

Fig. 5.3 represents a portion of Fig. 5.2 and it shows clearly that two sides, PS and PA, and an angle, $(\alpha - A)$, in the spherical triangle APS are known. Thus, the angle θ and the side ϕ can be quite easily calculated.

$$\cos\phi = \cos(90° - D)\cos(90° - \delta) + \sin(90° - D)\sin(90° - \delta)\cos(\alpha - A)$$
$$\sin\theta = \sin(90° - \delta)\sin(\alpha - A)/\sin\phi$$

Great circles which intersect at A are projected into the tangent plane as straight lines, and the angles between the great circles will be equal to the angles between the straight lines. Thus,

$$\angle PAS = \angle QAT = \theta$$

Also in triangle CAT

$$AT = AC \times \tan\phi$$

and it follows that

$$\xi' = AT \times \sin\theta = AC \times \tan\phi \times \sin\theta$$
$$\eta' = AT \times \cos\theta = AC \times \tan\phi \times \cos\theta$$

The geometry within the telescope is symmetrical to that on the sky, so we may also write

$$\xi = OC \times \tan\phi \times \sin\theta$$
$$\eta = OC \times \tan\phi \times \cos\theta$$

Since we express the standard coordinates in terms of the focal length of the telescope, OC, we can simplify these to

$$\xi = \tan\phi \times \sin\theta$$
$$\eta = \tan\phi \times \cos\theta$$

Having pointed out above that values of ϕ and θ can be calculated from the coordinates of A and S, we note without derivation that standard coordinates may be calculated from the equatorial coordinates of the plate center, A and D, and of the star, α and δ:

$$\xi = \cot\delta \times \sin(\alpha - A)/(\sin D + \cos D \times \cot\delta \times \cos(\alpha - A))$$
$$\eta = (\cos D - \cot\delta \times \sin D \times \cos(\alpha - A))/(\sin D + \cot\delta \times \cos D \times \cos(\alpha - A))$$

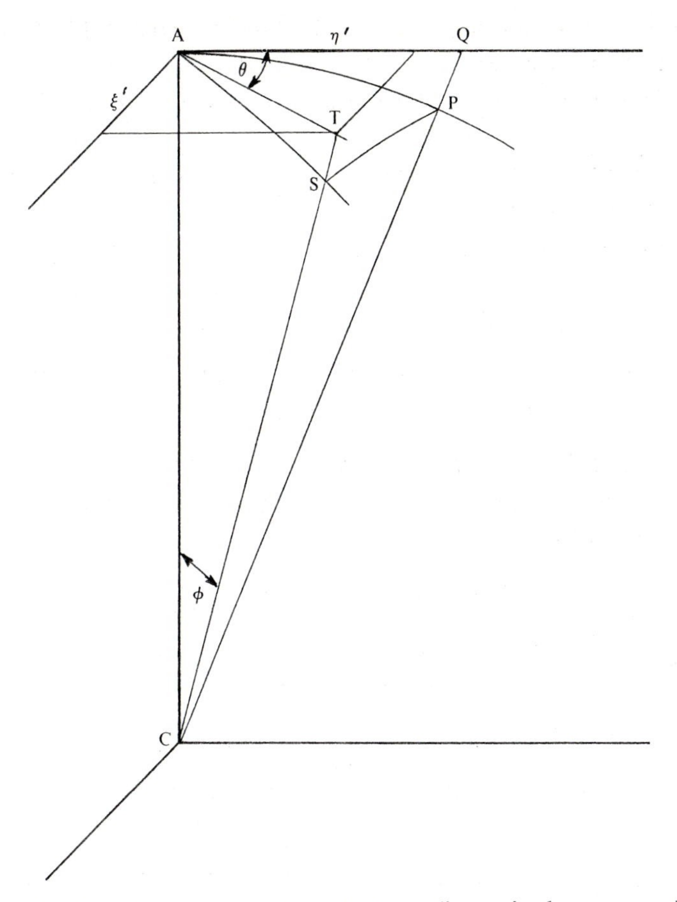

Fig. 5.2. The meaning of standard coordinates in the tangent plane.

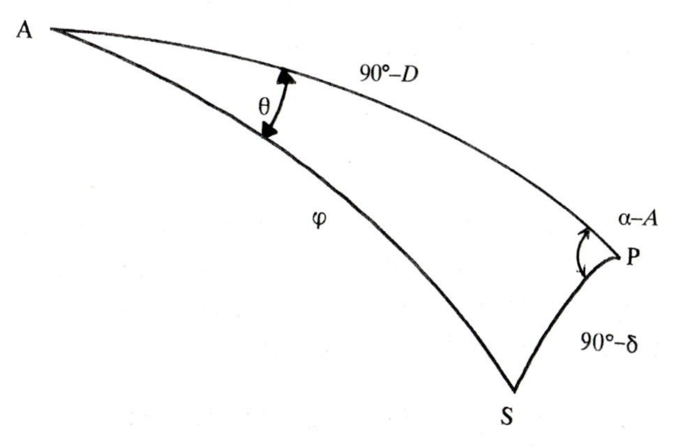

Fig. 5.3. The spherical triangle APS from Fig. 5.2.

It is also useful and necessary to have formulae by which α and δ may be calculated when the standard coordinates and the coordinates of the plate center are known. These formulae are written as follows:

$$\tan(\alpha - A) = \xi \times \sec D/(1 - \eta \times \tan D)$$
$$\cot \delta \times \cos(\alpha - A) = (1 - \eta \times \tan D)/(\eta + \tan D)$$

Measurements on the plate

Let us now see how the standard coordinates may be related to the measured positions of stars on a photographic plate. If we could locate the exact center of the plate, and if we could orient the plate so that its axes were parallel to the ξ' and η' axes on the tangent plane, then we could say that

$$\xi = x/\mathrm{OC} \text{ and } \eta = y/\mathrm{OC}$$

where x and y are distances on the plate measured along the ξ and η axes. x and y must, of course be expressed in the same units as the focal length, OC.

A number of complicating factors enter the problem to make the above relationships between x and y and ξ and η impossible to attain. We may list these factors as follows:

1. *Centering.* A and D may not be exactly known, and the corresponding point on the plate may not be easily found.
2. *Orientation.* The x and y axes on the plate may not be parallel to the ξ' and η' axes on the tangent plane.
3. *Perpendicularity.* The x and y axes may not be perpendicular.
4. *Tilt.* The plate may not have been perpendicular to the optical axis of the telescope during the exposure.
5. *Refraction.* Differential refraction can affect the positions of the images on the plate if the zenith distance is large and the field of view is wide.

Plate constants

All of the above factors combine with each other to cause the image of each star to be displaced by a small amount from its 'ideal' location on the plate. Furthermore, the amount of the displacement will depend on the location of the star on the plate. In other words, the displacements may be consistently larger in some portions of the plate than in others. Let us again assume that the measured quantities, x and y, are expressed in terms of the focal length of the telescope. Then we may note that ξ and η are related to x and y by the following:

$$\xi - x = ax + by + c$$
$$\eta - y = dx + ey + f$$

The quantities a through f are known as the plate constants, and they embody all of the errors introduced by the five factors described above. If the plate constants are known, then, these formulae may be used to find the standard coordinates of some object if its x–y coordinates on the plate have been measured. The standard coordinates thus determined lead in turn to the right ascension and declination of that object.

Due to the nature of the standard coordinates we may also write

$$\xi - x = a'\xi + b'\eta + c$$
$$\eta - y = d'\xi + e'\eta + f$$

If we consider now just the plate constants a, b and c, we realize that the plate must include at least three stars of known coordinates before we can solve for a, b, and c. For each of the three stars we may write an equation of the form just noted, since we can measure x and y and calculate ξ and η. We then have three equations and three unknowns, so we may solve for the plate constants, a, b and c. A similar procedure leads to the other three constants, d, e and f. If there are on the plate more than three stars of known position, then one may solve for the plate constants by the method of least squares.

Finding the unknown coordinates

Consider now the case of an asteroid for which coordinates are desired. On a photograph which includes the asteroid, measurements of three or more stars can give the six plate constants by the method described. The x and y coordinates of the asteroid may then be combined with the plate constants to give the standard coordinates of the asteroid. Finally, the standard coordinates may be used with the assumed coordinates, A and D, of the center of the plate to find α and δ, the coordinates of the asteroid.

There are a number of steps involved in this procedure, and it may seem complex. It is not a difficult problem to put into a computer, however, and those who expect to make even a few determinations of the positions of comets, asteroids or faint stars will find it worth their while to write the program.

Telescopes for astrometry

It was mentioned earlier that in photographic astrometry it was desirable to use a telescope with a wide field of view. The reason for this should now be clear. The wide field helps to insure that at least three stars with precise coordinates will be included on the photograph. Telescopes intended for astrometric use should, therefore, have some particular features and characteristics.

In the later years of the nineteenth century great strides were made in the application of photography to astronomy, and much effort was expended on the design of the best possible telescopes for the determination of precise positions. A wide field was desired, and so was minimal distortion around the edges of the field. Good correction for chromatic aberration was also sought in order to make the telescope as efficient as possible. A number of designs were tested, and ultimately several multi-element lens systems came into wide use. These included first of all the so-called 'standard astrograph', a telescope with an aperture of 13.4 inches and a focal length of 135 inches. A number of telescopes of this design were built, and they became the observational tools for the great Carte du Ciel program, an attempt to photograph the entire sky and catalog all stars brighter than fourteenth magnitude. This program was started in 1889 and has never been completed.

Telescopes intended for astrometric use are almost always refractors due to the fact that it is possible to make multi-element lenses which can produce photographs with good definition as far as three degrees from the optical axis. A long focal length is desirable also since this is what determines the plate scale. The greatest accuracy in measuring distances on a plate can be achieved when one centimeter on the plate represents only a small angle. Fig. 5.4 shows the relationships involved in calculating the plate scale.

We note right away that

$$\tan \phi = T'O/OC$$

Clearly, a long focal length results in a smaller value of ϕ for a given distance on the plate (one centimeter or one inch, for example). It is easy to calculate the plate scale in seconds of arc per millimeter or in minutes of arc per inch.

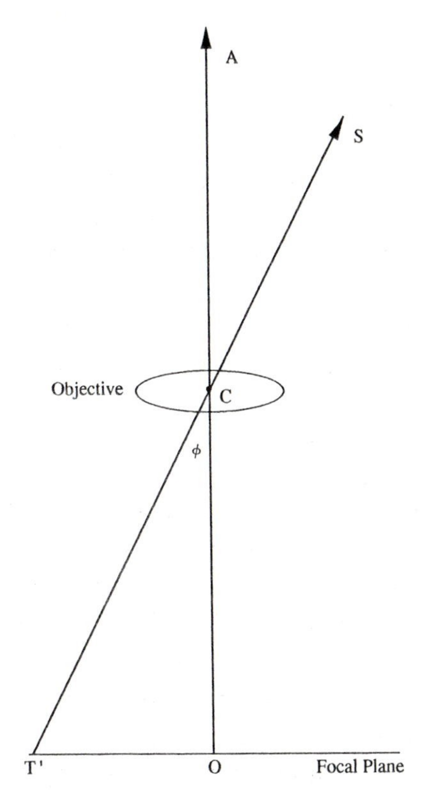

Fig. 5.4. Calculation of plate scale.

The CCD in astrometry

One of the great advances in astronomical imaging in the 1980s was the application of the charge-coupled device or CCD to problems in astronomy, and the applications included astrometry. The CCD is an array of very small photo-detectors or pixels. In some of the smaller examples there are more than one thousand rows and columns in an array only about 25 mm on a side. An image projected on to such an array causes a charge to accumulate in each pixel, and the amount of the charge depends upon the intensity of the light falling on that pixel. Row by row the charges may be transferred to a computer for storage, and the image can be reconstructed by the computer on command. When it is used in astrometry, the CCD acts both as the recording device and as the measuring instrument. The computer can be programmed to identify the images of stars and determine the x, y coordinates of those images. The reference by Monet at the end of this chapter describes the type of work that has been done.

A more detailed discussion of the CCD is included in Appendix 4.

QUESTIONS FOR REVIEW

1. Explain what is meant by the term 'standard coordinates'.
2. What must be known in order to calculate standard coordinates?
3. How are standard coordinates used to determine the unknown coordinates of an object such as a comet or an asteroid?
4. What are the advantages of using a two-dimensional array such as a CCD in astrometry?
5. Calculate the standard coordinates for a star at the position α=23 h 16.6 m, δ=+67° 50′. The nominal plate center is at α=23 h 30 m, δ=+68°.
6. Positions of several stars have been measured on a photograph, and the following plate constants have been determined:

a	−0.999968	*d*	−1.493850
b	−1.413165	*e*	−1.000471
c	2.336351	*f*	0.075562

An unknown object is located at x=4.6421, y=1.3177, and the center of the plate has the coordinates α=7 h, δ=+24°. Find the right ascension and declination of the unknown object.

Further reading

Green, R. M. (1985) *Spherical Astronomy*. Cambridge: Cambridge University Press. See Chapter 13 for a very lucid treatment of the process of going from measurements on a plate to equatorial coordinates.

Konig, A. (1962). Astrometry with astrographs. *Stars and Stellar Systems*, **11**, 461. See this paper for a more detailed treatment of the subject of the present chapter.

Monet, D. G. (1988). Recent advances in optical astrometry. *Annual Review of Astronomy and Astrophysics*, **26**, 413. In this paper Monet discusses the newest techniques in the observational side of astrometry. He reviews his own work with CCDs as well as that of many others.

Smart, W. M. (1977). *Textbook on Spherical Astronomy*, 6th edn. Cambridge: Cambridge University Press. Chapter XII with the misleading title, 'Astronomical photography', describes the basics of photographic astrometry.

van de Kamp, P. (1967). *Principles of Astrometry*. W. H. Freeman and Company. van de Kamp covers the basics very well in Chapter 5. In later chapters he describes special problems of astrometry with long-focus refractors. His research for many years dealt with very small variations in proper motion which might indicate the presence of planets.

6

Visual observations

In today's world astronomers seldom make extensive visual observations at the telescope. Instead they record their data photographically or electronically through an auxiliary instrument such as a camera, a CCD, a photometer or a spectrograph. With few exceptions, however, the final setting of the telescope is done visually so that the astronomer can be certain that the desired star is centered in the field. In the course of such work the visual appearance of a star can convey some important facts to the astronomer. We find that the appearance of the image depends upon the size of the telescope, the condition of the atmosphere and the characteristics of the eye. It is very helpful to be able to understand and recognize the effects produced by each of these factors.

The human eye

The astronomer's ultimate detector is, of course, the eye, and the eye is truly a remarkable organ. Let us describe its structure and the function of some of its important parts in the overall process of vision. Fig. 6.1 is a sketch of the right eye as it would appear looking down

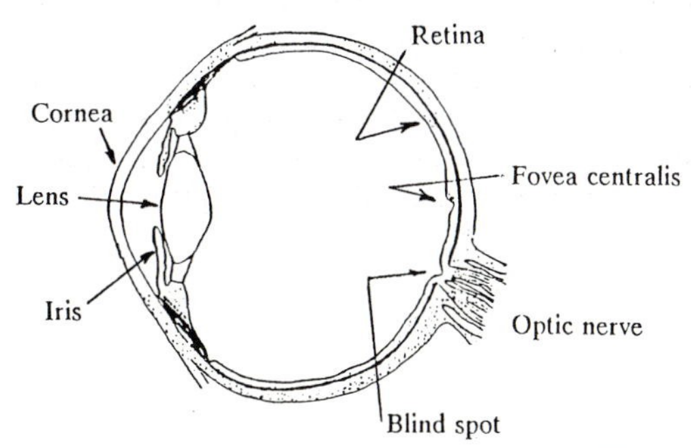

Fig. 6.1. Cross section of the human eye. This is the right eye seen from the top.

through the top of the head. The eye is essentially a spherical object which maintains its form by means of its tough outer layer, the sclera. The front, center portion of the sclera is the transparent cornea through which all light entering the eye must pass. Behind the cornea is the crystalline lens, and the two are separated by a small amount of clear liquid known as the aqueous humor. The eyeball is filled with a jelly-like substance, the vitreous humor, which helps it to maintain its shape.

Light is refracted at the outside surface of the cornea and at all of the interfaces within the eye. In a very real way, then, the optical characteristics of the eye are determined by the cornea, aqueous humor, crystalline lens and the vitreous humor. The focal length of the combination is about twenty-five millimeters, and this distance is the approximate diameter of the eyeball itself. Thus, we should expect an image to be formed on the inside surface of the spherical body of the eye. As the figure shows, the structure of the eye is such that rays may enter the eye from directions nearly 180 degrees apart. The inside lining of the eyeball is known as the retina, and it is in the retina that we find the very specialized cells which are the actual photodetectors. These will be described in more detail below.

If we consider for a moment just the optics of the eye, we realize that distant objects should be in focus on the retina, since the image distance should be equal to the focal length. Then if we try to look at nearer objects, the image distance should be greater, and the image should be focused behind the retina. By a process known as accommodation, however, the eye automatically corrects for a change in object distance by changing the focal length of the lens. This is done by muscles which surround the lens. When these muscles contract, the lens is squeezed in such a way that it becomes thicker, the curvature of its rear surface becomes greater, and nearer objects are brought into focus. As we become older the muscle which adjusts the lens becomes less able to do its job, and we find it increasingly difficult to focus on nearby objects. We compensate first by holding objects farther from the eye and later by wearing reading glasses.

The eye also possesses remarkable ability to adjust for differences in the level of illumination, and this is of considerable interest to astronomers. This process is known as adaptation, and it permits the eye to function through a range of brightness of ten billion to one. The ability to adapt to this remarkable range results from three mechanisms within the eye. The first and most obvious of these is a mechanical one made possible by the presence of an iris diaphragm located just in front of the lens. The iris is a sphincter muscle which can expand to an opening of about eight millimeters and contract to a diameter of about two millimeters depending upon the total illumination. The opening in the iris defines the hole that we commonly refer to as the pupil of the eye, and it is the color of the iris that determines the color of a person's eyes. In terms of area the extremes in the diameter of the iris offer a ratio of only sixteen to one – hardly enough to account for the great adaptive power of the eye. The balance of this adaptive ability is found in the retina, so let us turn now to the photoreceptive cells located there.

The surface of the retina is covered with two kinds of cells, the rods and the cones. Both types are sensitive to light, but each functions in a different manner. The rods respond to faint light more effectively than do the cones, and the cones serve better in bright light. Furthermore, it is only the cones that are capable of producing sensations of color. The distribution of the two types of cells is quite significant, for the cones are concentrated in the central part of the retina and the rods are found everywhere else. This accounts for the fact that when we wish to see a very faint object in a telescope, we often have to look slightly to one side of it so that the light can be focused on the rods rather than the cones. We often refer to this as averted vision. The nature of the rods gives us some insight into the reasons why it is hard to distinguish color in faint objects.

Two other important facts should be mentioned in connection with the distribution of the cones on the retina. It was mentioned above that the cones are concentrated in the central part of the retina. Going one step further now, we may add that there is a small region in which only cones are found and in which the cones are only about ten microns apart. The spacing of the cones determines in part the acuity of vision, so this is the area in which the greatest detail (the smallest angles) can be detected. This area, only about 0.25 mm in diameter is known as the fovea centralis, and it lies just outward of the optical axis of the eye. Taken together, the spacing of the cones and the diameter of the fovea dictate that the smallest detectable angle is about one minute of arc and the area of best definition is only about twenty-five minutes of arc in diameter.

The second point is that no cones or rods are found in the area at which the optic nerve leaves the eye. One may note in Fig. 6.1 that the retina is slightly depressed in this region. Since there are neither rods nor cones at this location, we have a blind spot, and light rays which fall here are not detected. The existence of the blind spot can be easily demonstrated by means of Fig. 6.2. The procedure is explained in the caption for the figure.

Returning now to the matter of the eye's ability to adapt to a great range in brightness, we must consider what actually happens when the rods and cones respond to light. In some truly remarkable studies scientists have found that when a photon strikes a rod or a cone, a measurable electric current is created. This current is passed on to the brain which recognizes that a stimulus has reached a particular part of the retina. When the level of illumination is high, as in daylight, the rods are not active, and the cones alone are sending stimuli to the brain. The sensitivity of the cones is only about one one-hundredth that of the rods, and when a subject goes from a well-lighted room into a nearly dark one, the cones are not able to detect the faint illumination. For a time the subject sees nothing, because the eye must switch over from cone vision to rod vision. This change does not occur all at once, so our night vision does not reach its maximum until large numbers of rods have become active.

There are actually three types of cones. They are distinguished by pigments within the cells, and these pigments determine the wavelength region to which the cone will respond. The wavelengths of the peak sensitivities are 4300, 5300 and 5600 ångströms, which means that they are in the blue, green and red. The brain processes signals from the three types of cones to give us our sense of color.

The response time for the rods of primates has been measured to be about 300 milliseconds, while that of the cones is about one quarter of that time. For this reason rapid motion can best be detected in bright light.

We should also point out that much of what we may describe as vision is not optical but is psychological. The eye is the detector, but the brain is the interpreter. What we 'see' depends quite a lot on how the brain interprets the optical stimuli. In astronomy in particular one's experience is a big help in trying to locate faint objects such as galaxies or planetary nebulae.

Fig. 6.2. A simple test for the blind spot. Hold the page vertically in front of you and close the left eye. With the right eye look directly at the +. Now move the page progressively toward you. You will see the square disappear when the page is about ten inches from your face. As you decrease the distance even more, the circle will disappear and the square will reappear. When the page is very close, you will again see both the square and the circle.

Astronomical seeing

The term 'seeing' has long been used in astronomy to describe certain effects of the earth's atmosphere on the visible images of the stars. We may describe these effects in terms of three modes of observation: the un-aided eye, a telescope with an aperture less than five inches (125 mm), or a telescope with an aperture larger than about twenty inches. Let us describe first the observable effect in each of the three cases. Then we can show how the observations have led to a simple theory which attempts to account for the effects.

Most casual observers are familiar with the idea that stars twinkle, and this twinkling or scintillation is usually seen as rapid fluctuations in the brightness of stars seen with the naked eye. When the stars are seen at large zenith distances (i.e., close to the horizon) they also seem to undergo rapid changes in color. On some nights the twinkling may be less noticeable than on other nights, and it is often less at high elevations. It is also frequently noticed that planets do not twinkle to the degree that stars do. Twinkling is just the naked eye manifestation of seeing.

In a small telescope one notices rapid movement of the star's image in the field of view along with changes in brightness, and an experienced observer can estimate the angle through which the image appears to move. On a night of average seeing the image may move randomly within an area about 2.5 arcsecs in diameter. When we speak here of 'small' telescopes, we refer to those with apertures of five inches or less.

The seeing effect is quite different in telescopes larger than twenty inches, for now one sees rapid changes in the size and shape of the image rather than changes in its position. Often the image appears to be a pulsating, formless blob in continual motion. It may momentarily break apart and then come together again. The average appearance of the image creates what is known as the 'seeing disc', and an experienced observer can estimate the size of this disc in arcsec. On a night of good seeing at a well-located observatory the seeing disc can be smaller than one arcsec. Seeing conditions would be considered to be very poor when the disc was as large as 10 arcsec. Optical theory requires that because of diffraction effects, the image of a point source should be a small disc surrounded by a series of progressively fainter rings. When conditions are especially good this pattern may actually be seen. When the seeing is poor, however, any semblance of the expected diffraction disc and rings will be impossible to recognize.

In Fig. 6.3 we have shown how the seeing differs in large and small telescopes. If a period could be assigned to the motion of the images, it would be the same in both cases. The differences in the observable effects of seeing are important clues to the factors that cause it.

telescopes. If a period could be assigned to the motion of the images, it would be the same in both cases. The differences in the observable effects of seeing are important clues to the factors that cause it.

A related observation may be made with a medium- to large-sized telescope on any clear night. The telescope should be pointed toward a bright star, and the eyepiece should then be removed. If the eye is placed near the focal plane, the observer can then look directly at the mirror or

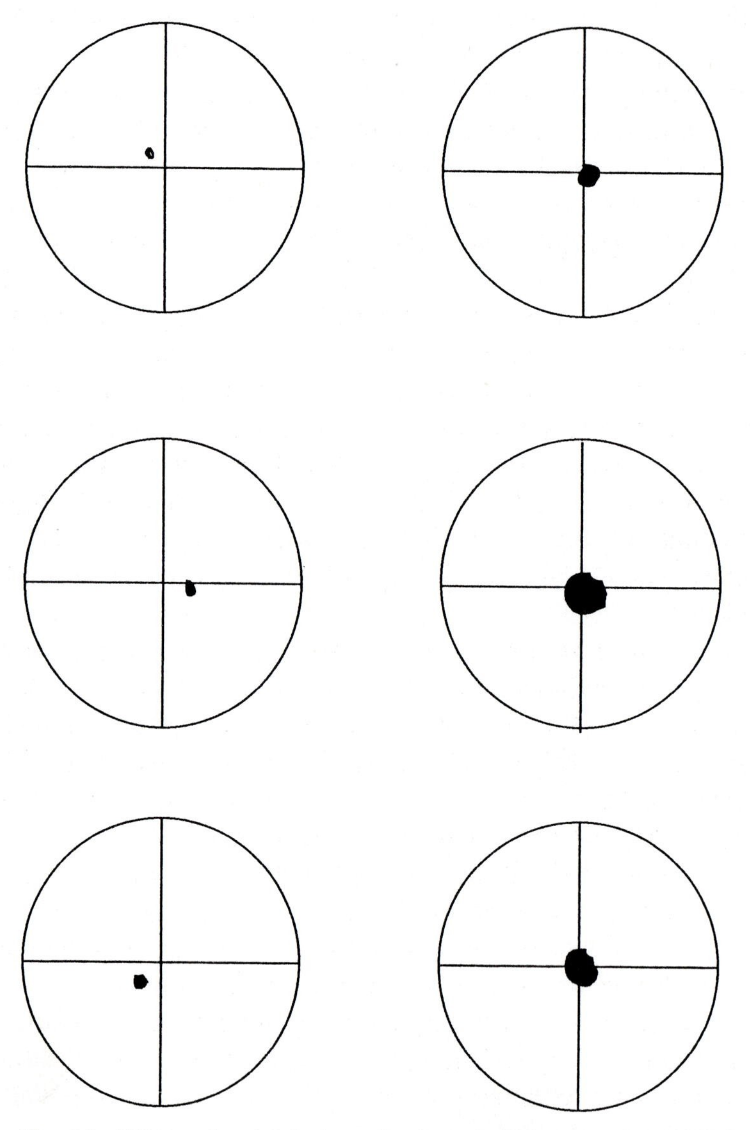

Fig. 6.3. Effects of seeing in a small telescope (left column) and in a large telescope (right column).

lens and see it illuminated by the star. One should then see a rapidly moving pattern of irregular dark bands on the objective. These observations suggest that the surface of the earth should be covered by an intricate pattern of light and shadow in constant motion. The distance between successive dark areas is in the range from 16 to 100 mm, and the pattern can have a transverse velocity roughly equal to that of the wind. This pattern may be compared to that seen on the bottom of a swimming pool when the sun is shining and the surface is covered with small waves.

Another aspect of the same effect is found in the shadow bands seen just before and after totality at the time of a solar eclipse. Shadow bands are patterns of light and dark which move across the surface of the earth at speeds up to tens of kilometers per hour. The period of the resulting fluctuations in brightness on the ground is comparable to that of the telescopic effects of seeing suggesting a common origin. Because of the sun's extended size in the sky, the shadow bands are not ordinarily visible.

From the observable effects which we have described here it is almost obvious that seeing is caused by small-scale variations in the index of refraction of the air above the observer. When the air is turbulent, there is extensive mixing of air of different temperatures coming from a number of heights above the ground, so we should expect to see the variations mentioned above. As the air moves above us the rays of starlight are refracted continuously in both amount and direction. From all of this a simple theory has been developed to help us to understand and predict the quality of the seeing.

Let us imagine that at some height above the earth the atmosphere consists of a layer of lens-like cells as suggested in Fig. 6.4. The cells could be volumes of air of differing index of refraction. Parallel rays entering these cells will then be refracted to produce patterns of light where rays are concentrated and patterns of shadow where rays are spread apart. Motion of the pattern on the ground is then explained by motion of this layer of cells above the ground. This concept can help to explain the patterns seen when we look directly at the mirror or lens of a telescope. The sizes of the patterns also give us a means of finding the average angular size of the cells above us in the atmosphere.

Now we are in a position to explain the differences seen when we use small and large telescopes. The aperture of a telescope defined here as small is not as large as the size of a cell projected on the ground. Therefore, as the air moves above us, rays of light entering the telescope come through only a part of one cell. The direction in which the rays enter the telescope is continually changing, so the star's image moves in the focal plane. In a large telescope the rays falling on the lens or mirror have passed through a number of adjacent cells. The rays are no longer parallel as they enter the telescope, and the telescope can only produce a large, blurred image.

Scintillation or twinkling results from the passage of the air mass above the observer and the resulting changes in intensity at ground level. The effect noticed for the planets is understood when we realize that the planet presents us with a disc that has a measurable angular diameter. If the angular diameter is great enough, the variations will be averaged through a column of air sufficiently large to eliminate the irregularities. In fact it is possible to notice a correlation between angular diameter and degree of twinkling. A planet such as Mars which is never more than 25 arcsec in diameter will often show some scintillation while Venus or Jupiter never show any when they are close enough to subtend considerably larger angles.

Careful studies have been made to relate the degree of scintillation to angular size, and these show that scintillation increases as angular size decreases until the angular size is only about 3 arcsec. This may be taken to mean that the angular size of an individual cell of turbulent air is about

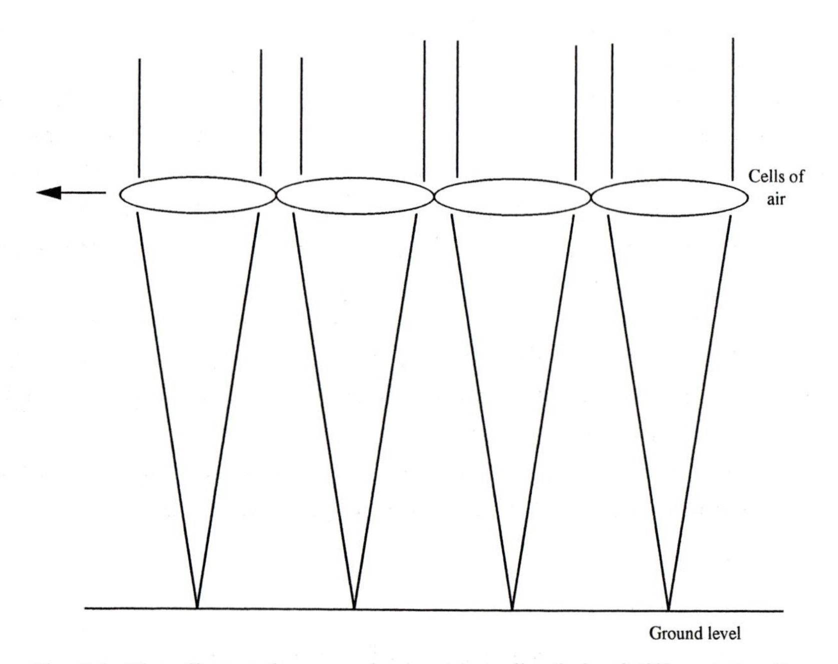

Fig. 6.4. The effect at the ground caused by cells of air of different density moving in the lower atmosphere.

3 arcsec. We said earlier that the patterns of light and dark on the earth had dimensions of the order of 100 mm. When we take these two facts together, we can calculate a distance for the turbulent layer that causes seeing. One hundred millimeters divided by the distance gives us the angular size of a cell as a small fraction of a radian. We may convert 3 arcsecs to a fraction of a radian by dividing by 206 265. We may equate these two and solve for *d*.

$$3/206\,265 = 100\text{ mm}/d$$
$$d = 100\text{ mm} \times 206\,265/3$$

The computed height, *d*, is about 7 km, and this is in good agreement with results obtained from balloon-borne telescopes.

Experience has also shown that there are other seeing problems which originate much closer to the earth's surface. The low-level effects become very important in the selection of the sites for new telescopes. Consider, for example, a ridge which runs north–south in a region in which the prevailing wind is from west to east. There will be a more or less constant flow of air over the ridge, and the air at the top of the ridge would be quite turbulent for a large part of the time. The ridge would not be a good site for a telescope. An isolated peak might be a better site because the moving masses of air could move around the peak. This might cause considerable turbulence on the lee side of the peak, but the summit might enjoy good seeing most of the time.

Sudden deterioration in seeing has sometimes been observed on still, windless nights at hilltop observatories. Cold air settles into the valleys and slowly pushes a layer of warmer air upward. When the boundary between the cold and warm layers reaches the height of the observer, the seeing suddenly becomes very bad. There is considerable turbulence at the interface, and this affects the seeing.

In the selection of sites for major observatories it is normal for tests of the seeing to be conducted over periods of several years. Only in this way can one have reliable data on the number of nights of good seeing to be expected in the course of a year.

The effects of seeing are, of course, eliminated from observations made using space telescopes in orbit around the earth.

Eyepieces

The final link in the chain of elements between the star and the eye is, of course, the eyepiece, and the quality of the observed image depends upon the characteristics of the eyepiece. In the simplest terms we may think of it as a single lens used as a magnifier. An image is formed in the focal plane by the objective, and the eyepiece is used to magnify that image as one might use a hand lens to magnify some small object. Most readers of this book will have already learned that the magnification of a telescope can be changed simply by changing the eyepiece, because magnification is given by the ratio of the focal length of the objective to that of the eyepiece. We increase or decrease the magnification simply by removing an eyepiece and replacing it with another one. There are limits, however, to the range of magnifications that can be used in practical observations. Too low a magnification can be wasteful of light, and too high a magnification can overemphasize all of the inherent defects in the image without showing any added detail. Consideration of the basic optical properties of eyepieces helps us to make the right choices and understand the limitations.

The eyepieces used in modern telescopes usually contain two or more lenses, and we shall discuss the reasons for this below. Let us first consider the case in which a simple, double-convex lens is used. The geometry is sketched in Fig. 6.5 where the objective is at O, and the eyepiece is at E. The image formed by O is indicated at I. The distances OI and EI are respectively the focal lengths of the objective and the eyepiece. We have drawn here some rays which indicate how a real image is formed by the objective and a virtual image is formed by the eye lens. Parallel rays from the object at infinity are refracted by the lens and converge to form part of the image. Diverging rays from the image are made parallel by the eyepiece, and these rays enter the eye. The angle at which these rays enter the eye defines the angular size of the virtual image. Magnification is then the ratio of angle a_E to angle a_O. The sizes of these angles are inversely proportional to the focal lengths of the two lenses, so we customarily write

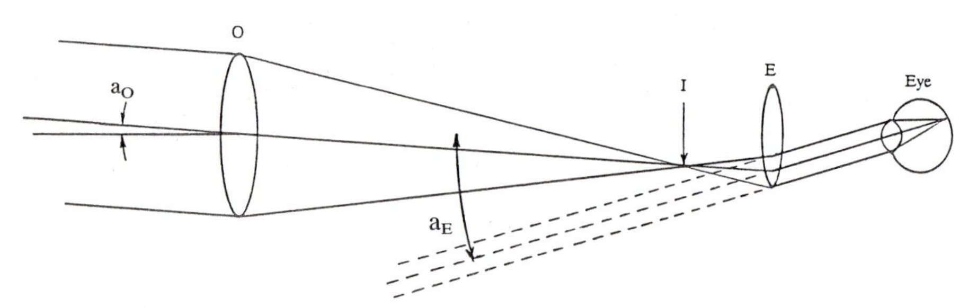

Fig. 6.5. Magnification results from the fact that the angular size of the virtual image, a(E), is larger than the true angular size, a(O).

$$\text{mag} = \frac{f_o}{f_e}$$

As drawn here, parallel rays enter the eye, and the virtual image is seen as if it were at infinity. When used in this manner, the eye is in its most relaxed position and does not tire easily. The location of the eye lens may actually be varied slightly without having a serious effect. If the distance EI is decreased, diverging rays will enter the eye, and the eye will have to adjust as if to focus on an object only a few feet away. This is shown in Fig. 6.6.

When we recall some of the simplest optical principles, we note that lens E will form an image of lens O. We have indicated this in Fig. 6.7 where the image of O is at O′. This image is referred to as the 'exit

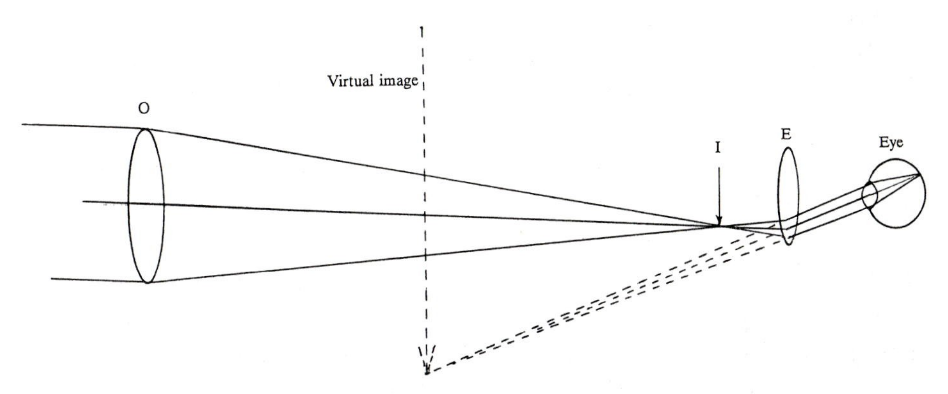

Fig. 6.6. The case when the eyepiece is too close to the image.

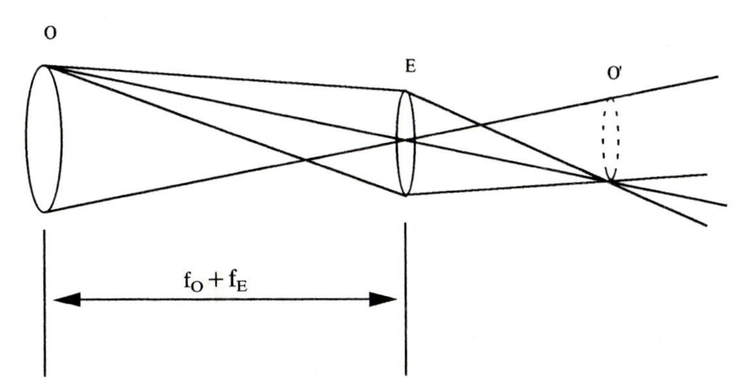

Fig. 6.7. Formation of the exit pupil which is the image of the objective. f_O and f_E are the focal lengths of the objective and of the eyepiece respectively.

pupil', and it is important for several reasons. First, we note that every ray entering the objective will have to pass through the exit pupil. This is where the beam emerging from the eyepiece will have its minimum diameter, so this is where the pupil of the observer's eye should be placed. If the exit pupil is larger than the pupil of the eye, some of the available light will be lost. Thus, in the design or selection of an eyepiece, care should be taken that the exit pupil is no larger than about 8 mm. To find the diameter of the exit pupil, first use the simple lens formula to calculate the image distance O′E. We recall the lens formula

$$1/F = 1/O + 1/I$$

and we substitute and solve for I, the image distance which is in this case the distance between the eyepiece and the exit pupil. We may illustrate this with an example in the case of a telescope of six-inch aperture and focal length of 90 inches. Consider that the eyepiece is a single lens with a focal length of three inches. The object distance is 93 inches, and we calculate that I is 3.1 inches. Then by a simple proportion

$$\mathrm{dia}(O')/\mathrm{dia}(O) = O'E/OE$$

Solving for the diameter of the exit pupil:

$$\mathrm{dia}(O')/6 = 3.1/93$$
$$\mathrm{dia}(O') = 0.2 \text{ in}$$

The exit pupil has a diameter of 0.2 in which is equivalent to 5.08 mm, so it can quite easily fit into the pupil of the eye. The magnification is, of course, 30×.

The distance O′E between the eyepiece and the exit pupil is known as the 'eye relief', and it is important that this should be large enough for comfortable viewing.

Let us consider an attempt to use a magnification of 300× with the telescope in the above example. We must use an eyepiece with a focal length of 0.3 in and the eye relief is now 0.301 in. The exit pupil is only 0.02 in or 0.51 mm in diameter. The exit pupil is now so small that only the center of the retina will be used, and the eye will not be able to produce its best results. In addition the observer's eye would have to be very close to the eyepiece.

An interesting and simple check of the exit pupil and eye relief may be performed in the daytime. The telescope is pointed toward the blue sky well away from the sun, and a card is held behind the eyepiece. At the correct distance a bright spot will be focused on the card. This is the image of the telescope's objective. Therefore, the diameter of the spot is the diameter of the exit pupil, and the distance of the card from the eyepiece is the eye relief.

From the above discussion it should be clear to the reader that with the proper assumptions about the exit pupil we should be able to calculate in a straightforward way the maximum and minimum useful magnifications for a particular telescope. This is greatly simplified if we first take note of some details which may be seen in Fig. 6.8. Here we see some rays which are parallel to the optical axis as they enter the objective and as they leave the eyepiece. These rays do not define the exit pupil, but we may assume that the eyepiece has been chosen so that the rays shown can enter the pupil of the eye. From the geometry here we see the following

$$D/d = F/f$$

Earlier we showed that the magnification is given by the ratio of the focal length of the objective to that of the eyepiece. This leads to

$$\text{mag} = F/f = D/d$$

In other words, we can express the magnification in terms of the diameters of the objective and the exit pupil. Now if we assume that the maximum diameter of the dark-adapted pupil is 8 mm we can say

$$\text{mag(min)} = D/0.8 \text{ cm}$$

D is measured in centimeters.

Similarly, if we assume that the exit pupil must be larger than 0.8 mm in order for the eye to achieve good definition, then

$$\text{mag(max)} = \text{D}/0.08 \text{ cm}$$

In the example used above this would give us maximum and minimum usable magnifications of 190× and 19×.

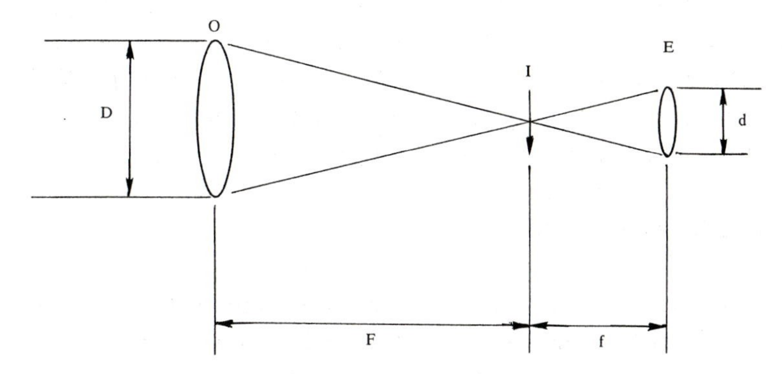

Fig. 6.8. Magnification is proportional to the ratio of lens diameters as well as to the ratio of focal lengths.

In general conditions the use of high powers is likely to be disappointing. When we magnify the good parts of the image, we also magnify the bad ones at the same time. On a night of poor seeing, for example, we usually achieve better results under lower powers. Also if the image quality is affected by lens aberrations, we magnify those and again make matters worse.

We have pointed out already that if too low a power is used, only part of the available light will be able to enter the eye, and the eyepiece will not be very efficient. The one situation when this may actually be advantageous is when one is observing the moon. The moon is usually so bright that one can well afford to lose some of the light.

Field lens

We have described so far the use of a double-convex lens as the eyepiece, but we mentioned earlier the fact that most eyepieces in use by astronomers are made with at least two individual lenses. Let us now outline the two principle reasons for the additional component or components.

In Fig. 6.9(*a*) we have again sketched an objective and an eyepiece with the image between them at I. We have drawn several off-axis rays which are refracted by O to form the extremity of the image. When we extend these rays beyond I, we see that some of them will not enter the eyepiece. The outer parts of the image will not be as efficiently illumin-

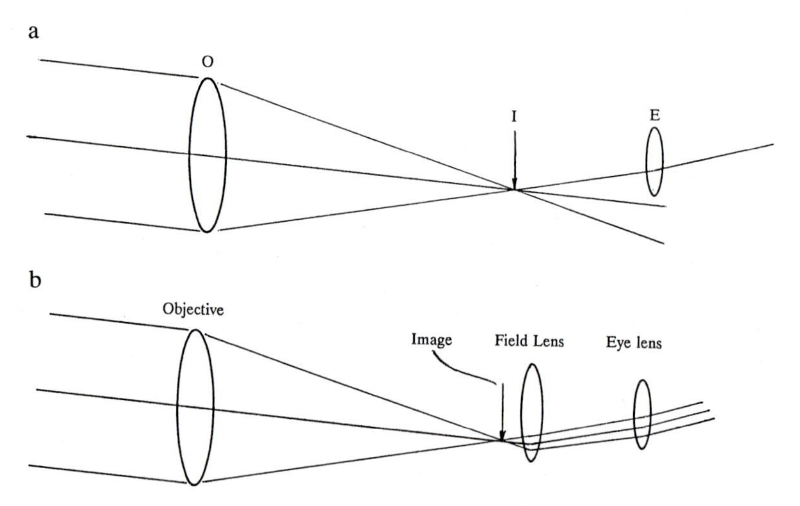

Fig. 6.9. (*a*) Some off-axis rays may miss the eyepiece. (*b*) The field lens directs off-axis rays to the eyepiece.

ated as the inner parts. The viewer will find that the angular size of the well-defined field is reduced. In order to correct this defect, one may place another lens in front of or behind the image. In Fig. 6.9(*b*) we have shown this lens behind the image, and one may note that rays which would otherwise have been lost are now refracted into the eye lens. The added lens is referred to as the 'field lens', because it actually increases and defines the angular size of the usable field of view.

The field lens also performs a second function in that it corrects for a type of chromatic aberration which would be introduced by the single lens used as an eyepiece. The problem in this case is known as lateral chromatic aberration, and it results from the fact that the magnification depends upon the wavelength. It can be shown that a compound eyepiece can be free from chromatic aberration if two conditions are met. First, both lenses must be made of the same kind of glass. More precisely, the refractive index must be the same for both lenses. Second, the separation between the lenses must be one half of the sum of the focal lengths of the two elements. Or if x is the distance between the lenses,

$$x=0.5(f_1+f_2)$$

In the design of a compound eyepiece one must know how to calculate the effective focal length of the two lenses used together. Without derivation we simply note that the appropriate formula is

$$1/f=1/f_1+1/f_2-x/(f_1\times f_2)$$

where x is again the distance between the two lenses, and f is the focal length of the combination.

Types of eyepieces

On the basis of the principles outlined above several types of eyepieces were designed many years ago and remain in common use today. We shall describe several of these and mention their advantages or disadvantages. The structure of each may be noted in Fig. 6.10.

1. *Huygens*: In this design the field lens is in front of the focal plane of the objective. This means that the image is located between the two lenses or inside of the unit. This type of eyepiece cannot be used as a magnifier. If crosswires are to be used, as in a finder, the crosswires must be placed inside. This can be done, but it is awkward, to say the least. At the observer's eye the angle subtended by a diameter of the field lens is approximately 40 degrees.

2. *Ramsden*: Here the field lens is behind the image plane of the

objective, so the eyepiece can be used as a magnifier outside of the telescope. With respect to the Ramsden we must point out that the separation of the two lenses does not conform to the criterion that was stated above for an eyepiece to be corrected for lateral chromatic aberration. The lenses are too close together, so this eyepiece is not entirely free of chromatic aberration. The angular field of view is about 35 degrees.

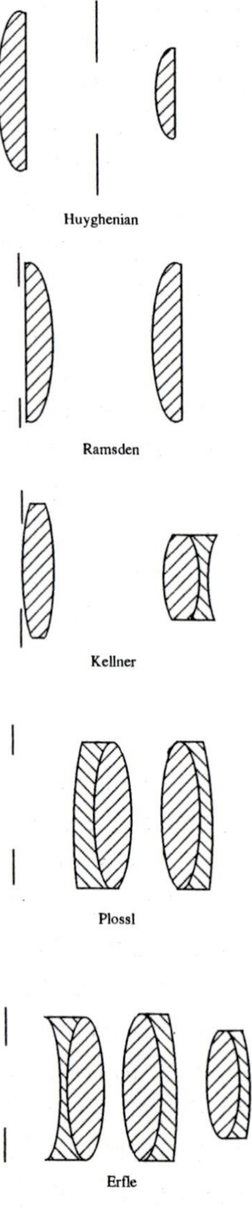

Fig. 6.10. Five designs for eyepieces.

3. *Kellner*: The distinguishing feature in this design is that the eye lens is an achromatic doublet. As in the objective of an achromatic refractor, one element is made of crown glass, and the other is made of flint glass. The Kellner is a versatile and widely used eyepiece. It is simple in its construction and therefore relatively inexpensive. The field of view is about 45 degrees.

It is hardly surprising to find that in recent times new designs have been developed in the ongoing quest for the most nearly perfect eyepiece. Let us continue by describing several eyepieces of the new generation.

4. *Plossl*: The parts of this design are sketched in Fig. 6.10, and we note the two identical doublets mounted close to each other. The elements made of Crown glass are facing each other. A field stop limits the angular field to about 40 degrees, and the eye relief is sufficient for comfortable use. Several makers have made small changes in design which resulted in considerable improvement in the functioning of these eyepieces. Many observers feel that this is one of the very finest designs in current use.

5. *Erfle*: This design may employ either three doublets or two doublets and a singlet. The principal advantage is that an angular field of 70 or 80 degrees may be obtained. This is offset by the fact that there is loss of light at each optical surface, so the overall efficiency is decreased. Eyepieces of this type were made in vast quantities during World War II for use in many items of military equipment, and many of these have found their way to the telescopes of both amateur and professional astronomers. The user of an Erfle eyepiece should be careful to note the size of the exit pupil to make sure that efficiency is reasonably large.

Field of view

In considering the view in an astronomical telescope, it is important that the user have some idea of the angular size of the area of sky seen in the eyepiece. Is it large or small compared to the moon, for example? Or is the optical view comparable to what might be displayed on a photograph? There are a number of ways to answer this question as one might imagine. The simplest method is to measure the time required for a star on the celestial equator to drift from one side of the field to the other when the telescope's drive has been turned off. This can be done with nothing more than a stopwatch. If it should take thirty seconds, for example, for a star to drift the diameter of the visible field, then the angular diameter of the field would be 7.5 minutes of arc. (Recall that four minutes of time are equivalent to an angle of one degree.) An angle of 7.5 minutes of arc is only one quarter of the angular diameter of the moon.

It is also quite simple to calculate the angular field of any combination of eyepiece and objective. This is done by simply dividing the angular field of the eyepiece by the magnification. For example, imagine that a Plossl eyepiece with a focal length of 25 mm was to be used with an objective of 250 cm focal length. An eyepiece of the Plossl design typically has an angular field of about 40 degrees. Thus, the combination would let us view an area of sky 40/100 degree or 24 minutes of arc in diameter. A word of caution is in order for those who might apply this method when using low powers and hoping to see a large area of the sky. Returning to the example cited here, we may easily calculate that in the focal plane of an objective of 250 cm focal length an angle of 24 arcmin is represented by a linear distance of roughly 1.75 cm. Now, since the field lens of the eyepiece is usually placed at or near the focal plane, it is obvious that the field lens must be at least 1.75 cm in diameter if the angular diameter of the field is to be the full 24 arcmin. The most common interchangeable eyepieces have an outside diameter of $1\frac{1}{4}$ in or 3.18 cm, so the field lens would be of adequate diameter in this example. If, however, we were to change the example so that the magnification was 50× instead of 100×, the rule discussed here would give an angular field with a diameter of 48 arcmin, and this would be larger than the field lens of the eyepiece.

The Barlow lens

There are a number of situations in which it is useful to observe with a magnification considerably higher than that which might normally be employed. Such might be the case, for example, if one was using a small telescope to view the planets. This can be accomplished quite nicely by means of the Barlow lens which is simply a negative lens placed just inside the focus of an objective. As may be seen in Fig. 6.11(*a*), the Barlow changes the angle at which rays converge, and in this way the effective focal length of the combination is increased. The magnification achieved with a particular eyepiece may be doubled through the use of a Barlow.

When parallel rays enter a negative lens as indicated in Fig. 6.11(*b*) the emerging rays appear to be diverging from a point behind the lens. This is point f in the figure. The focal length is then specified as the distance from f to the lens. When a Barlow lens is purchased, this focal length is specified by the manufacturer. Knowing the focal length of the Barlow one may calculate the effective focal length of an objective–Barlow combination using the following equation:

$$F=f_0 \times f_B/(f_B-d)$$

Here, and as indicated in Fig. 6.11(*a*), d is the distance between the original focal point of the objective and the Barlow. f_O and f_B are the focal lengths of the objective and of the Barlow. One should note in particular that the value of F can be varied simply by changing d. In practical use an observer would very likely wish to know just where the new focus would be located. The distance D between the Barlow and the new focal point may be found from

$$D=f_B \times d/(f_B-d)$$

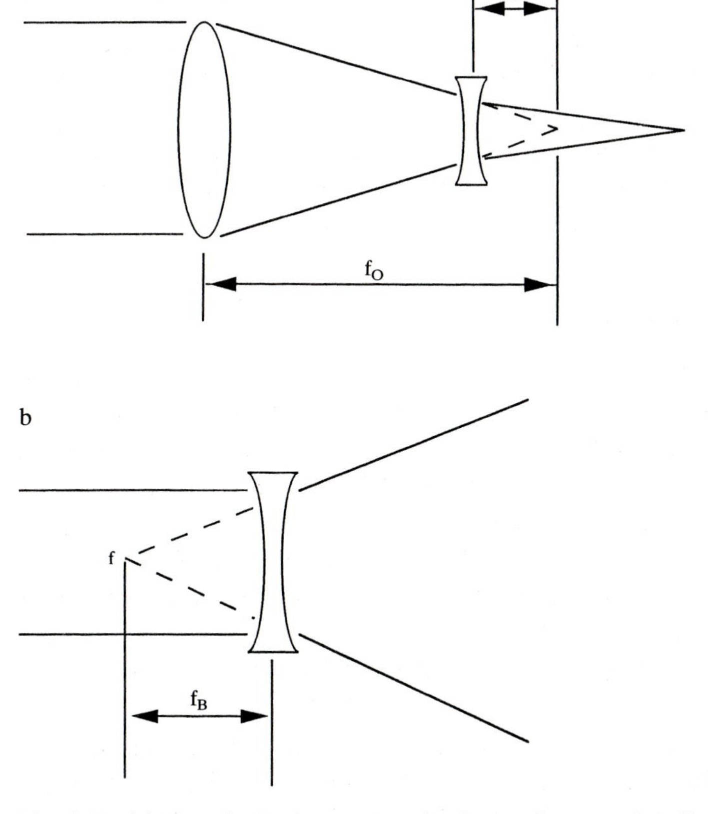

Fig. 6.11. (*a*) Use of a Barlow or negative lens to increase the effective focal length of a lens (*b*) The focal length of the Barlow lens.

Cross hairs and reticles

It is often quite useful to introduce into the light path some means of indicating the location of the center of the field of view. The most obvious case is, of course, that in which a finder telescope is used in conjunction with a larger telescope. The finder is typically a telescope of short focal length and wide field. If it has been carefully aligned so that its optical axis is parallel to that of the larger telescope, then an object centered in the finder will be centered in the larger one. The task of centering an object in the field is made simpler when two thin wires are arranged so that their intersection is at the center of the field. These wires are sometimes illuminated by means of a small lamp inside of the telescope. Cross wires must be extremely thin, and in a simple test one can easily see that even a human hair is really much too large for most applications. In former times it was quite common to use spider webs for crosswires, and this can still be done by a careful person.

A pattern more complex than crossed wires may sometimes be desired, and such devices are known as reticles. The term 'reticle' comes from a French word for 'net', and reminds us that in estimating size or position a pattern of intersecting lines could be quite useful. Today, reticles may be purchased in a great many patterns. They may be produced photographically with great precision. The most permanent ones are etched on glass.

The reticle or crosswires in an astronomical telescope should be placed in the focal plane of the objective. The eyepiece then magnifies the reticle at the same time that it magnifies the object under study. In a system which has not been properly adjusted, the observer can move the head slightly and see motion of the star with respect to the reticle. This parallax may be eliminated by adjusting the position of the reticle in the optical path. Parallax can be a serious problem when a microscope is being used to make precise positional measurements.

A specialized type of eyepiece known as the filar micrometer is used in the observation of visual binary stars. In this situation the observer wishes to measure the angular separation of the two stars and the orientation with respect to north of the line between them. These two quantities are known respectively as 'separation' and 'position angle'. The arrangement for doing this is seen in Fig. 6.12. The position angle is determined by rotating the entire assembly until the single central line passes through the images of both stars, and then reading the setting on a graduated scale. The separation is measured by adjusting the space between the two moveable lines until one passes through each of the

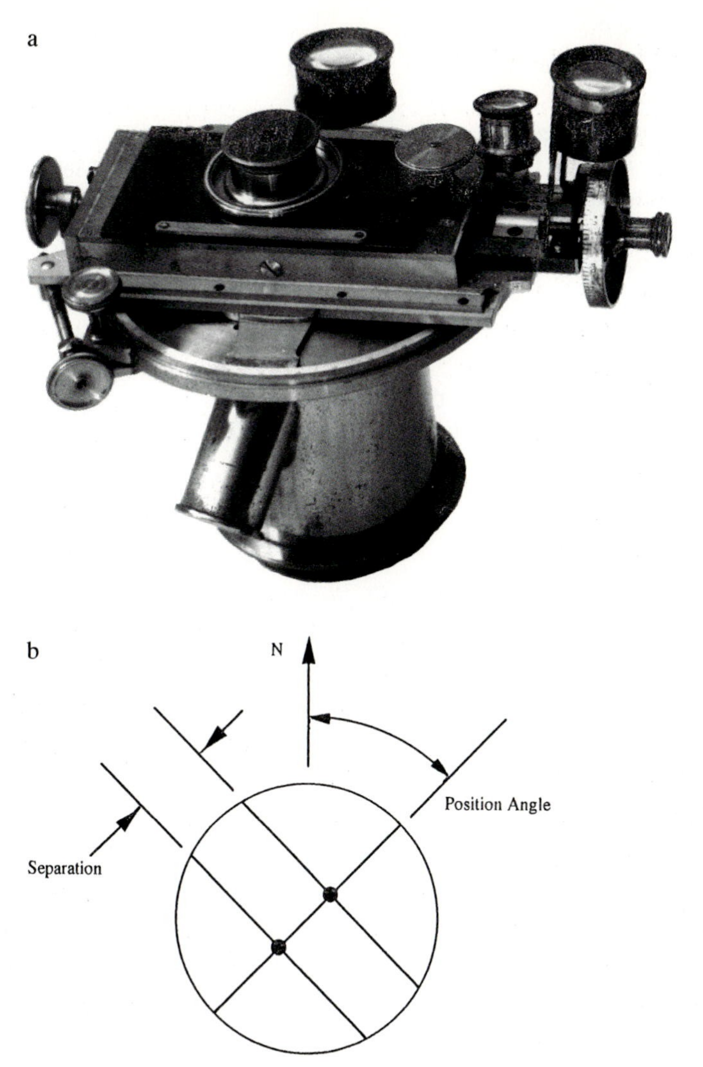

Fig. 6.12. (*a*) The filar micrometer used in the observation of visual binary stars. The knob at the right is used to adjust the separation. Position angle is adjusted by rotating the rectangular unit which carries the eyepiece. (*b*) The crosswires as viewed through the eyepiece.

images. The distance in scale units can be read from the instrument and then converted to an angular separation in the sky (see p. 78).

QUESTIONS FOR REVIEW

1. In what ways does the eye adapt to variations in brightness?
2. What are the roles played by the rods and cones in the process of vision?

3. What causes the effect which we refer to as 'astronomical seeing'? Why does this give observable effects which depend upon the size of the telescope being used?

4. Assume that a telescope with a six-inch aperture has a focal length of ninety inches. It is to be used with an eyepiece which has a focal length of one inch. The diameter of the field lens is 0.9 inch. What will be the diameter of the exit pupil? What will be the eye relief?

5. What will be the size of the piece of sky that can be seen with the eyepiece in question 4?

6. What factors govern the minimum magnification that can be efficiently used with a telescope?

7. Describe the applications of the Barlow lens.

8. In what way does the eye adjust itself to focus on objects which are close to it?

Further reading

Kuhn, R. (1975). Optical radiation receivers. In *Astronomy: A Handbook*, ed. G. Roth, p. 106. Sky Publishing Corporation. This is Chapter 3 in Roth's book, and it contains a concise treatment of visual observations. There is also some material on atmospheric effects.

Schnapf, J. L. and Baylor, D. A. (1987). How photoreceptor cells respond to light. *Scientific American*, **256,** 4, 40. In this article the authors present the newest results in the study of the way in which the eye responds to light.

Walker, G. (1987). *Astronomical Observations*. Cambridge: Cambridge University Press. In Chapter 4 Walker presents a more technical treatment of seeing and scintillation.

7

The magnitude system

Astronomers refer to the brightnesses of stars in terms of their magnitudes, with higher numbers indicating fainter stars. Thus, the brightest stars would be those of first magnitude, and those just barely visible to the naked eye on a clear, dark night of sixth magnitude. This scheme was introduced by Ptolemy in a catalog of stars which he included in his major work, Almagest, in the second century AD. The catalog lists nearly 1000 stars, and gives the position of each in the ecliptic system. It includes also Ptolemy's estimates of the magnitude, and this seems to be the first instance in which some scheme for specifying relative brightness had been used. With subsequent refinements this scheme has been in continual use ever since Ptolemy introduced it. Astronomers who have backgrounds in engineering may wring their hands over the more awkward aspects of the magnitude system, but it remains in worldwide use.

The practiced observer can learn to estimate the magnitudes of stars by memorizing the magnitudes of a few and using these as a frame of reference for the others. This was the general practice among astronomers from Ptolemy's time on through the years until the nineteenth century. Throughout this long period the human eye was really the only available detector, so precision was necessarily limited. Some observers who recorded estimates of magnitude were able to include notations indicating 'brighter than' or 'fainter than', but they could do little more than that.

Quantitative relationships

The general level of technology rose dramatically in the early nineteenth century as new methods of manufacture came into use. Machines and instruments of all types became more versatile and more precisely built, and astronomical telescopes shared in the advance. The same technology that produced better telescopes also began to produce the first auxiliary instruments that could assist the observer who was seeking more precise measurements at the telescope. Several of these new instruments were designed to permit astronomers to make better estimates of magnitudes, and observers began to list fractional magnitudes in their catalogs. As the observations proceeded from the bright stars to the fainter ones, two interesting points became clear. First, it was noticed that the original class designated as 'first magnitude' contained stars with too broad a range of brightness. Sirius, for example, was much too bright to be in the same class as Regulus. Second, the ratios of the brightness of stars in successive magnitude classes, while not uniform, was approximately 2.5/1. For example, a third magnitude star was about 2.5 times brighter than a fourth magnitude star. When the difference was two magnitudes, the ratio was 2.5 times 2.5 or 6.25, and so forth.

In about 1850 the astronomer, N. R. Pogson, at Oxford, England proposed the adoption of a precise magnitude scale that would formalize the loose relationship between magnitude and brightness. Pogson proposed that a magnitude difference of 5 should be exactly equivalent to a brightness ratio of 100/1. This may be said to be the formal definition of the magnitude scale as it has been used ever since. It is useful to write this definition in its mathematical form as follows

$$l_1/l_2=2.5119^{m_2-m_1}$$

where l_1 and l_2 are the measured brightnesses of two stars and m_1 and m_2 are their magnitudes. 2.5119 is the fifth root of 100, so whenever the difference in magnitude is 5, the brightness ratio will, in fact, be exactly 100/1. For computational purposes it is convenient to take the logarithm of each side and write

$$2.5\log(l_1/l_2)=m_2-m_1$$

When we examine either of these simple equations, two important points are noted. First, we can never determine the magnitude of a single star all by itself. We can only compare one star to another one and determine the difference between their magnitudes. Second, no zero point for the system is implied in this relationship. At some point astronomers must all agree that a certain star will be said to have a certain magnitude. After that, the magnitude of any other star may be found if the brightness of that star has been compared to that of the accepted standard star. The measurable quantity is the brightness, and the units in which this is measured are not important. Magnitude can only be calculated. The increase in precision of determinations of magnitude has followed the improvement in methods of measuring brightness, and magnitudes today can be known with errors as small as one thousandth of a magnitude.

Part of Pogson's original proposal was that the zero-point of the magnitude system should be based on average values for stars listed in early catalogs as being of sixth magnitude. With this base some of the brighter stars turned out to be brighter than first magnitude, so the scale had to be extended to zero magnitude and then to negative magnitudes. The magnitude of Capella, for example, is 0.1, and that of Sirius is −1.6. The system can also be extended to the brightest objects in the sky, so for Venus, the moon and the sun we find that the magnitudes are −4, −12.5 and −26.7. At the faint end, the scale is limited only by the sizes of our telescopes and the sensitivity of our detectors. Stars at least as faint as magnitude 23 have been recorded. It is impressive to note that a first magnitude star is more than six hundred million times as bright as a star of magnitude 23. We remind the reader that we are concerned here only with 'apparent' magnitudes. When we wish to compare the intrinsic properties of stars, we must calculate the 'absolute' magnitude from known values of distance and apparent magnitude.

In Table 7.1 we list for reference a few values of ratio of brightness and difference in magnitude.

As an exercise in the use of the magnitude formula the student should make some simple computations to see what the limiting magnitudes of telescopes of various sizes would be expected to be. One may start by assuming that the entrance pupil of the eye has a diameter of 8 mm and that the faintest star visible to the unaided eye under the best conditions has a magnitude of 6.0. We might ask what the magnitude would be for

Table 7.1. *Ratios of brightness*

Difference in magnitude	Ratio of brightness
1	2.51
2	6.30
3	15.84
4	39.81
5	100.00
10	10 000.00
15	1 000 000.00

the faintest star visible in a telescope of 25 mm aperture. The telescope functions much like a funnel in that it collects light in a large beam and compresses it into a beam small enough to fit into the eye. Thus, if we find the ratio of the areas of the telescope and the eye, we shall have the ratio of the brightnesses which produce equal responses in the retina. We may then convert the brightness ratio into a difference in magnitude.

$$(12.5\times12.5/4\times4)=l_2/l_1=156.25/16=9.77$$
$$2.5\times\log 9.77=2.47=m_2-m_1$$
$$m_2-6.0=2.47$$
$$m_2=2.47+6.0=8.47$$

Therefore, we should expect to see stars of nearly the ninth magnitude with a telescope of only 25 mm (≈1 inch) aperture.

During the nineteenth century astronomers went to great lengths to devise instruments with which they could make precise determinations of stellar magnitude at the telescope. These took many forms, but they are of historical interest only in today's world. Fig. 7.1 is a photograph of one of these, a wedge-type photometer.

Magnitude systems and color index

Photography became an important tool for astronomers in the latter half of the nineteenth century, and photographic photometry will be discussed in detail in the next chapter. Here, however, we must mention the fact that photography offered astronomers their first chance to make simple, quantitative measurements of the colors of stars. The earliest emulsions were most sensitive to blue light in contrast to the eye which is most sensitive to yellow light. The two curves in Fig. 7.2 show

Fig. 7.1. A wedge photometer for visual determination of magnitudes. The wedge is a filter which is more dense at one end than at the other. A measure of a star's brightness is obtained when the wedge is moved in front of the star until the increased density of the wedge causes the star to disappear. (Wellesley College photograph.)

the relative spectral sensitivities of the two detectors. The result of this is that when a photograph is compared to the visual appearance of the sky, the relative brightnesses of stars do not appear to be the same. A blue star will appear brighter on the photograph than it does to the eye, and a red star will appear brighter to the eye than it does on the photograph. It should be obvious, then, that the magnitudes of stars depend upon the means by which the stars have been observed. When we state the magnitude, we must also indicate how that magnitude was determined.

The photographic industry has for many years produced panchromatic emulsions which can very nearly reproduce the spectral sensitivity of the eye, so astronomers have been able to record the stars in a second region

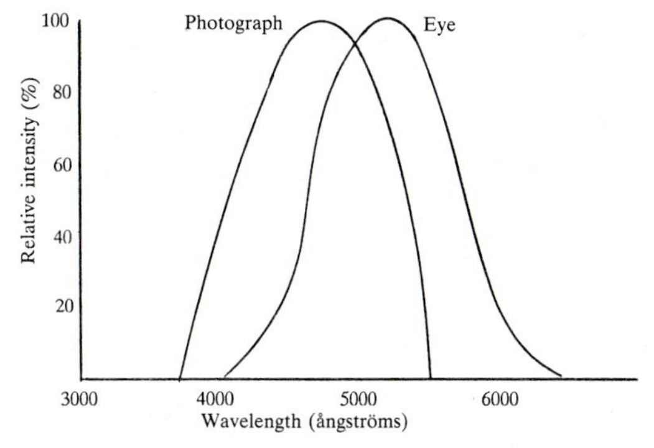

Fig. 7.2. Relative sensitivities of the eye and a common photographic emulsion.

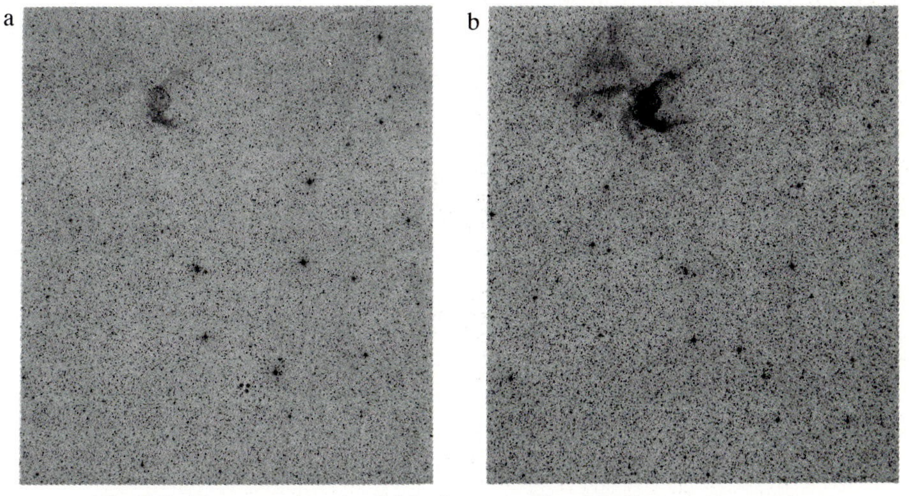

Fig. 7.3. Photographs of a field of stars made with filters and emulsions recording the blue light (*a*) and red light (*b*). © 1960 National Geographic Society/Palomar Sky Survey. Reproduced by permission of the California Institute of Technology.

of the spectrum. The two examples in Fig. 7.3 show clearly the difference achieved by using two different emulsion types.

Thus, with the ability to find magnitudes in more than one color system, astronomers are able to determine a 'color index' for any star. The color index is simply the difference between the magnitudes of a star in two color systems. For example, in the early days the term 'm_{pg}' was used to specify a magnitude found photographically, and the term 'm_v' was used to define a 'visual' magnitude. The difference, $m_{pg}-m_v$, is the color index. This quantity has the value 0 for a white star, and it can be 2 for a star which is very red. The color index for a blue star would, of course, be negative.

At the present time astronomers make use of a wide variety of photographic emulsions and photoelectric detectors. Each of these has its own spectral sensitivity, and each can be used with selected filters to isolate some range of wavelengths. Thus, stellar magnitudes may be determined in any system defined by a particular detector–filter combination. The possibilities are unlimited, but astronomers over the years have concentrated on just a few. In the early years the systems were chosen for practical reasons, but in more recent times they have been chosen for very specific astrophysical reasons. We shall discuss several of these in Chapters 8, 9 and 10.

International system

It was in the first years of this century that astronomers began to develop the techniques of photographic photometry. The need for standards to be used for reference was obvious and much effort was devoted to the selection and calibration of a suitable group of stars. At that time it seemed quite logical to choose as standards a few stars in the neighborhood of the north celestial pole, since most astronomers lived and worked in the northern hemisphere. In 1922 at the first meeting of the International Astronomical Union a list of standards was officially adopted and designated as 'The North Polar Sequence'. These stars define what is often referred to as the 'International System'. This work was largely the result of efforts by E. C. Pickering of the Harvard Observatory, and it was done with great care. Magnitudes in two colors were included for 135 stars, and these are specified as m_{pg} and m_v. A finding chart and a portion of the North Polar Sequence are included in Fig. 7.4.

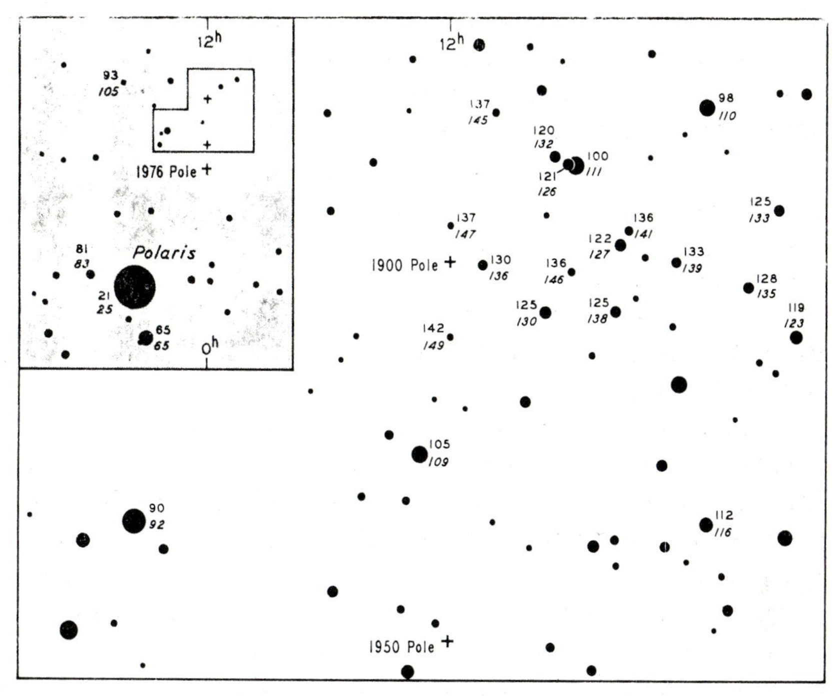

Fig. 7.4. The North Polar Sequence, a group of stars formerly used as magnitude standards. In the large portion of the diagram twenty stars have been identified by their magnitudes. The upper numbers are the photovisual magnitudes with the decimal point omitted. The lower numbers are the photographic magnitudes. The inset in the upper left indicates the location of the North Polar Sequence with respect to Polaris, and four more stars have been identified here. (Reproduced by permission of *Sky and Telescope* astronomy magazine, Cambridge, Mass.)

UBV system

With the introduction of photoelectric photometry in the 1950s it became possible to achieve very accurate magnitudes, and new standards were needed. At the same time it was recognized by H. L. Johnson and W. W. Morgan that observations in a third band could give astronomers much more information than was possible with just two colors. They built their three-color system around the 1P21 photomultiplier which will be described in Chapter 9. Their new band was centered at roughly 3600 Å in the ultraviolet and is designated simply as U. The other two bands very nearly match the wavelength ranges of the International System, and they are designated B and V for blue and visual. The spectral ranges and the central wavelengths of the bands were defined by the combination of the spectral sensitivity of the 1P21 and a set of filters which are listed in

Table 7.2. *Definition of the UBV system*

Band	Wavelength (Å)	Bandwidth (Å)	Filter
U	3650	680	Corning 9863
B	4440	980	Corning 5030 plus Schott GG-13
V	5480	890	Corning 3384

Table 7.2. The Corning filters are standard optical thickness, and the Schott filter is 2 mm thick. The two filters for the B band should be cemented together.

In order to duplicate the original instrumentation as closely as possible some other conditions should be met. The telescope should be a reflector with aluminized mirrors, and it should be 7000 feet above sea level. The photomultiplier should be refrigerated with CO_2. Very specific procedures were set forth by Johnson and Morgan for the reduction of the measurements made at the telescope to magnitudes. The procedures will be described in detail in Chapters 9 and 10.

Photographic magnitudes which are very close to the UBV system can be obtained if the filters and emulsions listed in Table 7.3 are used with a reflecting telescope. Glass lenses do not transmit in the ultraviolet, so only B and V can be determined with refractors. We shall see in Chapter 10 that many photometric systems have been introduced since the beginnings of UBV photometry. None of the others, however, have been as widely adopted as the UBV.

At the present time photoelectric standards have been established all over the sky. We have effective means of making transfers from photoelectric to photographic magnitudes, and so the North Polar Sequence is less widely used by professional astronomers than it was in earlier times.

Table 7.3. *Photographic definition of the UBV system*

Band	Emulsion	Filter
U	103aO	Schott UG2
B	103aO	Schott GG13
V	103aD	Schott GG495 (formerly GG11)

QUESTIONS FOR REVIEW

1. Write down an equation which defines the magnitude system.
2. Calculate the brightness ratio for two stars when the difference in magnitude is 3.78.
3. Calculate the magnitude of star A when it is compared to star B in the following example. The catalog magnitude of star B is 4.95, and star A is 32.9 times brighter than star B.
4. By what means are we able to specify the color of a star? What is the name given to the numerical specification of the color?
5. Imagine that two stars are so close to each other that they cannot be resolved in a six-inch telescope. If one star has a magnitude of 5.2 and the other has a magnitude of 8.4, what will be the magnitude of the two stars seen as one?

Further reading

Henden, A. A. and Kaitchuk, R. H. (1982). *Astronomical Photometry*. Van Nostrand Reinhold Company. See especially Section 1.2 beginning on page 5.

Kitchin, C. R. (1984). *Astrophysical Techniques*. Adam Hilger Limited. See Chapter 3 for another treatment of the background of the photometric system.

Miczaika, G. R. and Sinton, W. M. (1961). *Tools of the Astronomer*. Harvard University Press. See Chapter 5 for some basic material and for descriptions of early instruments for visual and photographic photometry.

Pannekoek, A. (1961). *A History of Astronomy*. Interscience Publishers. Consult the index for references to the origins and history of the magnitude system.

Struve, O. and Zebergs, V. (1962). *Astronomy of the Twentieth Century*. The Macmillan Company. Struve presents in Chapter V some interesting material on early methods used in photometry.

Walker, G. (1987). *Astronomical Observations*. Cambridge: Cambridge University Press. Refer to Chapter 1 for a concise treatment of the magnitude system.

8

Photographic photometry

It was not until the middle years of the nineteenth century that photographic practice reached what might be called the 'practical level'. The preparation and processing of plates could then be done in the field, and exposure times could be short enough to insure the widespread use of the new medium. Astronomers were quick to see the possibilities and to exploit them. The structure of the photographic emulsion (see Appendix 1) is such that photons can be accumulated in it until a detectable image has been produced. Thus, in astronomy, an exposure of many minutes can record stars that would be much too faint to be seen visually. Furthermore, the photograph records both the relative positions and the relative brightnesses of the stars. We discussed the question of positions in Chapter 5, so we are concerned here with the determination of stellar magnitudes from the photographic images.

When we examine a typical photograph of any region of the sky, we are immediately struck by the fact that the images of the stars do not all look alike. In Fig. 8.1, for example, the images of stars all look like small white dots in the photograph, but the images are not all the same size. Intuitively, we interpret the range in size to be an indication of a range in brightness. This is the correct interpretation, but the reasons behind it are not at once obvious. After all, the stars are point sources, and they should be photographed as points. When we photograph a landscape or the face of a friend, we expect each point in the scene to be faithfully reproduced in the picture, and this is what is normally achieved with a well-made camera. On a microscopic level we would find that the amount of detail in the common photograph is limited by the 'graininess' of the film, but this fact still does not explain why celestial point sources appear as dots of graded size in a photograph. In order to answer this question we take note of two facts.

First, the range of brightness in a typical scene is not very great. The

Fig. 8.1. Stars of the Pleiades and surrounding area (Wellesley College photograph).

brightest parts of a typical portrait are only a few times brighter than the dimmest parts. On the other hand, in a typical field of stars we might easily find images which are hundreds of times brighter than other images. Second, we must consider the actual microscopic structure of the image of a point source in the photographic emulsion. In Fig. 8.2(*a*) we have drawn, in an exaggerated way, a cross section of the emulsion and its glass backing. Photons focused on one point will enter the emulsion and sensitize any crystals of silver halide with which they collide. We should expect the exposed portion of the emulsion to appear as shown in Fig. 8.2(*b*). Here the blackened area indicates that in a small portion of the emulsion directly in the path of the light ray all of the crystals will very quickly be sensitized by the incoming photons. This small area is then said to be saturated and more photons will have no effect.

While most photons will go straight into the emulsion, a small number will be deflected or diverted while they are in the emulsion. The possible paths of some of these are shown in Fig. 8.2(*c*). Some of these scattered photons will be able to sensitize crystals along their new paths, so now we see that the actual image can be a conical volume of sensitized grains. If the exposure time is long or if the source is bright, the cone can become wider because there will be more photons scattered at larger angles. The net result is that on photographs of the sky the sizes of the images will in fact depend upon the brightness of the stars. The larger images are overexposed to a greater degree than are the smaller ones. This has been a somewhat idealized discussion, and there are other factors which are also relevant.

Atmospheric seeing is one such factor that has an important effect on the size of a stellar image. As we said in Chapter 6, seeing can cause rapid variations in the appearance of a stellar image. In a small telescope the image appears to be moving about over a small area, and in a large telescope the image may appear blurred and pulsating. In both cases the effect on a photograph is to cause the stellar image to become a spot of measurable diameter rather than a point. One should note here that seeing would tend to make all of the photographic images the same size, but now the density or blackness of each image would depend on the brightness of the star.

There are also some optical effects which contribute to the diameters of the photographic images. The diffraction disc and rings can never be eliminated in even the most carefully made system, so, again, point-like images are not to be expected. In reflecting systems the optical aberrations are such that it is quite normal for the images to appear as small spots in which, once again, the density depends on the brightness of the star.

The net result of all of these factors is that photographic images of stars are small spots of graded diameters and densities both of which depend upon the brightness of the stars. The problem for the photometrist is to devise some method of measuring these images and relating them to the scale of magnitudes. The density within an individual spot is not necessarily uniform, so it is not enough simply to measure the diameters of the images. One must try to measure both the diameter and the density at the same time since both of these quantities result from the total amount of light which formed the image. We shall describe the modern equipment by which this may be done, but let us first digress to describe more

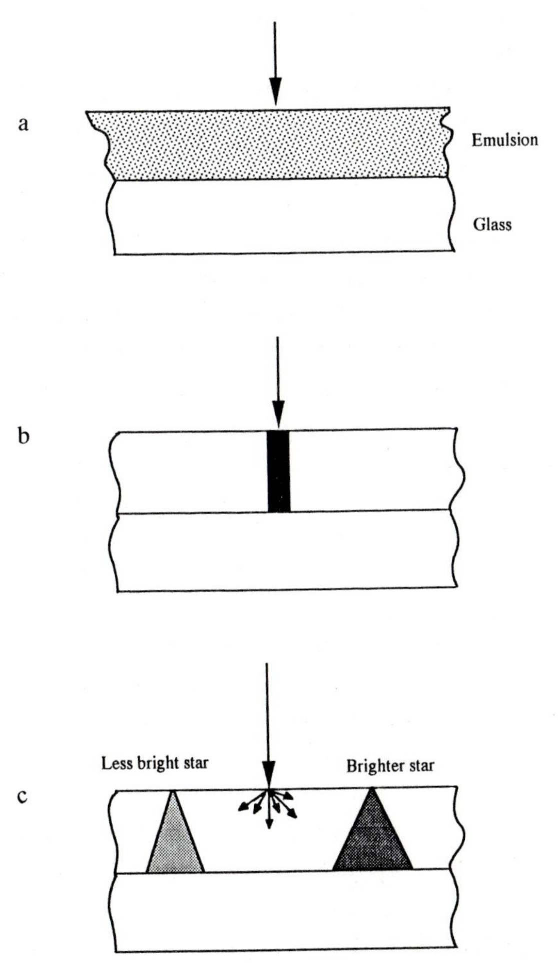

Fig. 8.2. Exaggerated structure of a photographic image of a star.

properties of photographic materials and some of the earlier, simpler alternatives.

The density of an exposed and developed emulsion depends upon the exposure, and the exposure may be defined as the product of the intensity of the incident light and the time during which the light falls on the emulsion. The response of a particular emulsion is described by the characteristic curve for that emulsion (see Appendix 1). A typical example of a characteristic curve is shown in Fig. 8.3. Here we note first that increases in total exposure result in increases in density. Along part of the curve which relates these two quantities the line is straight, and for exposures within a certain range density increases linearly with exposure. The lower curved portion indicates that very faint light may have no affect and that low total exposure results in underexposure. At the upper end we note that overexposure also results in a non-linear increase in density.

It is not difficult to design a simple instrument by which density of a developed emulsion can be measured, and in fact, a darkroom enlarger may be adapted to the purpose without serious modifications. The essentials of such a device are sketched in Fig. 8.4. Here a beam of light is caused to pass through the photograph and fall on a photo-detector of some sort. A mask or diaphragm is inserted to be sure that a uniform area of the photocell is used in all measurements. The current flowing in the external circuit may then be measured when the lamp is turned on. The system can be calibrated so that the measured current is a direct

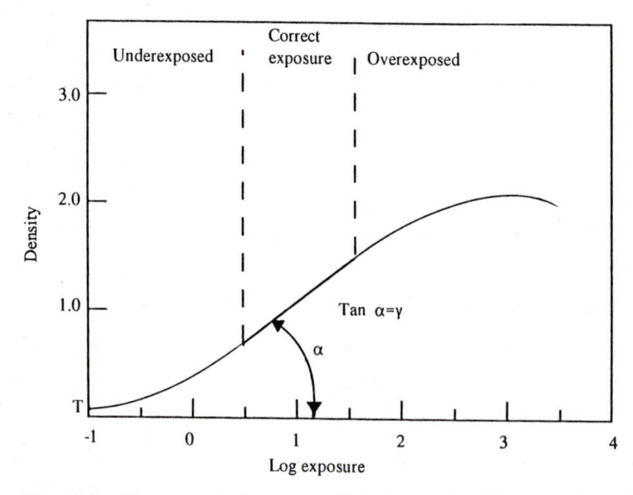

Fig. 8.3. Characteristic curve of a photographic emulsion. On the Density scale the letter, T, indicates the threshold density which is always present.

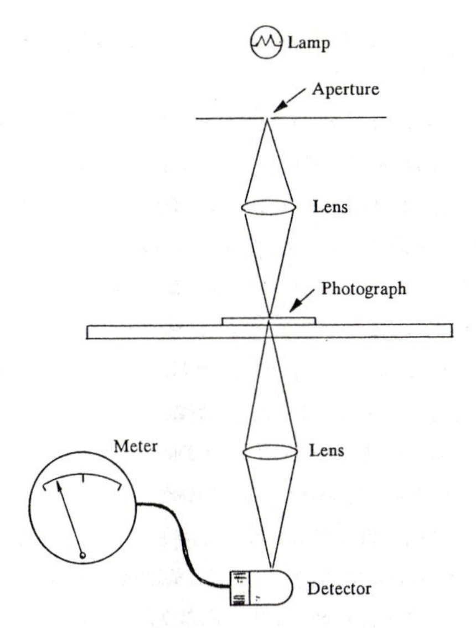

Fig. 8.4. Concept of a simple densitometer. A spot of light is focused on the emulsion. The light which is not blocked by the image of the star is measured by the photocell.

indicator of the density of the emulsion. The details of such instruments can, of course, vary considerably, but if care is taken to insure stability in all of the components, then relative and absolute densities can be determined with considerable precision.

Since density can be so easily measured, several astronomers designed special systems to take advantage of this. The first requirement was that the light should be spread out so that each star produced a small area of uniform density. The dimensions of all of the areas should always be the same. Thus, relative brightness could be determined from measures of the densities alone. The simplest way to do this is to move the plate to a position in which the images are out of focus. The images will then be of nearly uniform diameter, but their densities will be related to the stellar brightness. A sample of such extra-focal images is shown in Fig. 8.5.

It is interesting to note the lengths to which other workers went in their effort to obtain uniform areas in which to measure density. One rather complex device was known as the 'jiggle camera'. It was designed to move the plate continuously in the focal plane. The motion was random but was limited to a small square area. Here the developed images appeared as small squares of uniform dimensions with densities depending upon brightness. In another design small lenses were placed in the optical system near the focal plane, and these were focused in such a way

Fig. 8.5. Extra-focal images (Whitin Observatory photograph).

that each one formed an image of the objective on the plate. The arrangement of the lenses had to match the positions of the stars being photographed. In other words each star in the field was focused on a particular lens. For each field of stars a new plate had to be made to hold the small lenses. The final photograph recorded a pattern of images of the objective, and each image showed the objective uniformly illuminated by the light of just one of the stars to be studied. The originator of this idea was the French astronomer, C. Fabry. Because of its complexity this system never became widely used. We shall see in Chapter 9, however, that Fabry's use of an auxiliary lens to project an image of the objective on to the detector has found an important place in modern photoelectric photometry.

During an important period in astronomy, measurements on plates were made using a very simple scale of graded images. By varying the times of a series of exposures and shifting the position of the plate in the telescope between exposures, each star would produce a row of images differing by a fixed amount in terms of magnitude. The images for one star could be cut from the plate to make the comparison scale. This scale could then be compared to the images of stars on normal plates and values of magnitude could be read off. Some astronomers were able to obtain remarkable results with this simple arrangement.

The Cuffey iris photometer

To summarize the important problems in measuring stellar images on photographs, we remind the reader that as a result of seeing, photographic effects and optical problems, both the diameter and density of an image are affected by the brightness of a star. When we examine a photograph such as the one in Fig. 8.1, for example, we immediately relate the brightness of a star to the size of its image. Thus, the most nearly ideal photometer should be able to measure both the diameter and the density of an image at the same time. James Cuffey, then at the University of San Diego, designed and built such an instrument in 1950, and his principles have been the basis of the design of most plate-measuring photometers ever since. Fig. 8.6 is a photograph of a Cuffey iris photometer, and Fig. 8.7 is a sketch depicting the principal elements of the system.

In studying Fig. 8.7 one should notice first that light from the lamp goes along two different paths to the detector which in this case is a photomultiplier tube (RCA 931-A). One beam is the measuring beam, and the other is the monitoring beam. The 'chopper' is a segmented disc which spins in the light path and transmits first one beam and then the

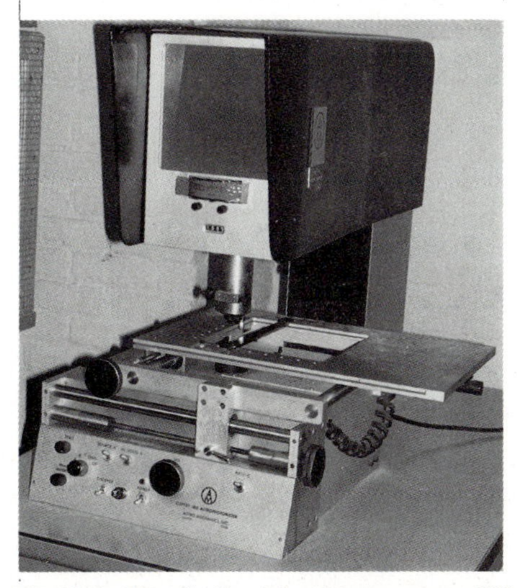

Fig. 8.6. The Cuffey iris photometer at Whitin Observatory, Wellesley College (Wellesley College photograph).

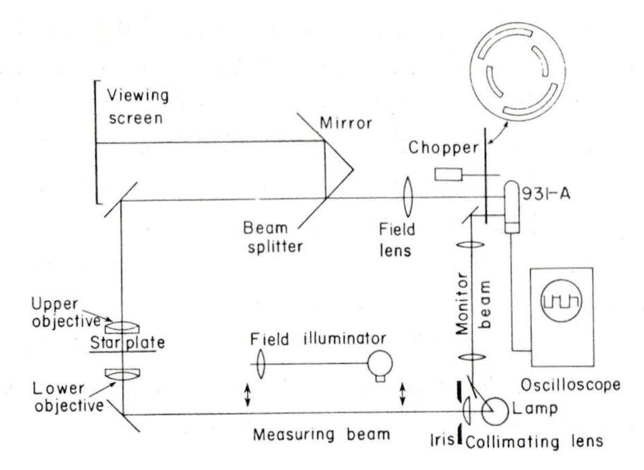

Fig. 8.7. Schematic diagram of the operating parts of the iris photometer. Reprinted by permission of the publisher from *Tools of the Astronomer* by G. R. Miczaika and William M. Sinton, Cambridge, Mass: Harvard University Press. Copyright © 1961 by the President and Fellows of Harvard College.

other. The output of the detector may be displayed on an oscilloscope as indicated in Fig. 8.8. Here the higher level indicates the intensity of light in the measuring beam and the lower level that in the monitor. An adjustable iris is placed in the light path, and a system of lenses projects

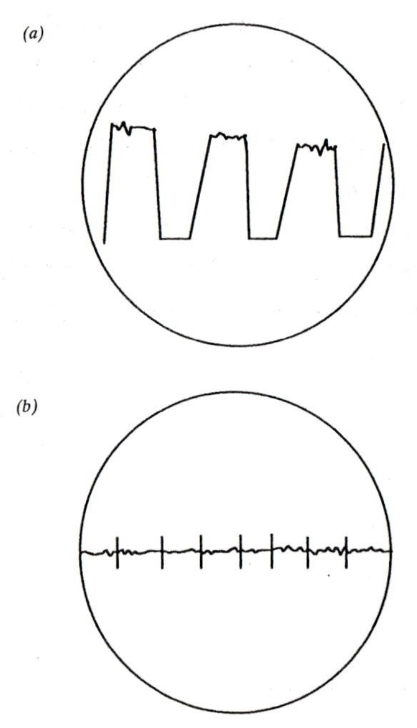

Fig. 8.8. Output from the photometer when the two beams are (*a*) unbalanced and (*b*) balanced.

an image of the iris on to the plate being measured. The plate may be moved around until the image of a star is centered in the projected iris. The image now blocks some of the light passing through the iris, and so the amount reaching the detector is reduced. The operator may now adjust the diameter of the iris until at some point the amount of light in both beams will be equal. This condition is indicated when the trace on the oscilloscope shows a straight line (Fig. 8.8(*b*)). The operator then records a number related to the diameter of the iris.

Clearly, the technique outlined above provides a measure of the combined effects of the diameter and the density of the image. The next step is to relate these measurements to the magnitude system.

Finding the scale and zero-point

Let us suppose for a moment that we have photographed a star field in which accurate magnitudes are already known for a number of stars. We measure the images of the stars using equipment and methods such as those described above. Now let us plot the settings of the iris against the magnitudes of the stars as indicated in Fig. 8.9. The points here fall along a straight line in this simplified example. If we also have an iris reading for a star of unknown magnitude, then we could find that magnitude either by reading it from the graph or by computing it from the equation of the best line through the points. Methods of finding the equation of a line representing a set of points such as these are discussed in Appendix 3. We may note also that the solutions described there give a measure of the precision of the computed result.

The line representing the points in Fig. 8.9 is, of course, defined by its slope and its intercept, and in the case of magnitudes these quantities are respectively the scale and the zero-point of the magnitude system on the

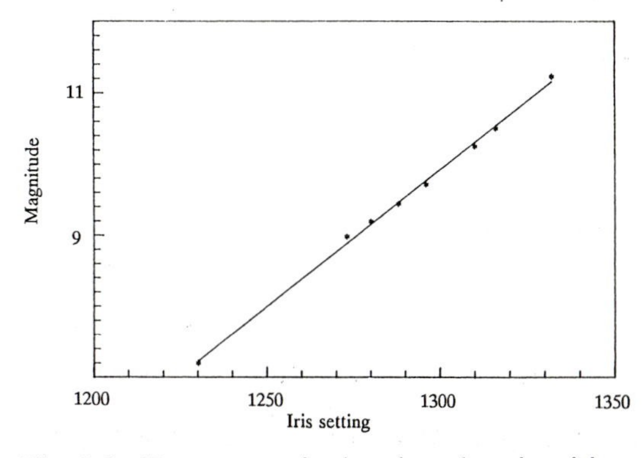

Fig. 8.9. Known magnitudes plotted against iris readings for a group of stars.

plate being measured. We need a minimum of two points to define the line, so we must know the magnitudes of at least two stars before we may begin. Our confidence in the computed magnitude of a star will increase to the degree that we are able to increase the number of comparison stars for which we have well-known magnitudes. Later in this chapter we shall outline a method by which the scale may be determined from photographs, but an accurate zero-point requires that there must be at least one star of known magnitude on the plate.

Example: In order to illustrate the application of the procedures described above, let us analyze the data in Table 8.1. The star field is the one shown in Fig. 8.10 which is the AAVSO[1] chart for the variable star RX Piscium. The magnitudes in the table have been taken from the chart, and the magnitudes may be used to identify stars on the chart. The numbers on the chart are the magnitudes with the decimal point omitted.

This field was photographed by students at Wellesley College, and the plate was measured on the iris photometer there. The measurements are listed in the table under the column marked 'Iris'.

In Fig. 8.11 magnitude has been plotted against iris reading, and a line has been fitted through the points. The equation of the best-fitting line is

$$Y=0.01492148X-14.70978$$

or

$$\text{mag}=0.01492148\ \text{iris}-14.70978$$

This equation was found by means of the method of least squares which is described in Appendix 3.

We may now determine the magnitude of any other star on the photograph by substituting the iris reading for that star into this equation.

Table 8.1. *Data from a photograph of RX Piscium*

Iris	Magnitude
1770	11.7
1764	11.4
1793.7	12.0
1660	10.0
1665	10.3
1675.3	10.3
1839	12.8
1858.3	13.1

[1]The AAVSO is The American Association of Variable Star Observers. See Chapter 14.

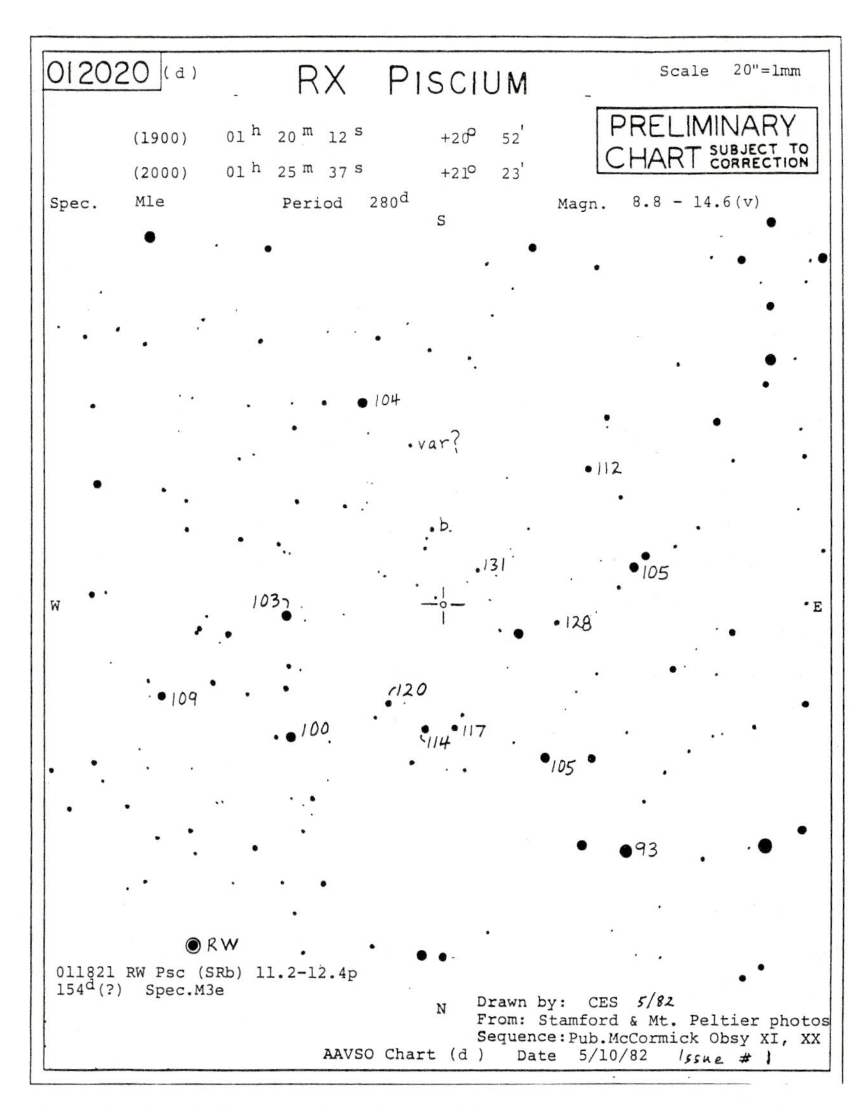

Fig. 8.10. Star field centered on the variable star, RX Piscium. This has been reduced from the original, so the scale is not what is indicated in this illustration. (Courtesy of AAVSO.)

The numerical solution from which the equation for the line was determined also gives us the standard deviation of the magnitudes. In the present example we note that the standard deviation, σ, is ±0.12 mag. This is not a particularly good result in terms of what can sometimes be achieved photographically, but it is not unexpected when we look again at Fig. 8.11. There is noticeable scatter on both sides of the straight line

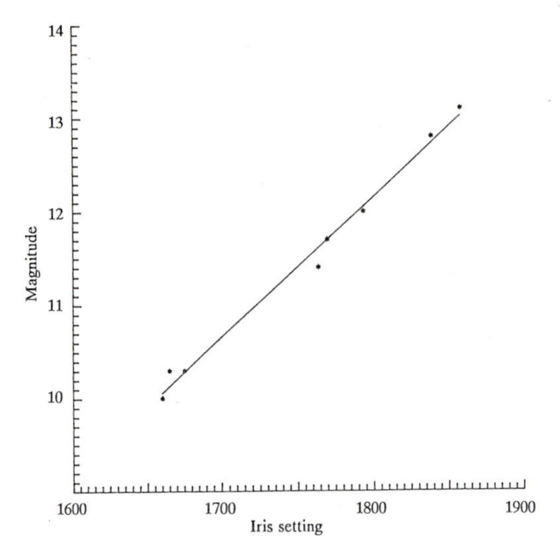

Fig. 8.11. Computer plot of magnitude versus iris reading from a photograph of RX Piscium.

fitted through the points. Scatter along this line may be due to some or all of the following causes:

1. errors in the 'known' magnitudes;
2. errors in making the measurements;
3. photographic effects.

Errors in measuring can result when the image is poorly centered in the projected iris or when the iris is either too large or too small. The iris photometer usually includes a set of filters for changing the intensity of the measuring beam. In a properly adjusted instrument, balance between the two beams will be achieved when the iris does not block the edges of the image and does not extend far beyond the limits of the image. Care and experience are needed to insure the best results.

In some cases the stars of known magnitude may cover a relatively large range of magnitude and the graph of magnitude versus iris reading may be a curve rather than a straight line. This curve may resemble that in Fig. 8.12. The general form of this one is similar to that of the characteristic curve of a photographic emulsion (see Appendix 1), and this need not be surprising. At its upper end the curve bends toward the horizontal because the brightest stars are overexposed. At the lower end the curve bends the opposite way to approach the horizontal, and this time it is due to underexposure. One should avoid the extremes if a numerical solution is to be based on the fit of a straight line to the points. On the other hand,

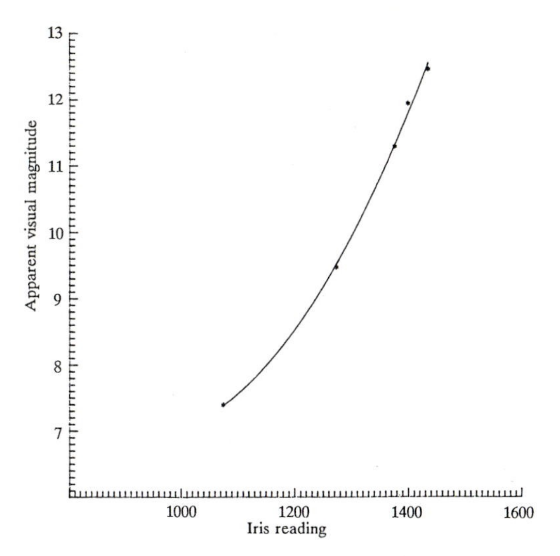

Fig. 8.12. Plotted magnitudes and iris readings for a photograph of W Piscium. Clearly, a curve fits this data better than a straight line would.

more of the points might be used if a second-order curve (Appendix 3) was to be fitted to the data.

Methods of finding the magnitude scale

The steps outlined in the previous section might have frequent application in the study of variable stars. A group of unchanging comparison stars might be selected and used to establish the scale and zero-point of a long series of plates. A graphical or a numerical solution would be required for each plate, since differences in exposure and processing cannot be avoided. For best results the magnitudes of the comparison stars should be as precise as possible, and this means that they should be determined photoelectrically by the methods to be described in the next chapter.

There are, however, several photographic methods by which the magnitude scale may be determined, and we shall describe two of these. In each method multiple images are obtained on the plate for each star. The images differ by amounts which are equivalent to a known difference in magnitude.

In the first method we simply make a double exposure with a slight change in position of the plate or pointing of the telescope between exposures. If the exposure times are properly chosen we obtain pairs of images which have the appearance of two stars differing in magnitude by some pre-selected amount. In preparation for this, the ratio of the two required exposure times may be found from the basic equation by which the magnitude system is defined. As we shall see in the next paragraph, however, a small empirical adjustment in that equation is required.

In a developed photographic emulsion density depends upon the intensity of light (I) and the time of exposure (t), and we might write

$$\text{density}=k(I\times t)$$

We should be able to obtain the same density by using a less-bright source and a longer time, or

$$I_1\times t_1=I_2\times t_2$$

It has been found, however, that this expression does not hold. With fainter light we need a longer exposure than would be predicted by the direct relationship. This is known as 'failure of the reciprocity law' (see p. 300). Experiments have indicated that t should be raised to the power, p, and p is approximately equal to 0.8. We could write this as follows

$$\log (I_1/I_2)=p \log (t_2/t_1)$$

When we substitute this in the magnitude equation we have

$$2.5\, p \log (t_2/t_1)=m_2-m_1$$

for a single source exposed for two different times. If we choose m_2-m_1 to be 1.0 and let $p=0.8$ we may solve this for t_2/t_1. We find that the desired ratio of exposure times is 3.16, so an exposure time of $3.16t$ gives images which are the equivalent of those formed by stars one magnitude brighter. If we want an interval of only half a magnitude, then the ratio of the times is 1.77.

This method has the disadvantage that the two exposures cannot be made at the same time. Errors will be introduced if there is any change in the condition of the sky from one exposure to the next.

A second method of obtaining graded images involves the use of a device known as an 'objective grating'. In essence the objective grating is a very coarse diffraction grating placed in front of the lens of the telescope. The theory of this will not be described here, so we shall simply say that the grating produces a central image with a series of spectra on either side of it. A sample exposure is shown in Fig. 8.13. The dispersion of these spectra is so low that they look just like stellar images. The actual grating consists of a series of parallel bars mounted on a frame. The spacing between the bars is equal to the diameter of the bars.

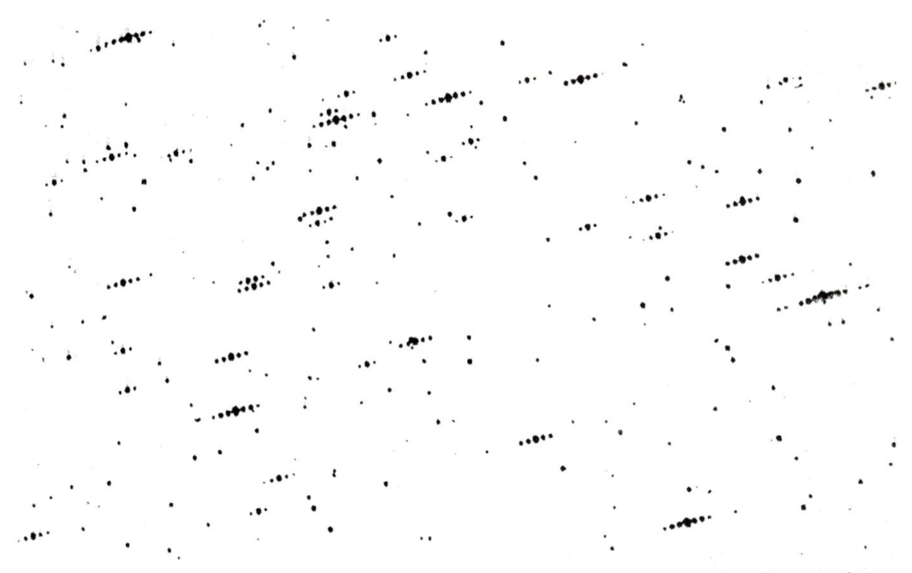

Fig. 8.13. A field of stars photographed through a coarse diffraction grating.

Obviously, only half of the incident light can come through the grating. The theory predicts that half of the transmitted light will go into the central image, and the remainder will be distributed in the other images. Furthermore, the ratio of the intensity of light producing the central image to that forming the first image on either side is 2.46. Putting this into the magnitude equation, we see that the equivalent difference in magnitude for the two images is 0.977 or very nearly 1.0. Thus, it has been customary to refer to this type of grating as a 'one magnitude' grating.

Let us now see how graded images produced either with the grating or by two exposures can be used to find the magnitude scale. One must select about six stars well spaced in brightness and measure the central and first-order images of each. The iris readings from the central or zero-order image of each star may now be plotted against those of the first-order. The points should fall along a curved path such as that shown in Fig. 8.14. A line should also be drawn through the diagram at an angle of forty-five degrees to the axes. Select one star, number 4 in this case, near the center of the range of magnitude. The iris readings which located this point were 4_0 on the vertical scale and 4_1 on the horizontal scale. Now by means of the diagonal line transfer point 4_1 to the vertical scale. Since the images which produced point 4 were different by the equivalent of one magnitude, we now have two points on the vertical scale, 4_0 and 4_1, which may be said to be different from each other by one magnitude. We find

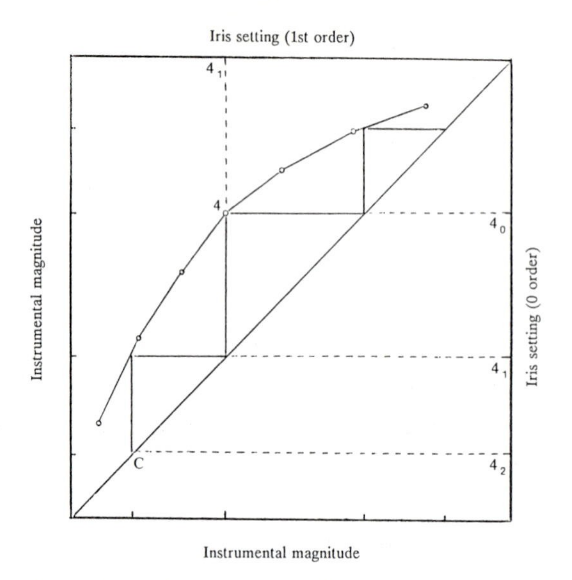

Fig. 8.14. A diagram for finding the magnitude scale from graded images. The small circles represent iris readings for first-order images plotted against readings for zero-order images. The vertical and horizontal lines define intervals of one magnitude.

another point one magnitude fainter than 4_1 by extending the horizontal line from point 4_1 to the curve and drawing a vertical line from this intersection to the diagonal. This defines point C, and a horizontal line through C defines point 4_2 one magnitude fainter than point 4_1. Continuing in this manner with a series of vertical and horizontal lines, we may find other one-magnitude intervals progressively fainter or brighter than 4_0. We note that on the vertical scale the intervals representing one magnitude are not constant.

The final curve may now be drawn by plotting iris reading on one scale and magnitude on the other. We choose some arbitrary zero point on the magnitude scale, and we transfer the iris readings directly from Fig. 8.14 to Fig. 8.15. The resulting curve gives us the relationship between iris readings and magnitude over the range of the stars used in the sample. Then, of course, if the magnitude of any one of the stars is known, the magnitudes of the others may be calculated. One may enter this diagram with the iris readings for other stars and find their magnitudes. We have described these procedures in terms of a graphical analysis of the data, but we remind the reader that this could just as well be done numerically.

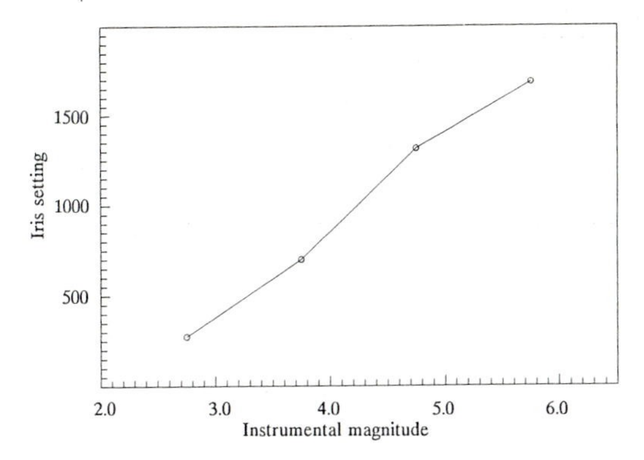

Fig. 8.15. Calibration curve relating iris readings to instrumental magnitudes.

Precision of photographic magnitudes

It is generally accepted that when all parts of the process are carried out with the utmost care the photographic method yields magnitudes which are reliable to about 0.01 magnitude. Sources of error may enter at every stage from the manufacture of the plates to the measuring. A measure of the precision may be obtained in the course of the numerical solution to find the best line when iris readings are plotted against known magnitudes. This may be thought of as the scatter on either side of the line in Fig. 8.15. The smaller the scatter, the better are the results. Today astronomy calls for photometry which can produce magnitudes which are accurate to within 0.001 magnitude. Results with this precision are routinely determined with the photoelectric equipment which is to be described in the next chapter.

As an interesting exercise one may return to the equation by which we define the magnitude system and see what sort of brightness ratio is equivalent to a difference in magnitude of 0.01. The result is that $l_1/l_2=1.00925$. Surely, it is a tribute to photographic science to realize that results accurate to one part in one hundred are attainable.

QUESTIONS FOR REVIEW

1. Describe the factors which affect the diameter and density of a star's image on a photograph.
2. Using the slope and intercept calculated in the example on p. 122, calculate the magnitude of a star if its iris reading was 1721.0.
3. Repeat the steps in Question 2 for a case in which the iris reading was 1891.5. What concern might you feel for the accuracy of this magnitude?

4. What is meant by 'failure of the reciprocity law', and how can the magnitude equation be modified to account for it?

Further reading

Binnendijk, L. (1960). *Properties of Double Stars.* University of Pennsylvania Press. In spite of its misleading title, this book contains some fine material on photometry. See Chapter V.

Miczaika, G. R. and Sinton, W. M. (1961). *Tools of the Astronomer*. Harvard University Press. The methods of measuring stellar images for photographic photometry are introduced in Chapter 5.

Stock, J. and Williams, A. D. (1962). Photographic photometry. *Stars and Stellar Systems*, **II,** 374. See this chapter for some important material on ways to insure reliable results in photographic photometry.

9

Photoelectric photometry

In the astronomical world of today most of the serious photometry is done photoelectrically, and this is due principally to the increased accuracy obtainable in the results. Another factor is the great versatility of modern photoelectric systems, for astronomers are now able to observe at wavelengths from the ultraviolet through the visible and infrared almost to the boundary with the radio spectrum at millimeter wavelengths. These great advantages were the result of the development of a variety of photodetectors sensitive at wavelengths spanning this entire broad range. Of equal importance, however, were developments in electronics which followed the discovery of the transistor in 1946. Solid state and digital electronic systems are now stable, reliable, compact, versatile and inexpensive. Without them photometry would be more difficult, more complex and more time-consuming.

The key element in any photoelectric system is the actual detector or 'photocell', and the currently used photocells are of three very different types. All three have been important in astronomy. First are the photo-emissive types in which photons impinging on certain metallic surfaces in a vacuum cause the emission of electrons from the surface. The sensitive surface is referred to as the 'cathode', and the electrons emitted from it are drawn to an 'anode' which is maintained at a positive voltage by an external power source. The rate of flow of electrons from the cathode to the anode is a measure of the intensity of the light. Devices of this sort have been in use in practical ways almost since the discovery of the photoelectric effect by Hertz in 1887, but they were not widely used until the 1930s. The early cells were not very sensitive, and the equipment for detecting the weak output of the cells was bulky and hard to use. Second are the photo-conductive cells in which the electrical resistance of materials such as selenium and lead sulfide changes in the presence of light. These were the first detectors used in photoelectric astronomy, and today they are used for observations in the infrared. Third are the photodiodes which are semiconductor devices related to the transistor. There were no widespread applications of these in astronomy until after about 1980, but they are being more widely used every year. The discussion which follows will be centered on the photo-emissive detectors, since these have historically had the broadest use among astronomers.

Complete photoelectric systems may take many forms, but they always incorporate certain fundamental elements. The heart of each is, of course, the actual light-sensitive detector of the types mentioned above. From this detector electrons flow through appropriate circuitry to the second element, an amplifier which enhances the very weak signal coming from the detector. The final element is a display or recording unit. These elements are sketched in Fig. 9.1. We must note here that each of these elements can be found in many forms. The amplifier, for example, may be replaced by a counter, or the recorder may be replaced by a small

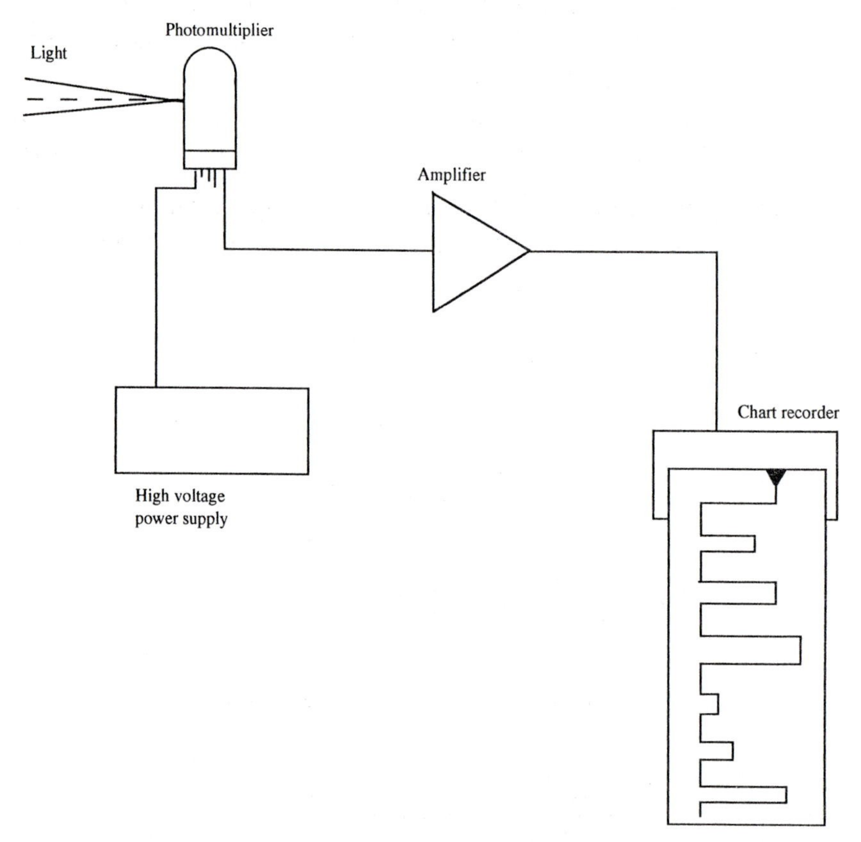

Fig. 9.1. Major elements in a simple photoelectric photometer. Today chart recorders have been replaced by computers. (See Fig. 9.7.)

computer. Our discussion of the elements in photoelectric systems will be oriented more toward the user than toward the person who is attempting to design and build a new photometer.

The photometer's construction

In considering a practical photometer that can be mounted on a telescope we recognize that the light focused by the objective must fall on the sensitive area of the detector. Beyond that, however, we find that many practical problems are involved, and we must actually design a rather complex instrument to take care of them. The complete instrument must function in the following ways:

1. support the detector in its proper position relative to the optical system of the telescope;
2. permit the introduction of filters into the light-path so that observations may be made at selected regions of the spectrum;
3. eliminate as much as possible of the background light of the sky near the star being observed;
4. permit rapid acquisition and identification of a star to be observed;
5. make use of a large percentage of the area of the detector.

In Fig. 9.2 we have drawn a sketch of a typical photometer which has provision for these five functions. The figure indicates that the converg-

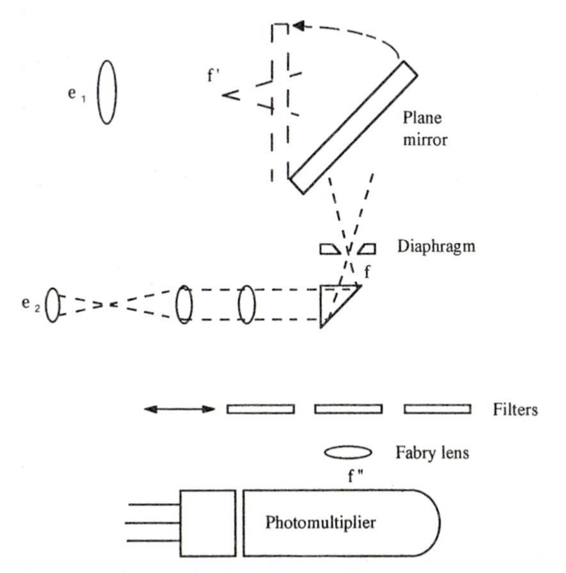

Fig. 9.2. The arrangement of the principal parts in a photoelectric photometer.

ing rays from the objective enter the photometer from the top. The first element in the beam is a mirror mounted on a hinge so that it can be moved into or out of the light path. With the mirror in the position shown here, light will be reflected to the side and brought to a focus at f′. The eyepiece, e_1, shown here permits the observer to view a wide field and to pick out the star to be observed. The observer may then adjust the telescope to bring that star to the center of the field. The hinged mirror is then moved out of the way, and the light is brought to a focus at f.

We have also indicated that in the focal plane there is a small aperture or diaphragm. This has been introduced to block out the sky surrounding the star. A second moveable mirror (or a prism) is mounted behind this aperture, and now if the observer looks through the second eyepiece, e_2, only the selected star will be seen. The magnification here is much greater than with e_1 and the field of view is much smaller. The observer can now adjust the telescope to bring the star into the center of this aperture. On nights of poor seeing or when measuring a bright star, the edges of the image may appear to be blocked by the aperture, so a slide containing a set of graduated apertures is required. The observer selects the smallest one which can still permit all of the light of the star to pass through. When several stars are being observed in sequence, the same aperture should be used for all of them. After the desired star has been centered in the aperture, the second mirror may be moved out of the light path.

When the sky is very dark, it may be impossible for the observer to see the edges of the diaphragm. He or she will not be able to know whether or not the star is centered. For this reason a small, dim light is usually placed on the sky side of the diaphragm. When the lamp is on, some of its light will be scattered by the edges of the diaphragm making it visible to the observer. A switch in the lamp's circuit can usually be arranged to turn the lamp off whenever the second mirror is moved out of the beam.

Behind the second mirror a filter has been indicated in Fig. 9.2. The filter combined with the spectral sensitivity of the detector makes it possible for the user to isolate a specific range of wavelengths. In the typical photometer provision is made for the rapid change from one filter to another. This is usually accomplished by mounting several filters on a slide or on a wheel.

With the most commonly used detectors we can never be certain that the sensitivity of the detector is uniform over its entire surface, so we cannot know where the image of the star should be focused for best results. We eliminate this problem by making use of the maximum possible area of the cathode. In Fig. 9.2 a small lens is shown behind the filter. This is referred to as the Fabry lens, and it functions in the manner described for Fabry lenses on p. 118. The light of a star on the telescope's axis will illuminate the objective more or less uniformly, so the image at f″ is a uniformly illuminated spot which is the image of the objective. Recalling the discussion of eyepieces in Chapter 6, one should note that the spot described here is just the exit pupil of the objective–Fabry lens system. By properly choosing the focal length of the Fabry lens, one can be certain that the image at f″ is small enough to fit on to the detector. By this simple expedient, we insure that the same area of the detector is always used, and we eliminate concern for small irregularities in sensitivity.

Altogether, then, the photometer becomes a complex instrument, but each of the elements described here serves a necessary function. Let us now turn to some of the detectors which are the heart of any photometric system.

Photomultipliers

In the long list of technological items which have been adopted by astronomers in the last forty years the photomultiplier stands out conspicuously. This device is a small, rugged detector of great sensitivity. The first photomultipliers were developed in the years between 1940 and 1950, and their early applications were in the sound systems of motion picture projectors. Many other applications for them arose during World War II just as advances in electronic amplifiers and recorders were being made. Photomultipliers have since taken on many forms, and two of these are illustrated in Fig. 9.3. By 1950 astronomers had begun to use photomultipliers for their photoelectric observations, and in almost all cases they chose one particular tube. This was the 1P21 manufactured by RCA, and its interior parts are shown in Fig. 9.3(*a*). The cathode (or photocathode) is the surface which is actually struck by the incoming photons. This surface has been coated with a compound containing an alkali metal such as caesium, and an electron is frequently emitted from this surface when a photon hits it. An external power source supplies a potential difference between the cathode and the next element, the first dynode. This potential causes electrons from the cathode to be directed to the first dynode. Whenever an electron strikes the surface of a dynode, several electrons are emitted from its surface. These secondary electrons are directed to the next dynode where each of them causes the release of more electrons. This sequence is repeated at each dynode, so that a very large number of electrons are emitted from the last dynode in the chain. If each incoming electron causes the emission of five secondary electrons, then at the end of the nine stages in the 1P21 nearly two million electrons will be emitted for each electron from the cathode. These electrons are attracted to the final collector, the anode, and a current flows in the external circuitry. The great power of the photomultiplier arises because of the very high amplification which takes place in the chain of dynodes.

Some photomultipliers are built with their windows in the end of the tube rather than in the side as in the 1P21. In others the photoemissive surface is actually a coating on the inside of the window (Fig. 9.3(*b*)). Some have fourteen or twenty-one stages or dynodes giving internal amplifications of hundreds of millions.

As mentioned above, an external source is required to provide the voltages between the successive stages. This is a DC power supply that must be quite stable and should provide about one hundred volts per stage. Recommended voltages are, of course, supplied by each manufacturer.

It is normal for an occasional electron to be emitted spontaneously from the cathode or from a dynode even when no light is entering the

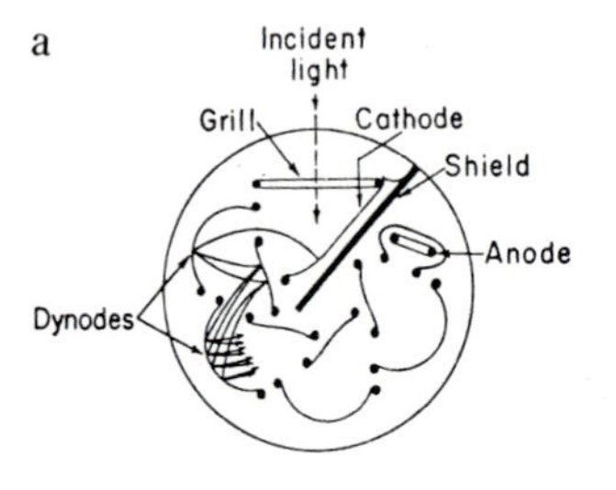

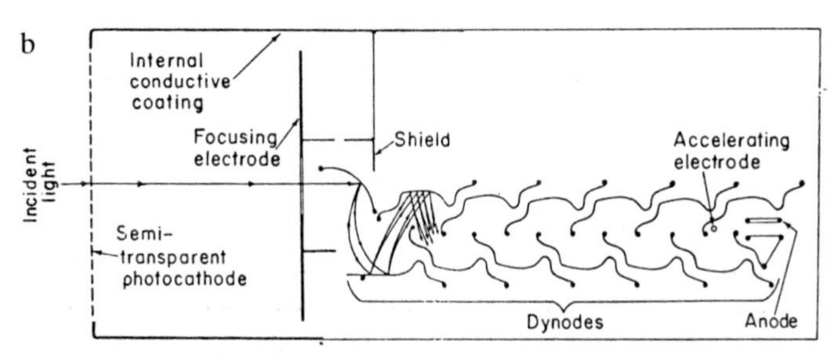

Fig. 9.3. (*a*) The internal arrangement of the 1P21. (*b*) The internal arrangement of an end-window photomultiplier, the RCA 6810. (Courtesy of Burle Industries, Inc.)

tube. These electrons are multiplied in the chain of dynodes, and are the source of what is known as the dark current. This unavoidable and unwanted flow of electrons appears in the output as an irregular signal sometimes referred to as 'noise'. A weak signal can sometimes be lost altogether if the ratio of signal to noise is not high enough.

Fortunately, there is a simple measure which can reduce the dark current dramatically. One need only cool the photomultiplier to approximately the temperature of solid carbon dioxide (−80°C) to see a substantial reduction in dark current. The record reproduced in Fig. 9.4 shows this vividly. In a period of five minutes the level of the dark current dropped from sixty-two units to less than two units. Dark current may also be reduced by making the area of the cathode as small as possible. The most common practice is to surround the photomultiplier with an insulated container in which crushed dry ice can be placed. A typical cold box of this sort is shown in Fig. 9.5. In the age of solid state electronics,

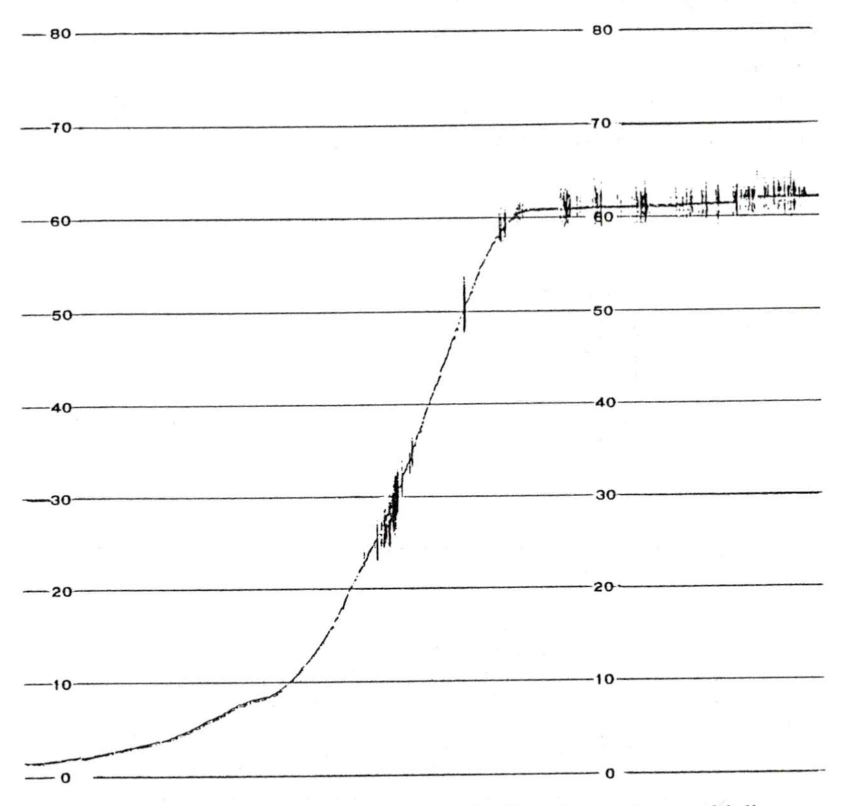

Fig. 9.4. Decrease in dark current as a result of cooling a photomultiplier with dry ice, solid CO_2. The time during which the signal dropped from sixty units to approximately two units was five minutes.

Fig. 9.5. A cold box shown as the lowest unit in this photograph serves as the housing for a 1P21 photomultiplier. (Photograph by the author.)

thermoelectric coolers have been developed which are capable of providing suitable temperatures. When provision for cooling is provided, the Fabry lens may serve as a window admitting light to the closed chamber in which the photomultiplier is located. In humid weather dew could form on the chilled lens, so a heating element is placed around it to keep it slightly warmer than its surroundings.

Considerable research over the years has been directed toward the improvement of the efficiency of photocathodes. By using more complex caesium compounds in the photoemissive surfaces great improvements in sensitivity have been achieved. In addition the range of detectable wavelengths has been extended. One of these photoemissive coatings is designated S-20, and its sensitivity at various wavelengths is shown in Fig. 9.6. Here we see both high sensitivity in the ultraviolet and sensitivity far beyond the visible in the red. The material used in the S-20 coating is referred to as a trialkali compound and it contains sodium, potassium and antimony in addition to caesium.

By properly combining filters with the spectral sensitivity of the cathode, one can isolate selected regions of the spectrum for particular purposes. This was done in 1953 by H. L. Johnson in an attempt to define a photometric system which would take advantage of the many features of the 1P21 and still be compatible with the earlier photographic system. By means of filters he isolated three specific bands – one each in the

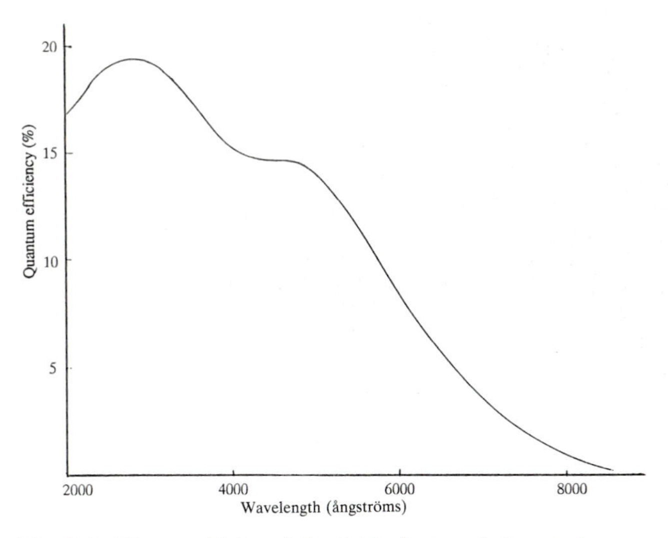

Fig. 9.6. The sensitivity of the S-20 photoemissive surface.

ultraviolet, the blue and the visual portions of the spectrum. The bands in the blue and visual were essentially the same as those in the earlier photographic system. The band in the ultraviolet was chosen to include a region near the series limit in the spectrum of hydrogen for very specific astrophysical reasons. Johnson's system has come to be known simply as the UBV system, and it has gained world-wide acceptance. Other combinations of filter and detector have since come into wide use as we shall see later in the next chapter.

Those who are just beginning their use of photomultipliers should be made aware of the fact that these tubes can be ruined for astronomical use if they are ever exposed to strong light when the high voltage is turned on. Even when the voltage is not turned on it is recommended that the cathode should never be exposed to direct sunlight.

Photodiodes as detectors

One of the most widely used devices which followed the invention of the transistor by Shockley, Bardeen and Brattain in 1947 was the solid-state diode. The name comes from earlier times when a diode was a vacuum tube containing two electrical elements (the anode and cathode in a photoemissive cell, for example). For the most part, diodes were used as rectifiers – devices which would permit the flow of electrons in one direction but not in the other. All early radios and television sets used diode tubes to convert alternating current (AC) to direct current (DC). Solid-state diodes serve the same purpose. They are made of two different types of silicon crystals bonded together, and they can be as small as the head of a match. In addition, with variations in their structure diodes can be made to emit light. These are the light-emitting diodes or LEDs that we often see in digital clocks, counters, cash registers and other displays. (LEDs should not be confused with liquid crystal displays or LCDs.) Other variations of the basic diode cause a voltage difference to appear between their two ends when the diode is exposed to light, and these have become known as 'photodiodes'. They are frequently used to turn lights on or off and to perform other useful functions.

Recent research has resulted in photodiodes which are sensitive enough to find applications in astronomy. Now when these devices are coupled with modern low-noise amplifiers they can be used in systems which rival the performance of photomultipliers. In some areas the photodiodes have the advantage. Let us list a few comparisons.

Photodiodes are sensitive to radiation in the entire range from below 3000 Å to above 11 000 Å. One detector could, therefore, be used with appropriate filters in many wavelength bands.

The quantum efficiency, a measure of the rate at which a detector converts photons to some useful form such as current or voltage, is two or three times better for photodiodes than for photomultipliers.

A high voltage is not needed for the photodiode, so another area of expense and space is eliminated.

The cost of the photodiode is low compared to that of the photomultiplier.

On the other side, the photodiode cannot be used in pulse-counting systems, and no amplification takes place inside of it.

We may note also that the photodiode is not normally used with either a sky-limiting aperture or with a Fabry lens. This, of course, simplifies construction of the complete photometer. At the present time some very successful photometers use photodiodes. An example is shown in Fig. 9.7. In some of these a digital display is obtained by means of a voltage-to-frequency converter. This is an electronic device in which the frequency of an oscillating signal varies with variations in an input voltage. A counter then counts the number of cycles which occur in some preset period of time.

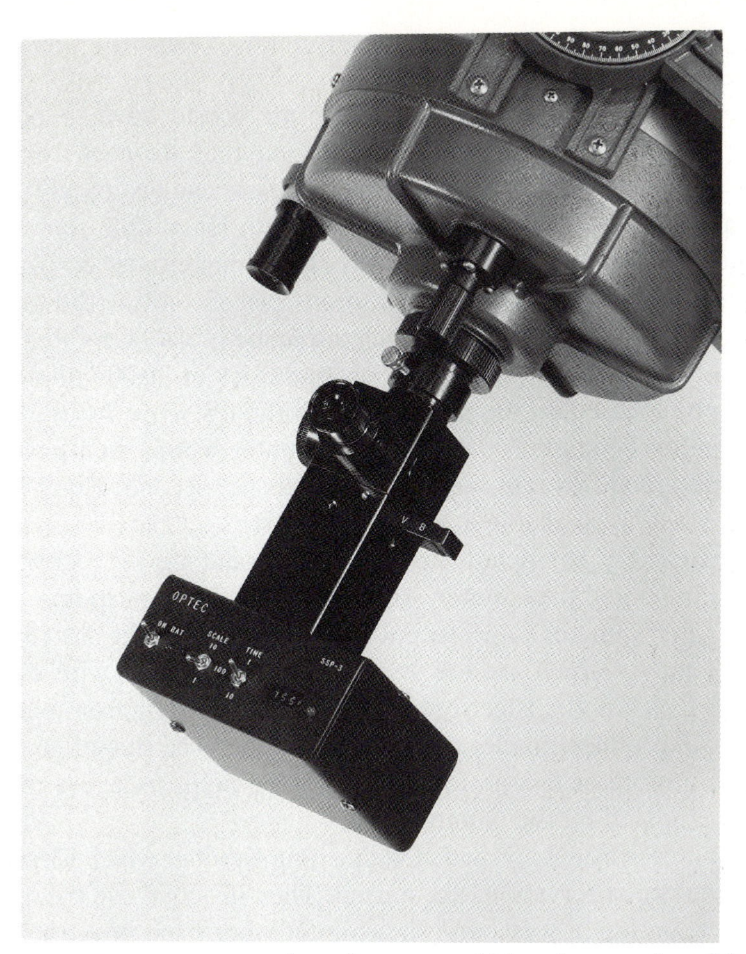

Fig. 9.7. A compact, modern photometer which makes use of a solid state detector. (Courtesy of Optec, Inc.)

Pulse counting

Today most photometric observations are made with equipment designed to count electrical pulses generated by the photomultiplier. As we said above, each photoelectron which is emitted from the cathode produces something on the order of a million electrons at the anode. Thus, the output of the photomultiplier consists of a series of pulses – one for each photon which stimulates the cathode. In a strong light these pulses come so rapidly that we detect a continuing flow of electrons or a direct current. Most of the time, however, astronomers work with faint stars, and the pulses are less frequent. Electronic counters are readily available with counting rates sufficiently high to perform very well in photometric systems.

To get an idea of what the counting rate would have to be let us consider the number of photons per second that we could expect to detect with a twenty-four inch telescope. The eye can just barely detect a sixth magnitude star, and the flux from a sixth magnitude star is about two hundred photons per second. From a tenth magnitude star we should expect to receive only about five photons per second at the eye. The collecting area of the twenty-four inch telescope is 5800 times that of the eye, so we could expect something on the order of 30 000 photons per second to be focused on the detector. With the telescope focused on the star of tenth magnitude, our electronic counter would then have to be able to operate at this rate, i.e. 30 000 counts per second. Equipment of this sort is readily available at reasonable cost.

It is customary to include another electronic device between the photomultiplier and the counter. This addition has two purposes. First, it acts as a 'discriminator', and second, it is a 'shaper'. The discriminator eliminates pulses which are too small to have originated with electrons emitted at the cathode. Electrons emitted spontaneously from one of the dynodes will produce smaller pulses at the anode. The shaper gives all of the pulses a more or less uniform shape which helps to insure that they will not be missed by the counter.

The counter is usually provided with a built-in timer which is set by the operator before observations are started. Then in actual use the operator starts the counting process, and the counter stops itself after the pre-set period – ten seconds, for example. The number displayed on the counter is then the number of pulses detected in the desired period. If the counting period is short and the light is faint, then the number of pulses counted per period may not be constant. This is due to the statistics of the emission process at the cathode. The duration of the counting period should be long enough to insure uniform counts if all other factors remain constant. The user should also be careful to avoid sources which are so bright that pulses come too rapidly to be counted.

In some systems a printer is connected to the counter so that the output is recorded at the end of each counting period. Fig. 9.8 is a block diagram showing the arrangement of the electronic parts in a pulse-counting system. We have indicated also that it is possible to include a clock in the system so that the time of each observation can be recorded on the printout.

The widespread use of microcomputers has opened up new possibili-

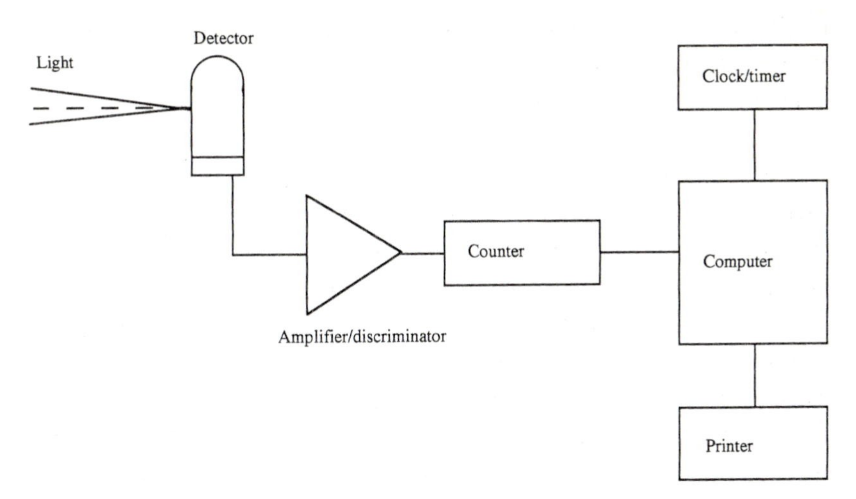

Fig. 9.8. Principal parts of a pulse-counting system.

ties in the design of photometric systems, for the computer itself can operate as the counter, the clock and the timer. The computer can be programmed to give cues to the operator and to record the data in any desired format.

Magnitudes with the CCD

The most advanced techniques in photometry make use of the charge-coupled device, or CCD. The construction and use of the CCD are described in Appendix 4, and we note that the CCD has been used for the determination of precise positions of stars. Here let us show how this two-dimensional electronic detector can be used in photometry. As described in Appendix 4, the CCD is basically a two-dimensional array of photodetectors. An electrical charge accumulates on each detector or pixel in proportion to the number of photons which strike that pixel. The charges on all of the pixels can be transferred to a computer for storage or for the reconstruction of an image.

When used in photometry, three distinct advantages of the CCD are available. First, the quantum efficiency of the CCD is greater than that of the photographic emulsion. That is, sixty per cent of the photons which strike a pixel will contribute to the stored charge in the CCD as opposed to only one or two per cent for a photographic emulsion. Second, the computer stores a digital record of the charge on each pixel. Therefore, a measure of the brightness of a star may be obtained by adding up the charges on all of the pixels which are part of the image of a star. Third, the brightnesses of all stars imaged on the array are determined simultaneously. Fourth, the brightnesses of stars can be measured over a range of 12 magnitudes. On photographs, a bright star may be overexposed when an exposure time is long enough to record faint stars.

The CCD is not without its disadvantages, of course, and the first of these is its size. An array only 25 mm on a side may contain more than 640 000 pixels. It is difficult to manufacture such an array and be certain that so many pixels function properly. Beyond this is the need for storage space in a computer, and this increases dramatically as the size of the array increases. Some means must also be found by which the computer can discriminate between pixels which are part of a stellar image and those which are not. This is done as part of the reduction procedure.

Amplifier/recorder systems

In the earliest photoelectric systems the signal at the anode of the photomultiplier was led to an amplifier and then to a chart recorder. This arrangement was quite satisfactory after astronomers and electrical engineers had learned to build stable, reliable DC amplifiers. Equipment of this sort had a tendency to drift or change its characteristics with time and with changes in temperature. Now, however, such problems have been overcome through better circuits and components.

It was customary in the 1950s to preserve the data by means of a chart recorder. This was essentially a sensitive voltmeter in which a pen moved across a chart as the voltage changed. The chart was driven at a constant speed and was usually left running throughout an entire observing session. The time could be written on the chart at the beginning of an evening, and could be checked at the end. To minimize the problem of finding the average height of this noisy signal, a current integrator was added to the circuitry. After each observation a short horizontal line representing the average signal was drawn on the chart by the pen. This saved considerable time.

These early systems had the advantage of simplicity, and the astronomers using them usually could diagnose problems quickly. But there were disadvantages as well. Sometimes in cold weather, for example, the pens would not write, or the roll of chart paper would have to be replaced at the most critical moment in an evening's program. The record always contained some noise of the type represented by the width of the trace when the light of a star was being measured. The operator had to adjust the sensitivity of the amplifier when changing from one star to another or when changing from one filter to the next. Amplifier gain settings and filter identifications had to be written down for later use in the numerical reductions.

With practice, however, astronomers did learn to work quickly and efficiently with amplifiers and recorders, and they acquired accurate data in this manner for many years.

Observing procedures

Over the years certain practices have developed for the actual sequence of operations in using photometric equipment. The routines at the telescope are derived from two guiding principles: first, we wish to be certain that what we record is just the light from the star under study, and second, we must make comparisons between stars so that differences in magnitude may later be calculated.

When we center a star in an aperture and record its brightness with a photometer, we actually record the brightness of both the star and the sky, for the background sky is never totally dark. Therefore, we must subtract the brightness of the sky if we are to have a good measurement of the brightness of the star by itself. The general practice is to move the telescope away from the star by a small amount and record the brightness of the sky. Then re-center the star in the aperture and record the brightness of (star+sky). A second observation of sky is often made, so that the average of the two can be subtracted from (star+sky). If the sky is very dark, the sky brightness may contribute an insignificant amount to the total, but the observer should always make some observations to be sure. A portion of a chart showing observations of sky and star is shown in Fig. 9.9. When observations are being made through several filters, then measurements of sky brightness should, of course, be made through each filter. Experience will provide the observer with an idea of how often to make a sky reading, but it is safe to say that the sky should be measured each time the telescope is moved through any significant angle on the sky. Only at remote, very dark locations can one expect the background to be reasonably uniform, and the sky is usually brighter near the horizon.

When we recall again the equation which defines the magnitude scale, we note that we measure ratios of brightness, and we calculate a difference in magnitude only by comparing two stars. This should be kept in mind in planning a series of observations. If we wish to determine the

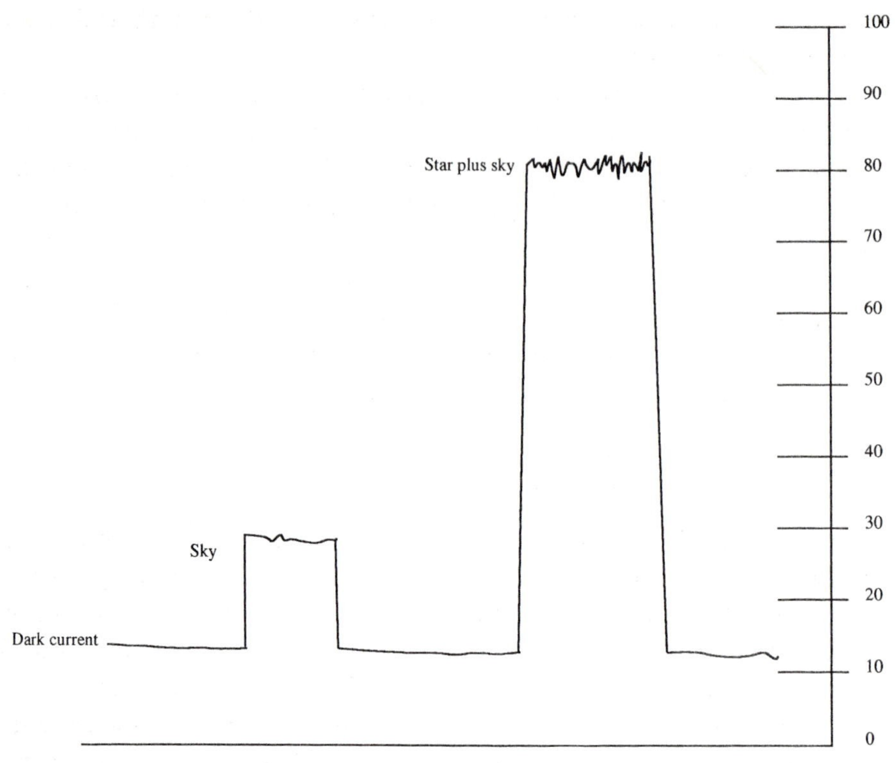

Fig. 9.9. Sample observations of star and sky (hypothetical).

magnitude of a variable star at some particular time, for example, we must observe sequentially both the variable and a comparison star which is chosen for reference. Care should be taken that the comparison star is constant in its brightness. When possible, the comparison star should have very nearly the same brightness and color as the variable star. It is also very helpful if the comparison star is a star of known magnitude.

As mentioned earlier, it is important to record the time at which each observation is made. We may add that the time is often one of the most precisely known quantities in an observational program. When data are recorded by a computer, the computer can be programmed to record the time from its own internal clock.

In a practical sense there are two types of observational programs. These may be described as 'differential' and 'standard'. In the first case the goal is simply to determine the difference in magnitude between two stars. One of these may have a known magnitude and the other an unknown magnitude. Or one star might be constant and the other variable. In these cases we seek only to find the difference in magnitude between two stars, and both the observations and their numerical treatment can be quite simple.

In using the term 'standard' to describe an observational program, we mean that the observer is seeking to produce magnitudes which fit quite closely the scale and zero-point of some standardized system such as the UBV. In this case the program must include observations of several stars for which precise magnitudes are already known, and the numerical reductions to the standard system become complex. These will be described later in this chapter. Programs involving stars which are not close together on the celestial sphere also require corrections for the effects of the earth's atmosphere. These, too, will be described below.

Reduction of the observations

In considering the reduction of differential observations to magnitude, let us consider data acquired with a pulse-counting photometer, and let us assume that the counts obtained from the sky have been subtracted from those obtained from (star+sky). Clearly the number of counts obtained for a star in a given period of time is a measure of the brightness of the star. We need only to substitute counts for brightness in the magnitude equation. The ratio of the counts for two stars leads us directly to the difference in magnitude.

As an example, imagine that for stars A and B which are close to each other in the sky, we obtain in ten seconds counts of 5274 for A and 4029 for B. Then

$$m_B - m_A = 2.5 \log (5274/4029)$$
$$m_B - m_A = 0.292$$

Such simple calculations are easily done with a hand calculator, and this simplicity is one of the great advantages of the pulse counter as a device for data acquisition. An observer would undoubtedly make a series of observations of the two stars to see how closely the results could be repeated. When necessary, the counting period or 'integration time' could be increased in order to provide a better average of the number of counts received per second. The reader should remember that counts are treated in the same way whether they were recorded by a true pulse-counting system or by a voltage-to-frequency converter and counter.

In older systems in which the data were recorded by means of an amplifier and chart recorder, the procedure was basically the same but was made more complex by the limitations of both the amplifier and the chart recorder.

Atmospheric extinction

Even on the clearest of days or nights the earth's atmosphere is not completely transparent. Particles of dust, droplets of water and molecules scatter and absorb light as it passes through the air, and the longer the path in the atmosphere, the greater is the effect. We notice this quite vividly near sunset when we see that the sun is considerably dimmer than it was at noon. When the sun is low in the sky, its light must travel through more atmosphere to reach our eye than when it is high (see Fig. 9.10). The sun also appears redder at sunset, and this is an important indication that the absorption and scattering in our atmosphere are wavelength dependent. The shorter wavelengths are affected more than the longer ones, so at sunset the blue light has, in a sense, been removed from the beam. What is left then appears redder than normal. The blue light does eventually reach the surface of the earth, but when it does so it is not coming from the direction of the sun. It comes from all parts of the sky and gives the sky its blue color. Now, if our atmosphere has such an obvious effect on the light from the sun, it must affect the light from the stars in the same manner. When we make accurate photoelectric observations, we can readily see that starlight is also dimmed and reddened by the time it reaches the surface of the earth. This effect is referred to as 'atmospheric extinction'. The process of correcting for extinction gives us accurate values for a star's magnitude and color as they would appear to an observer above the earth's atmosphere.

The effect of extinction on the brightness of a star depends upon the

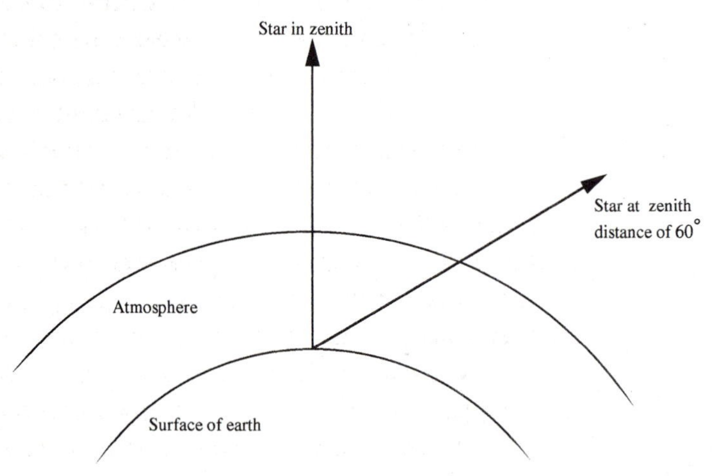

Fig. 9.10. Path length through the atmosphere varies with zenith distance.

altitude of the star above the horizon, so we should be able to notice the effect if we make a series of observations when a star is at different altitudes. We might select a star with declination equal to our latitude so that the star would pass through our zenith. Then if we began measuring the brightness when the star was low and followed it to its meridian passage, we should see a noticeable brightening. If we continued to follow it after meridian passage, we should see the star become dimmer. Observations of this sort provide the data for the analysis of extinction, so we may now proceed to the method.

We are concerned initially with the differences in path length through the atmosphere for objects seen at different zenith distances. This is seen by referring again to Fig. 9.10 above. The minimum path length is for a star in the zenith, so let us call this thickness of atmosphere 'one air-mass'. Then the air-mass for a star at a greater zenith distance will be greater than one. We may find an approximate path length simply by saying

$$\text{air-mass} = \sec z$$

where z is the zenith distance. This expression is approximate only because it does not take into account the curvature of the earth. A more exact expression is

$$\text{air-mass} = \sec z - 0.0018167(\sec z - 1) - 0.002875(\sec z - 1)^2 - 0.0008083(\sec z - 1)^3$$

We must point out here that for values of z less than 60 degrees the two expressions give very nearly identical values for the air mass. In addition, as a practical matter it is usually difficult to observe a star at a zenith distance greater than 60 degrees. In Table 9.1 we list values of air-mass for several values of zenith distance so that the reader may see the way in which air mass increases. For convenience we may repeat here an expression by which one may calculate the zenith distance using the methods of Chapter 4.

$$\sec z = 1/(\sin(\text{lat}) \sin(\text{dec}) + \cos(\text{lat}) \cos(\text{dec}) \cos(\text{HA}))$$

Let us now return to the hypothetical observations of a star as it

Table 9.1. *Air-mass and zenith distance*

z	30	60	70	75	79
sec z	1.155	2.000	2.924	3.864	5.241
Air-mass	1.154	1.995	2.904	3.816	5.120

approaches and passes through the zenith. From the measurements of brightness (i.e. counts) we compute magnitudes on some instrumental system, and we plot these against air-mass as in Fig. 9.11(*a*). The plotted points define a line, and we find the slope and intercept of this line. We shall use the value of the slope to define a quantity, k, the 'coefficient of extinction'. The observer who is just beginning to make photoelectric observations should take the time to compute extinction coefficients for several stars on several nights. It will usually be found that the coefficient is not constant from night to night and may vary during the course of one night.

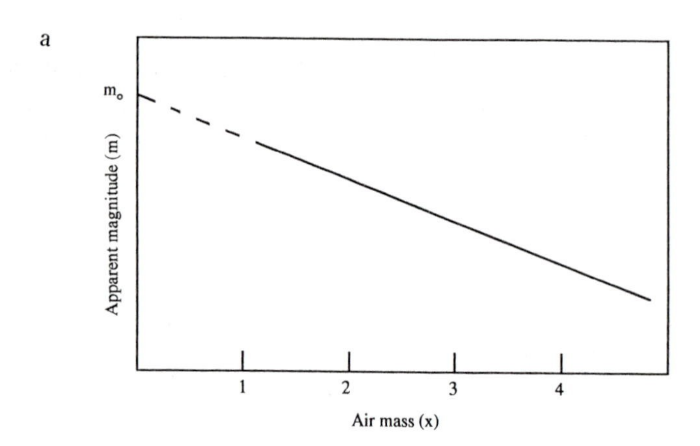

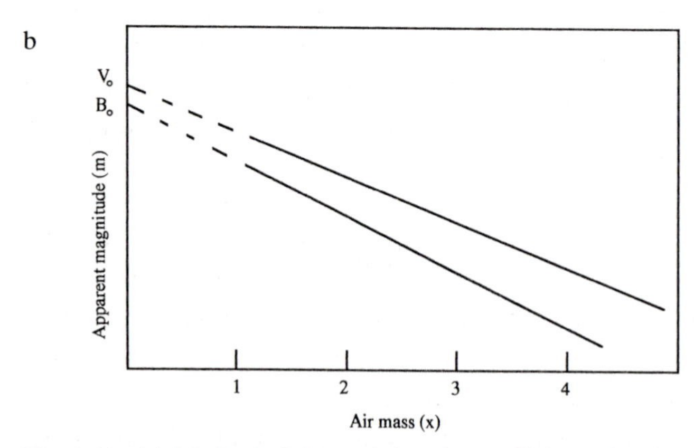

Fig. 9.11. (*a*) Method of determining the coefficient of extinction. The slope of the solid line is the coefficient of extinction. The magnitude when the air mass is zero is the magnitude outside of the atmosphere. (*b*) Dependence of extinction on wavelength.

The magnitude of a star may now be corrected for extinction by

$$m_0 = m - k \sec(z)$$

Here m_0 is the magnitude of the star as it would be seen from above the atmosphere, and it is, of course, the intercept of the line with the magnitude axis.

Earlier in this section we recalled the well-known fact that extinction depends upon the wavelength, and that extinction is greater for blue light than for red. Thus, we expect that in making observations through several filters, we should obtain a different coefficient of extinction at each range of wavelengths defined by a particular filter–detector combination. It is interesting to note in Fig. 9.11(*b*) the manner in which extinction varies with wavelength.

Color index was defined in Chapter 7 as being simply the difference between magnitudes determined at two specific wavelengths, or (B−V), for example. Thus, the true color index outside the atmosphere would be given by $B_0 - V_0$. Let us now make some substitutions and write

$$C_0 = B_0 - V_0 = B - k_B X - V + k_V X$$
$$C_0 = B - V - X(k_B - k_V)$$

X in these equations represents *air-mass*, which was defined earlier. If we let k_c be the difference between the two coefficients, then

$$C_0 = C - k_C X$$

From this it should be quite clear that a color coefficient k_C can be computed by the same method used for the coefficients k_B and k_V.

Another important factor is to be noted in Fig. 9.12. Here we have plotted observed color index against air-mass for two stars: a red one and a blue one. Clearly we are faced with the strange fact that the values of k_C are not the same in the two cases. Some means of correcting for this must be applied. Let us first, however, try to understand how this effect can take place.

We said earlier that when we combine the spectral sensitivity of a photocathode with the spectral transmissivity of a filter, we define a certain range of wavelengths. This range in turn defines a color system in which we may make routine observations. Now we must point out that the transmissivity of the filters actually depends upon the color of the incident light. This seemingly anomalous condition arises because the filters do not transmit a sharply defined band of wavelengths but rather a region poorly defined at long and short wavelength ends. It is the combination of the transmission curve for the filter with the spectral energy distribution of the star that determines the distribution of energy in the transmitted beam. If the peak in the curve for the star is shifted to the left or to the right, then the peak in the curve for the transmitted light will also shift. So the effective wavelength at which the filter transmits depends upon the color of the incident light. Thus, the two values of k_C in Fig. 9.12 are not the same, and we are presented with the need for another correction which we may refer to as the second-order correction. If we choose to think of the coefficient of extinction, k, as being composed of two parts, we may say that

$$k=k'+k''C$$

Substituting in the earlier equation gives

$$m_0=m-k'X-k''CX$$

In the same way the expression for the colors becomes

$$C_0=C-h'X-h''CX$$

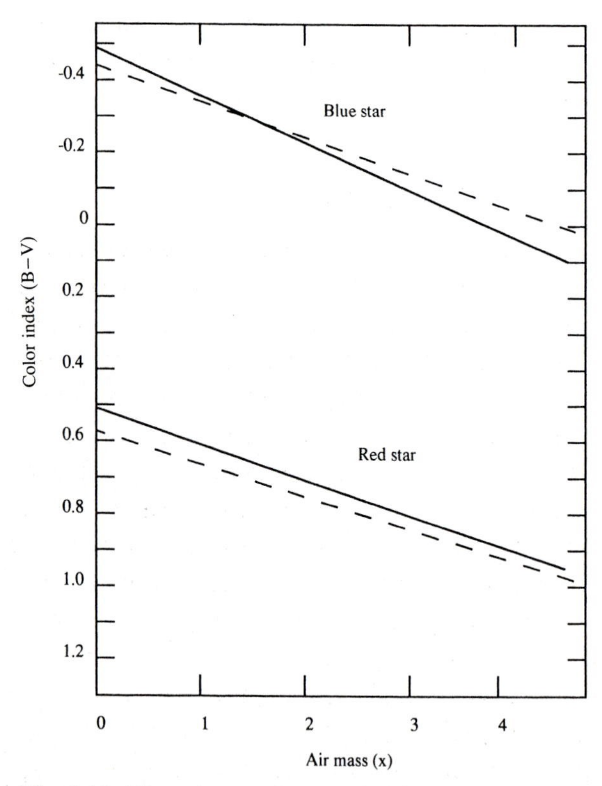

Fig. 9.12. The color coefficient of extinction depends upon the color of the star. The solid lines are for B magnitudes, and the dashed lines are for V magnitudes.

Here h' and h'' are the color extinction coefficients which are sometimes designated as k'_C and k''_C. The second order correction is quite small in the visual system (V) and in systems at longer wavelengths. In these cases it can usually be ignored. In the blue (B) k'' might have a value of about 0.03. This means that the second-order term could amount to only a few thousandths of a magnitude. One should always strive for the greatest possible precision, but there are undoubtedly times when the observations do not warrant this correction. When it does become necessary to use the color terms, the coefficients k'' and h'' may be determined from a series of observations of a close pair of stars of differing color. (Albireo, for example.) From each pair of observations differences $\Delta(m)$ and $k''\Delta(CX)$ may be found. When $\Delta(m)$ is plotted against $\Delta(CX)$ for a number of values of X, k'' may be found from the slope of the line. In the same way h'' may be found from the relationship between $\Delta(C)$ and $\Delta(CX)$.

The first-order coefficients, k' and h', may be found by now plotting $(m-k''_C X)$ against X and $(C-h''_C X)$ against X. This is, of course, very similar to the method described earlier for the determination of k and h when second-order corrections were not being considered. Again, the second-order coefficients are usually quite small.

Conversion to a standard system

We wrote earlier that each complete photometer and its filter/detector system will be slightly different from all other systems. Thus, the observer is always faced with the problem of transforming observations made on the local instrumental system to one of the standardized systems such as the UBV. Again we may recall that there will be differences in scale, zero-point and color response between the two. If a series of stars of known magnitude and color are observed, then their magnitudes and colors on the instrumental system may be compared to the standards by means of a diagram such as that in Fig. 9.13. The ideal would, of course, be for the line relating the two systems to have a slope of 45 degrees and pass through the origin. The ideal is never met, and we are more likely to find that the slope is not 45 degrees and that the line does not pass through the origin. In addition, we could expect a different line for each value of color. In actual practice a series of standard stars at various zenith distances are observed. Magnitudes and colors from these observations are used to find the coefficients of the transformation equations. These equations may then be used to convert

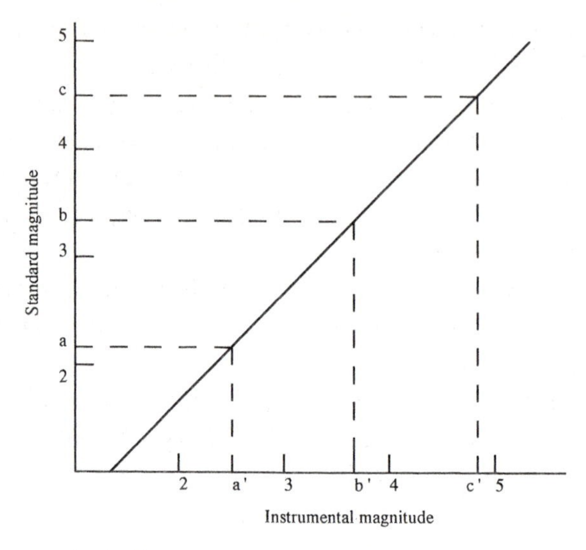

Fig. 9.13. Hypothetical relationship between standard and instrumental magnitudes for stars of the same color.

magnitudes and colors of any star to magnitudes and colors on the standard system. We are assuming at this point that the direct observations have been completely corrected for extinction.

Those readers who find themselves faced with the necessity of making precise transfers to a standard system are advised to consult the lengthy article by Robert H. Hardie in *Stars and Stellar Systems*, beginning on page 178. That article is quite complete, and describes methods by which all of the parameters of a transformation may be determined from a minimal number of observations. Equally useful are the details spelled out in *Astronomical Photometry* by Henden and Kaitchuck.

Photoelectric standards

The UBV system originated with the work of H. L. Johnson and W. W. Morgan published in the *Astrophysical Journal*, Volume 117, 1953 beginning on page 313. That article contains a list of 290 standard stars which are nicely spaced in right ascension. Most of the stars in that list are brighter than fifth magnitude.

Johnson and D. L. Harris published another list of 108 stars in the *Astrophysical Journal*, Volume 120, page 196, 1954. This list may be said to be the real definition of the UBV system.

An extensive and very useful catalog of magnitudes and colors of 1325 stars on the UBV system was published as the *Arizona-Tonantzintla Catalog* in 1965. This may be found in *Sky and Telescope*, Vol. 30, July 1965 beginning on page 24. This list also includes observations in the R and I bands, that is, the red and infrared about which more will be said in the next chapter. The limiting magnitude here is 5.00, but some stars of sixth magnitude are also to be found. The catalog is complete from the north pole to a declination of −30°.

Standards as faint as magnitude 17 may be found in Publications of the U.S. Navy Observatory, Vol. XX, Part VII. The stars here are all located in some of the Selected Areas.

More standards in the Selected Areas were published by Landolt in the *Astronomical Journal*, Vol. 78, 1973, page 959. This list contains stars with V magnitudes in the range from 10.5 to 12.5.

Automated observatories

With the advances in computers and in electronic control systems during the 1980s, astronomers began to give serious thought to photometric equipment which could be remotely controlled. There are a number of very practical reasons why this could be advantageous. The best sites in terms of weather and good seeing are often on remote mountain tops where dark skies still exist. Beyond this a computer-controlled telescope can schedule the observations of several observers to make the most efficient use of the hours of darkness.

In the United States a small group led by Russell Genet and Louís Boyd has pioneered the development of automated or robotic observatories and has been operating a successful observatory since 1983. They have developed single purpose telescopes and have designed controls which insure that the telescope will find the correct star.

QUESTIONS FOR REVIEW

1. Draw a sketch of a complete photoelectric photometer and explain the function of each part.
2. Why is the Fabry lens not needed when the detector is a photodiode rather than a photomultiplier?
3. What are some of the advantages of the photodiode as the detector in a photoelectric photometer? What are some of the disadvantages?
4. Imagine that the observation of a star plus sky produces 1877 counts in ten seconds. The expected precision of the observations is 0.05 mag. What would be

the maximum number of counts on the sky alone that would permit the observer to ignore the contribution of the sky?

5. Two stars are observed with a pulse-counting photometer. One gives 2043 counts in a ten-second integration period. The other gives 644 counts in an equal period. The brighter star has a magnitude of 6.37. What is the magnitude of the fainter one?

6. 4489 counts are recorded for a star when it is passing through an observer's zenith. Two hours later the star's zenith distance is 76°, and 4376 counts are recorded in an equal time interval. What is the approximate value of the coefficient of extinction?

7. What are the advantages to the observer of making differential observations?

8. What advantages does the charge-coupled device have over the photomultiplier and the photodiode in photometry?

9. A star is observed through the B and V filters with a pulse-counting photometer. The number of counts through the B filter is 2592, and the number through the V filter is 3087. What is the color index of the star?

10. Explain how you would select and make use of stars from lists of photometric standards.

Further reading

Binnendijk, L. (1960). *Properties of Double Stars*. University of Pennsylvania Press. See especially the material on extinction beginning on page 239.

Eccles, M. J., Sim, M. E. and Tritton, K. P. (1983). *Low Light Level Detectors in Astronomy*. Cambridge: Cambridge University Press. True to its title this book covers everything from photography to CCDs in its eight chapters.

Hall, D. S. and Genet, R. M. (1982). *Photoelectric Photometry of Variable Stars*. International Amateur–Professional Photoelectric Photometry. Most of this book is devoted to the material that must be understood by anyone who is assembling the equipment and selecting stars for a program of observations of variable stars.

Hall, D. S., Genet, R. M. and Thurston, B. L. (1986). *Automatic Photoelectric Telescopes*. The Fairborn Press. The twenty-two papers in this small volume describe the work that has been done up to 1986 on developing fully automated telescopes dedicated solely to photometry.

Hardie, R. H. (1962). Photoelectric reductions. In *Stars and Stellar Systems*, vol. II, Astronomical Techniques. University of Chicago Press. This is the classic reference on the reduction of photoelectric data.

Henden, A. A. and Kaitchuk, R. H. (1982). *Astronomical Photometry*. Van Nostrand Reinhold Company, Inc. See Chapters 6, 7 and 9 for another look at the topics covered in the present chapter.

Johnson, H. L. (1962). Photoelectric photometers and amplifiers. In *Stars and Stellar Systems*, vol. II, Astronomical Techniques. University of Chicago Press. This chapter will give the reader a fine review of the construction of the photometer, but the material on amplifiers and recorders has been superceded.

Miczaika, G. R. and Sinton, W. M. (1961). *Tools of the Astronomer*. Harvard University Press. Basic material is good, but modern developments are not covered.

Wolpert, R. C. and Genet, R. M. (1983). *Advances in Photoelectric Photometry*, vol. 1. Fairborn Observatory. This book contains a number of recommendations for observational programs and some reports on the kinds of equipment in use by amateur astronomers.

10

Applications and extensions of photometry

In previous chapters we have commented on the widespread use of the UBV system, but we have not had much to say about the applications that have caused it to become so widely adopted. In this chapter we shall first demonstrate that three-color observations can give us much more than just the apparent magnitude and color of a star. We shall show that spectral types and absolute magnitudes can be predicted and that we can often know how much the light of distant stars has been dimmed and reddened by interstellar dust.

The outstanding success of UBV photometry led later astronomers to design several other filter-detector combinations to search for very specific kinds of stellar data. Several such systems were used, for example, to assign spectral types and luminosity classes without the need for spectrograms. Specific combinations of filters and detectors made it possible for astronomers to extend their observations into the infrared part of the spectrum. Results here have been surprising and significant. We shall describe several of these additional photometric systems and their applications only briefly. Our emphasis will be on the UBV system which has been the system most accessible to those who are starting out in photometry.

Applications of UBV photometry

In both Chapters 8 and 9 we have mentioned the (B−V) color index. Now we remind the reader that the color of a star is a result of the temperature of the star. The appearance of the spectrum of a star also depends upon the star's temperature, so color and spectral class are closely related to each other. We should, therefore, produce a reasonable copy of the H–R diagram by plotting absolute magnitude, M_V, against (B−V), and this has been done in Fig. 10.1 where only main-

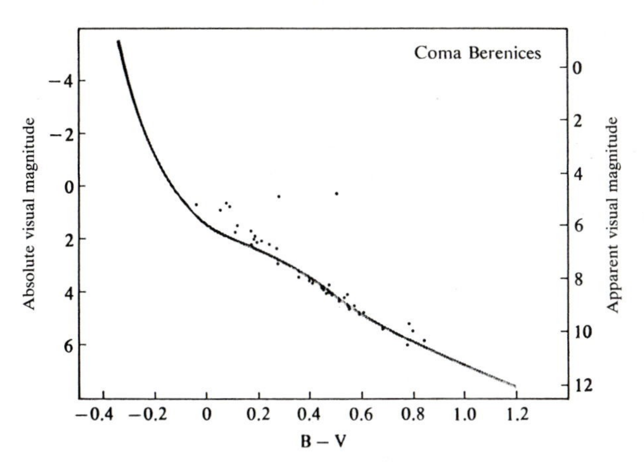

Fig. 10.1. The plot of absolute visual magnitude, M_V, versus the color index. The dots represent stars in the Coma Berenices cluster, and the apparent magnitude scale for that cluster is shown on the right.

sequence stars have been included. There are no surprises here. The normal main sequence is well defined.

Absolute magnitude is a quantity that we normally calculate after parallax has been measured for a star, and parallax is a difficult quantity to find. Thus, the diagram in Fig. 10.1 is not likely to be very useful. On the other hand, if we observe the stars in a cluster such as the Hyades, we are looking at stars which are all at nearly the same distance from us. In Fig. 10.2 we show visual apparent magnitude, M_V, plotted against (B−V) for the stars of the Hyades. Here again the main sequence is obvious for

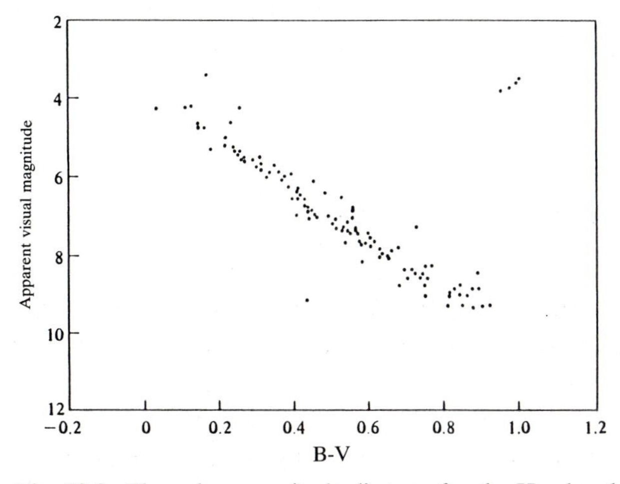

Fig. 10.2. The color–magnitude diagram for the Hyades cluster.

members of the cluster. This very simple color–magnitude diagram tells us two important things. First, if we superimpose this on the H–R diagram and match the two main sequences, we may determine the difference (m–M) which is commonly known as the distance modulus. We recall the definition of absolute magnitude:

$$M = m + 5 - 5\log R$$

where R is the distance to a star in parsecs. When we have found $m-M$ from the color–magnitude diagram for the cluster we may easily determine R. Distances to both galactic and globular clusters have been determined in this way. Second, the color–magnitude diagram is a reliable indicator of which stars are members of the cluster as opposed to being foreground or background stars. Foreground stars may appear above the main sequence, and background stars will appear below it. There can, of course, be ambiguity between foreground stars which would be on the main sequence of the HR diagram and giants or supergiants on the color–magnitude diagram.

The addition of the third filter, U, makes possible the creation of two other differences in magnitude, and one of these is (U–B). When we plot this new color index against the familiar (B−V) for a group of main-sequence stars, we have the color–color diagram shown in Fig. 10.3. Spectral types and absolute magnitudes were previously known for the stars in this sample, and these have been indicated along the curved central line in this diagram. Here we see a unique relation between the two color indices and the true characteristics of stars. The power of three-

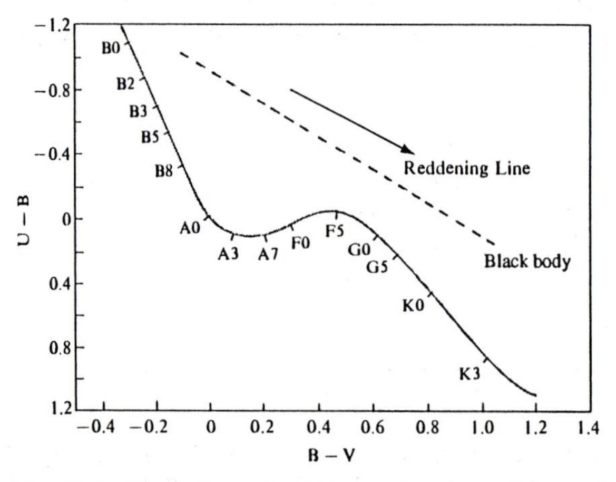

Fig. 10.3. The color–color diagram for stars of the main sequence.

color photometry should be quite clear, because each range of absolute magnitude and spectral type has a unique position.

A straight, dashed line has also been drawn on Fig. 10.3. This is the path along which we would expect to find objects which radiated exactly according to Planck's Law. The temperature of such a theoretically perfect radiator determines that body's distribution of energy with wavelength, its Planck curve or spectral energy distribution. We could measure or calculate values for (U – B) and for (B – V) for such an object, and it is these values which define the dashed line in the figure. Since the stars fall along the curved line, it is quite clear that stars do not radiate as perfect radiators. There are two principal reasons for the deviations, and they may be understood with reference to Fig. 10.4. Here, the solid line shows the observed distribution of energy with wavelength for a star of spectral class A. At longer wavelengths this line approaches the Planck curve for an object at 10000 K. The bands covered by the UBV system are also shown. From the placement of the U and B bands we note that (U−B) is a measure of the size of the Balmer discontinuity which has been marked in the figure. To the left of this point the absorption lines of hydrogen are so numerous and close together that the star's energy curve is pressed downward. To the right of the Balmer discontinuity the star's energy distribution approaches the Planck curve. Since the Balmer lines (and the Balmer discontinuity) are strongest in the spectra of stars of spectral class A, the deviation in Fig. 10.3 is greatest for the A-type stars. Proceeding toward the F-type stars the Balmer discontinuity becomes less, and as expected, the curve bends back toward the line representing perfect radiators.

The appearance of spectra for all types of stars are described in detail in Chapter 12, and there the reader will find that the lines of metals are

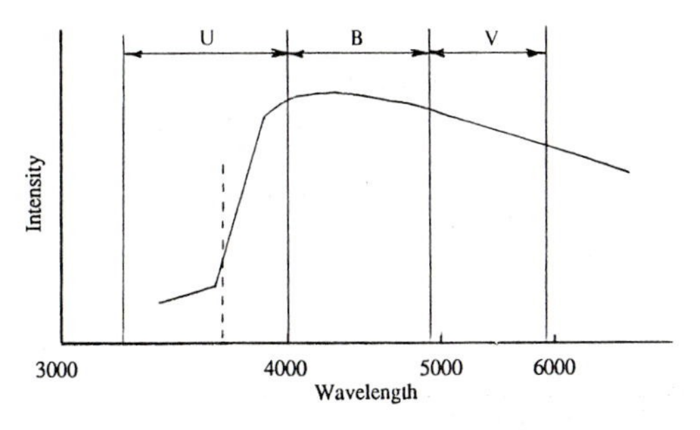

Fig. 10.4. The spectral energy distribution of a star of spectral type A shown with the bands which define the UBV system. The dashed vertical line has been drawn at the wavelength of the Balmer discontinuity.

the most conspicuous features in the spectra of the cooler stars, that is, those in spectral classes F through K. These lines become more numerous in the progressively cooler stars, so the ultraviolet end of the spectrum is again depressed with respect to the Planck curve. This is reflected in Fig. 10.3 where the curved line makes its second bend away from the dashed line. Thus, our theory of the manner in which spectra are affected by temperature is nicely supported by the color–color diagram.

The above discussion of the curved locus of points representing main-sequence stars in the color–color diagram surely implies that a star's position on the line is an indication of that star's spectral type. Segments of the line have been marked for spectral types from B through K. We must emphasize here the fact that the above applies only to main-sequence stars. Giants, supergiants and white dwarfs radiate more nearly like black bodies, so their positions on the color–color plot follow a line very much like that for the ideal black bodies.

In Fig. 10.3 there is a solid line marked 'Reddening line', and this line represents the direction in which a star's location on this diagram will be shifted if its light has been affected by interstellar absorption. When we measure U, B and V for a star in the direction of the Milky Way we can expect that the star's light will have passed through significant amounts of interstellar dust on its way to us. The effect is the same as it is in the earth's atmosphere, for the photons of short wavelength will be scattered more than those of longer wavelength. On Fig. 10.3 a reddened star's location will not be on the curved line but will have been moved along a line parallel to the reddening line. This displacement can provide a clue to the amount of interstellar dust along the line of sight to the star. Because of the curved line the correction for interstellar extinction is ambiguous for many stars. A reddened B star might be displaced to the location of the F5 stars on the diagram. Or an A7 star which has been reddened might be mistaken for an unreddened star of spectral type K. Methods of eliminating the ambiguity are described in references by Sharpless and by Henden and Kaitchuck.

The observer should correct for the effects of interstellar absorption in all cases in which the distance to an object is a concern. We can feel certain that the light of all stars in or near the direction of the Milky Way has been dimmed and reddened in this manner.

Extensions into the infrared

Astronomical interest in the red part of the spectrum may be said to go back as far as William Herschel's detection of invisible radiation from the sun in 1799, and a system of red magnitudes was in use in the early 1900s. A limited number of standards in the red were included in the original definition of the International System in 1922 (see p. 108). Over the years this interest was spurred in part by the fact that a number of stars are obviously red in color. As photoelectric methods began to develop in the years between 1930 and 1950, some of the earliest detectors were photoconductive cells made of lead sulfide or of selenium. These are of the photo-conductive type in which the electrical resistance of the material depends upon the amount of light falling on it. Even though these cells are sensitive in the red, they are not very efficient and did not see wide usage. Progress in photometry in the red was slow until the development of photomultipliers using the S20 photocathode. These provided extended sensitivity in the red and infrared and made possible the first precise magnitude standards in longer wavelengths.

It was in the early 1950s that Gerald Kron and his associates at Lick Observatory published a series of papers which defined the R and I magnitudes (for red and infrared) centered on 6800 Å and 8250 Å respectively. They established standards conveniently located all around the celestial sphere, and these standards reach stars as faint as eleventh magnitude. A list of eighty-eight primary standards may be found in a paper by H. L. Johnson in *Stars and Stellar Systems*.

Some of the early observations at long wavelengths suggested that many stars were so red that their peak emission was at wavelengths considerably longer than those of the R and I systems. In addition, astronomers had learned that surprises were often found in unexplored regions of the spectrum, so they began serious efforts to extend their observations even further into the infrared. Here they knew that they would have to face serious problems due to the nature of the earth's atmosphere. A number of molecules (CO_2, O_3, N_2O, H_2O) absorb across very broad spectral ranges in the far infrared, and the absorption is almost continuous. Fortunately, there are a series of narrow 'windows' or wavelength bands of good transparency in which photometric observations can be made. The wavelengths of these windows are listed in Table 10.1, and it is of necessity that these define the far infrared photometric system. It is too bad that a system could not have been designed around specific astrophysical criteria as was the case in the choice of the position of the U band in the UBV system.

Observations at these very long wavelengths are by no means simple. It is not just a matter of placing a suitable detector (usually lead sulfide) at the focus of a telescope. The complications arise from the fact that so many things inside and outside of the observatory dome radiate at the wavelengths that the observer is trying to detect from the stars. For example, the very molecules that limit the transparency of the atmosphere are also radiating in the same region, and the telescope, observer and dome are radiating as black-bodies with their emission peaks in the infrared. In an optical analogy, one can imagine trying to make normal photometric observations in the daytime with a glowing telescope. In order to see just how much radiation is coming from the star alone, the usual practice is to switch the telescope rapidly back and forth onto the star and off of it. In effect the observer makes a series of observations of star-plus-sky and sky alone. The difference is a measure of the radiation from the star by itself. Radiation from the detector's immediate surroundings and noise from the detector are lessened by keeping the detector at the temperature of liquid helium or liquid nitrogen.

Infrared photometers must also be equipped with filters which transmit only in a desired range of wavelengths just as in the case of optical photometry. Multi-layer interference filters can be produced to transmit only the wavelengths which match the atmospheric windows which define each band in the system.

The complexity of the equipment and the need for extreme cooling mean that infrared astronomy is a specialty that is not as readily available as observations in the UBV bands. This type of astronomy is carried on

Table 10.1. *Wavelength bands in the infrared*

Band	Effective
J	12 500
K	22 000
L	34 000
M	50 000
N	102 000

mostly at observatories dedicated just to infrared work or at observatories which are large enough to be able to make a major commitment to it.

Many of the problems caused by the earth's atmosphere are reduced if the observatory can be located as high as possible above sea level, so infrared observatories have been located in high, dry locations. Great success has also been achieved with telescopes carried aloft by a large helium-filled balloons. (See references at the end of this chapter.)

The ultimate location for an infrared telescope is, of course, in orbit altogether outside the earth's atmosphere, and a very successful program began in January 1983 with the launch of the Infrared Astronomical Satellite (IRAS), a joint effort of the Netherlands, the United Kingdom and the United States. For nearly a year this spacecraft collected observations and transmitted them to the ground. Using a telescope of 60 cm (23.5 in) observations were made at four wavelengths and covered 96% of the sky. The four bands were at 12, 22, 60 and 100 μm, so the IRAS data is at wavelengths much longer than those used in the JKLMN system. A summary of the analysis of the IRAS data may be found in articles by Robinson and by Schorn listed at the end of this chapter.

Narrow and intermediate band photometry

When we described the UBV system in Chapter 8, we mentioned the reasons why the three bands are located where they are in the spectrum. The B and V bands are essentially at the locations defined in the International System of 1922, and these were located by the sensitivity of early photographic emulsions in the case of B and the sensitivity of the human eye in the case of V. The location of the U band was chosen specifically to be at wavelengths shorter than 3650 Å, the wavelength of the Balmer discontinuity. This meant that the quantity (U−B) was a measure of the size of the Balmer discontinuity. In other words, the location of the U band was chosen for specific astrophysical reasons. We mentioned earlier in this chapter that the locations of bands in the far infrared, J, K, L, M and N, were dictated not by important features in the spectra, but by transparent regions in the earth's atmosphere. Let us now look at systems defined completely by the nature of the tasks for which they were intended.

One such is the Hβ narrow band system developed by David Crawford, then at Yerkes and McDonald Observatories. The details of this system may be found by consulting the references. Here we shall give some background information and only a brief description of the principal points.

The most easily visible features in stellar spectra will be described in detail in Chapter 12, but we may mention here that the absorption lines of hydrogen are conspicuous in the spectra of the hotter stars and are detectable in the spectra of almost all stars. It had been noticed in the early 1930s that the width of the Hβ line could be correlated with the spectral class. If a quantity known as the 'equivalent width' is determined from measurements of the Hβ line, one can see a well-defined relationship between spectral class and equivalent width. The principal application of equivalent widths, it should be noted, is in the determination of the relative abundances of elements in stellar atmospheres.

Beginning with work in 1958 Crawford introduced a photometric system which could give accurate spectral types for hot stars from two simple measurements. He made use of two narrow band interference filters centered on the wavelength of the Hβ line (4861 Å). Transmission profiles of the two filters are shown in Fig. 10.5, and we note that one filter transmits through a wider range of wavelengths than the other. Observations are made through both filters and a difference in magnitude is determined (see p. 104). This difference is defined as the β parameter or the β index. From Fig. 10.6 we may see just how effective this index is in specifying the absolute magnitude of a star.

When we recall that the two filters transmit only in very narrow ranges of wavelengths, we realize that we cannot expect to study faint stars with this system unless we are able to use a large telescope. We repeat also that the system works only for those stars in which the hydrogen lines are strong. On the positive side we note that the β index does not suffer from the effects of extinction in the earth's atmosphere, and is not affected by the presence of interstellar dust.

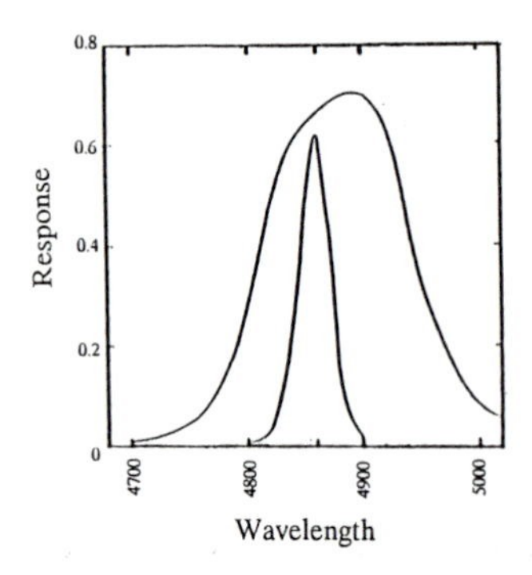

Fig. 10.5. The transmission bands for the Hβ system.

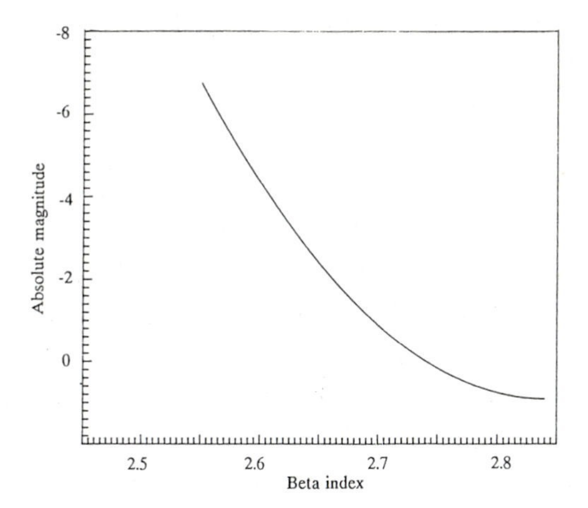

Fig. 10.6. The relationship between the Hβ index and absolute magnitude. (Plotted from published data.)

Crawford and his associates also produced another important photometric system in which a set of filters define four intermediate bands. The four are designated u, v, b and y for ultraviolet, violet, blue and yellow, and their transmission curves are contrasted with those of the UBV filters in Fig. 10.7. The classic discussion of uvby photometry is found in the reference by Stromgren at the end of this chapter. Using differences

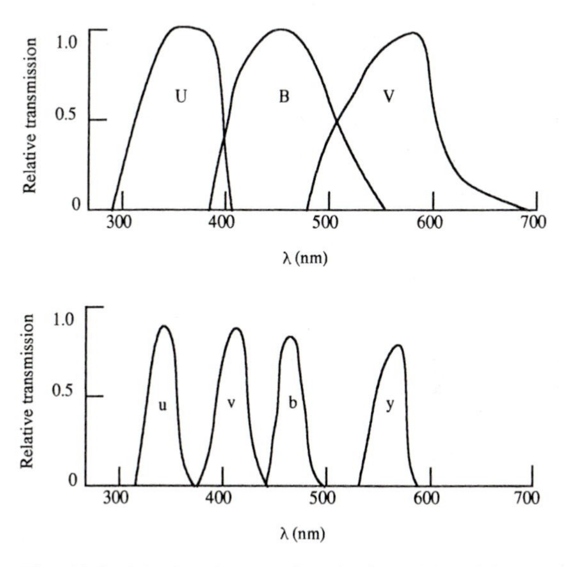

Fig. 10.7. The bandpasses for the broad band UBV system compared to those of the uvby intermediate band system.

between intensities measured through adjacent filters, Stromgren defines two indices as follows:

$$c_1 = (\mathrm{u}-\mathrm{v})-(\mathrm{v}-\mathrm{b})$$

$$m_1 = (\mathrm{v}-\mathrm{b})-(\mathrm{b}-\mathrm{y})$$

The first of these is a measure of the height of the Balmer discontinuity, and the second is a measure of the amount by which the star's continuum has been depressed by the numerous lines of the metals. These two indices must be corrected for interstellar absorption, but this can be done in an unambiguous way.

The relationship between c_1 and m_1 after correction for interstellar absorption is shown in Stromgren's diagram in Fig. 10.8. Main sequence stars fall betwen the two curved lines, and giants lie in the three regions above the curves. Quite clearly this diagram provides a straight forward method of determining spectral type and luminosity class from fairly simple observations.

Quite a few other photometric systems have been developed over the years, and each has its own applications. Some are for general use, while others provide methods for looking for stars with special characteristics. Many of these are described in Golay's book referenced below.

Those observers who are just getting started in photometry can find plenty of work that is well-suited to the UBV system, and should be encouraged to become experienced in it before moving on to the more specialized systems. UBV photometry has stood the test of time, and after thirty-five years it is still the most widely used system.

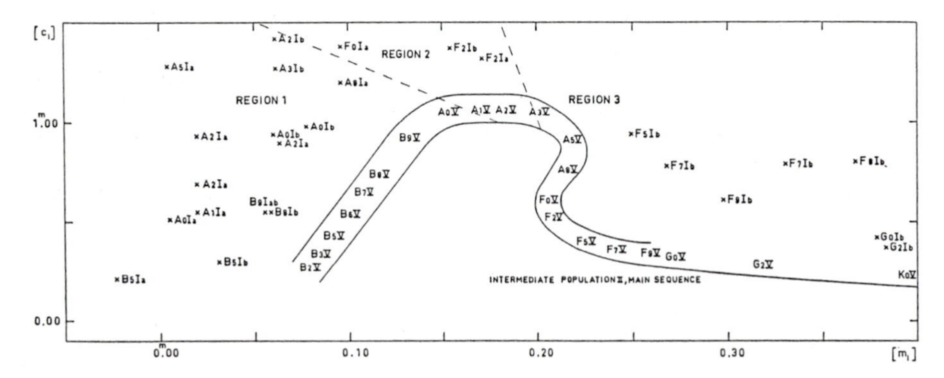

Fig. 10.8. The relation between the m_1 and c_1 indices of the four-color system. (Reproduced with permission from the *Annual Review of Astronomy and Astrophysics*, Vol. 4, © 1966 by Annual Reviews.)

QUESTIONS FOR REVIEW

1. Why should we expect the color–magnitude diagram for a galactic cluster to resemble the H–R diagram? How is this used to find the distance to the cluster?

2. Make a photocopy of Fig. 10.3 and sketch on it the locus of stars which were members of a reddened galactic cluster. How would the amount of reddening be estimated?

3. Why were astronomers interested in extending their observations into the infrared? What made this possible?

4. What is the Hβ index, and what does it measure?

5. What are some applications of Hβ photometry that might make an observer want to use it?

6. What is the main disadvantage of both the narrow-band and the intermediate-band systems?

Further reading

Becker, W. (1963). Applications of multicolor photometry. In *Stars And Stellar Systems*, vol. III, Basic Astronomical Data, p. 241. Chicago: University of Chicago Press.

Golay, M. (1974). *Introduction to Astronomical Photometry*. Reidel Publishing Company. This book is an in-depth summary of the applications of multi-color photometry. See especially Chapters 2 through 5.

Habing, H. J., and Neugebauer, G. (1984). The infrared sky. *Scientific American*, **251**, 5, 49. This is a clear and comprehensive article by two recognized authorities in infrared astronomy.

Johnson, H. L. (1963). Photometric systems. In *Stars And Stellar Systems*, vol. III, Basic Astronomical Data, p. 204. This chapter contains precise definitions of the UBV and several other photometric systems.

Kitchin, C. R. (1984). *Astrophysical Techniques*, Adam Hilger Ltd. Refer to the latter parts of Chapter 3 for parallel material.

Low, F. and Rieke, G. H. (1974). The instrumentation and techniques of infrared photometry. *Methods of Experimental Physics*, **12**, 415. These writers were among the real pioneers of infrared photometry.

Robinson, L. J. (1984). The Frigid World of IRAS – I. *Sky And Telescope*, **69**, 1, 4. Schorn, R. A. (1984). The Frigid World of IRAS – II. *Sky And Telescope*, **69**, 2, 119. These two articles summarize the equipment and the scientific results of the Infrared Astronomy Satellite program.

Stromgren, B. (1966). Spectral classification through photoelectric narrow-band photometry. *Annual Review of Astronomy and Astrophysics*, **4**, 433.

11

The spectrograph

It is safe to say that the spectrograph, a relatively simple instrument, brought about a virtual revolution in astronomy. Although Newton had examined the spectrum of sunlight and Fraunhofer had seen the spectra of a few stars, the spectroscope was not extensively used on telescopes until the latter half of the nineteenth century. Beginning in about 1860 William Huggins in England and Fr. Angelo Secchi in Rome did their first experiments on the light from the moon, the planets and the brighter stars. The spectrograph was slowly refined and improved, and eventually it made possible a series of new understandings of the nature of the sun and stars. First came the identification of a few absorption lines in solar and stellar spectra; then came recognition of several distinct classes of spectra. By the end of the century the construction of spectrographs had been refined to the point that radial velocities could be measured with confidence. Today we have a remarkable understanding of the physical processes which occur in the atmospheres of the sun and the stars.

Over the years there have been a number of major developments which have increased the efficiency of the spectrograph, and there have been changes in the means by which the light is dispersed. The general principles of the spectrograph are not complicated, however, and will be outlined below. We shall discuss first the prism and then the grating as dispersive elements. This will be followed by a description of the practical considerations in the design and use of a spectrograph. The chapter will end with descriptions of some specialized types of spectrographs and some methods of increasing the efficiency. It has been the general practice to use the term spectroscope when referring to an instrument used visually and to use the term spectrograph when referring to an instrument which records the spectrum on a photograph or by some other means.

The major parts of the spectrograph are shown schematically in Fig. 11.1 in their proper relationship to each other. Here the dispersive element is indicated simply as a rectangle. A source of light to be studied is shown at the left, and we note that this source is almost always a narrow slit. On a telescope the slit is illuminated by placing it in the focal plane of the objective and allowing the image of a star to fall on it. Rays from the slit are made parallel by means of a lens, the collimator, between the slit and the disperser. Parallel rays emerge from the disperser and are brought to a focus by the camera at the right side of the diagram. Since light of different colors is dispersed by different amounts, parallel beams in each color enter the camera lens from slightly different directions, and the camera forms an image of the source in each color. It should be noted here that in modern spectrographs one or more of the three elements, collimator, disperser and camera, may be made with reflecting rather than refracting surfaces. If the slit is illuminated by the light of a glowing gas such as hydrogen, then in the focal plane of the camera there will be a series of images of the slit – one in each color present in the light of hydrogen. We speak of these images as 'spectrum lines' since the slit itself looks like a line. If the slit is made wider, then the images formed by the camera are also made wider. If the slit is illuminated by white light, then the images of the slit run together to form a continuous spectrum.

Dispersion by a prism

Newton, Huggins, Secchi and the other early experimenters all used prisms as a dispersing element, so we shall discuss the prism in some detail. The ability of a prism to disperse light depends, of course, on the fact that in transparent materials the index of refraction varies with the wavelength of the light. When light passes through a piece of glass with

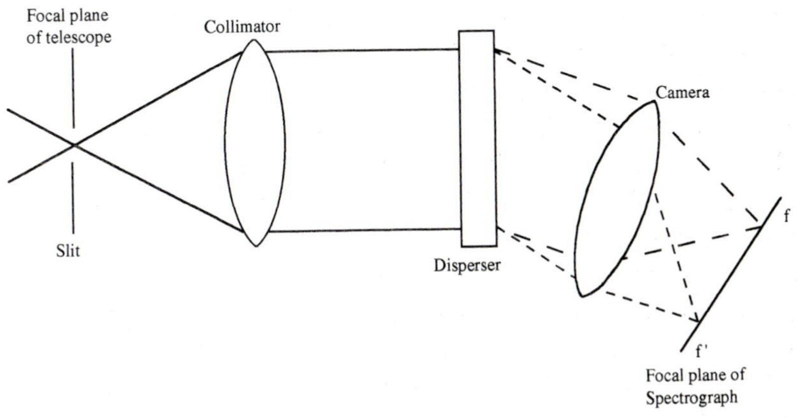

Fig. 11.1. The essential parts of a spectrograph.

parallel sides, there is some dispersion as light goes from air to glass. This dispersion is reversed when the light comes out of the glass and continues in its original direction. In a prism, however, the angle of incidence on the second face (that is, glass to air) is such that the original refraction is not reversed. Beams of different wavelength emerge in different directions and can be brought to a focus in different locations by the camera.

In order to see clearly how the prism functions, let us try to compute the refraction at the incident and emerging faces of a prism for light of several wavelengths. The procedure requires that we first describe several properties of the prism. In Fig. 11.2 the triangle abc represents the end view of a prism made of some material with a refractive index of n. Imagine that a narrow beam of monochromatic light is allowed to fall on the left side of the prism so that its angle of incidence is ϕ. The beam will emerge from the right side of the prism having been diverted through the angle δ defined in the diagram. If the angle ϕ is large, δ will also be large. Now if ϕ is slowly reduced, δ will become smaller for a time and then become larger. We note the angle of minimum deviation which we shall call δ_m. It is interesting and significant that when the condition of minimum deviation has been achieved, the angles ϕ and ϕ' are equal. This condition is shown in Fig. 11.2. One should keep in mind that it was stated above that we had assumed that the incident beam of light was monochromatic. That is, the beam contains light of only one wavelength. The angle of minimum deviation will be different for each wavelength because of the fact that the index of refraction, n, depends upon the wavelength.

Many good books on optics contain derivations of the formula for minimum deviation, so we shall not include such a proof here. We simply state that the angle of minimum deviation may be found from

$$\sin\tfrac{1}{2}(\delta_m+A)=n\sin(A/2)$$

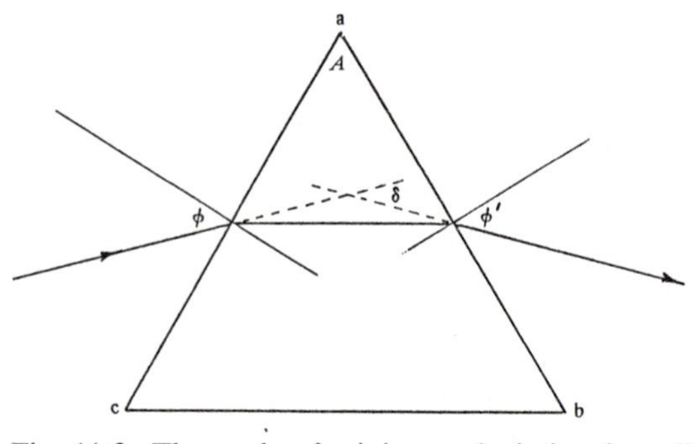

Fig. 11.2. The angle of minimum deviation for a light ray refracted by a prism.

where A is the vertex angle of the prism and n is the index of refraction.

We wish now to show how the direction of the emerging ray can be determined for any wavelength. From earlier statements it should be clear that one must know the index of refraction of the material at hand for a variety of wavelengths. In Fig. 11.3 we have shown a general case in which a ray passes through the prism at some angle other than δ_m. From this we may note several important points. In the four-sided figure, abcd, there are right angles at b and d. This means that the sum of angles A and C is 180 degrees. In triangle bcd

$$\phi_1+\phi_2+C=180$$

Then it follows that

$$A=\phi_1+\phi_2$$

Referring again to Fig. 11.3 we may note that δ is an external angle to triangle sbd. This means that δ is equal to the sum of the two non-adjacent interior angles. Therefore,

$$\delta=(\phi-\phi_1)+(\phi'-\phi_2)$$

Rearranging, we have

$$\delta=\phi+\phi'-(\phi_1+\phi_2)$$
$$\delta=\phi+\phi'-A$$

Let us assume now that we wish to calculate the refracted directions for a number of wavelengths. We must first select an angle of incidence, ϕ, at which the beam will enter the prism. We shall want ϕ to be the angle of minimum deviation for some central wavelength in the range being studied. Then, in order to find ϕ we must begin by calculating δ_m from the formula given above:

$$\sin\tfrac{1}{2}(\delta_m+A)=n\,\sin(A/2)$$

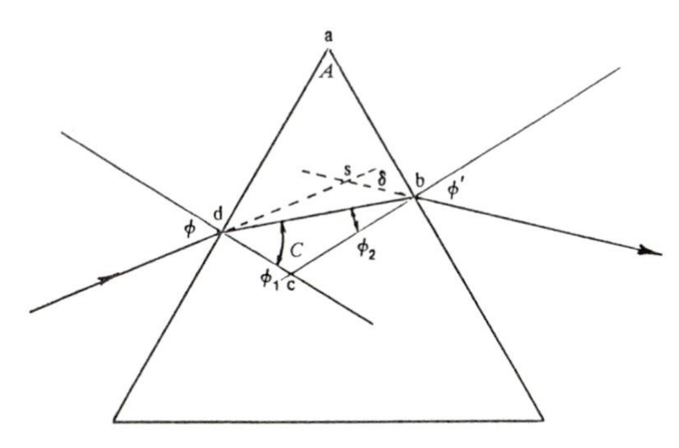

Fig. 11.3. The general case for dispersion in a prism.

For the case of minimum deviation $\phi=\phi'$, so we note that

$$\delta_m=2\phi-A \text{ and } \phi=\tfrac{1}{2}(\delta_m+A)$$

We may substitute ϕ for $\frac{1}{2}(\delta_m+A)$ in the earlier expression, and we have

$$\sin\phi=n\sin(A/2)$$

Thus, from the shapes of the prism and a knowledge of n for some selected wavelength, we may calculate ϕ, the angle of incidence of the incoming ray.

Now one may compute the deviation at each of the two surfaces of the prism for light of several wavelengths provided that the index of refraction is known for each of those wavelengths. The angle of incidence at the first surface (that is, at d) is simply ϕ, and the value of ϕ_1 is found from Snell's Law:

$$\sin\phi=n\sin\phi_1$$

At the second surface the angle of incidence is ϕ_2, and ϕ_2 may be found from the earlier statement that $A=\phi_1+\phi_2$, since A and ϕ_1 are now assumed to be known. The final step is to find ϕ' from another application of Snell's Law

$$\sin\phi'=n\sin\phi_2$$

We may illustrate these steps with an example using five wavelengths. Let us assume that the prism is made of high-dispersion crown glass and that the angle of the prism is sixty degrees (i.e., $A=60$ degrees). We assume, further, that the wavelength of minimum deviation is 4860 Å. The index of refraction for each of the five wavelengths is given in Table 11.1. As above, we find ϕ from

$$\sin\phi=n\sin(A/2)$$

and in this example ϕ is equal to 49.774 degrees. We may then calculate ϕ_1, ϕ_2 and ϕ'. Results for these five wavelengths are listed in Table 11.2. We confirm from this that light of shorter wavelengths is refracted more than light of longer wavelengths. Thus, if a collimated beam of light enters the prism of Fig. 11.3 on the left, parallel beams in each color or

Table 11.1. *Index of refraction*

Wavelength	3610	4340	4860	5890	6560
n	1.546	1.533	1.527	1.520	1.517

Table 11.2. *Variation of Φ' with wavelength*

Wavelength (λ)	n	ϕ_1	ϕ_2	ϕ'
3610	1.546	29.594	30.406	51.486
4340	1.533	29.871	30.129	50.308
4860	1.527	30.000	30.000	49.774
5890	1.520	30.153	29.847	49.156
6560	1.517	30.218	29.782	48.894

wavelength will emerge in slightly different directions at the right. Each of these parallel beams enters the camera lens and is brought to a focus somewhere in the plane ff′ of Fig. 11.1. We shall return to the camera lens after the next section in which we shall describe the grating as a disperser.

Dispersion by a grating

In something of an oversimplification the diffraction grating may be thought of as a plate on which a very large number of grooves have been cut, and by 'large number' we mean that the number of grooves per millimeter may range from several hundred to several thousand. The grating may be designed to disperse light which passes through it (a transmission grating) or light which is reflected from it (a reflection grating), but in both cases light will be dispersed in a direction at right angles to the direction in which the grooves run. The dispersion by the grating is different from that of the prism in three respects. First, the dispersion is uniform over moderate ranges of wavelength, and second, the red light is bent more than the blue. Third, several spectra of decreasing brightness are formed on either side of a central image. These are referred to as the first-order, second-order, third-order etc., and the central image is referred to as the zero-order.

For the sake of our introductory discussion let us consider a transmission grating as sketched in Fig. 11.4. Again, a slit acts as the source, and a collimated beam enters the disperser. The camera lens forms an image of the slit in each color present in the original beam and forms the multiple spectra on either side of the zero-order image. In this sketch the grating is shown as a series of open spaces and opaque strips of equal width, and gratings can be made in this way. We shall see below, however, that large increases in efficiency are possible when the grating consists of non-symmetrical, but specially shaped grooves.

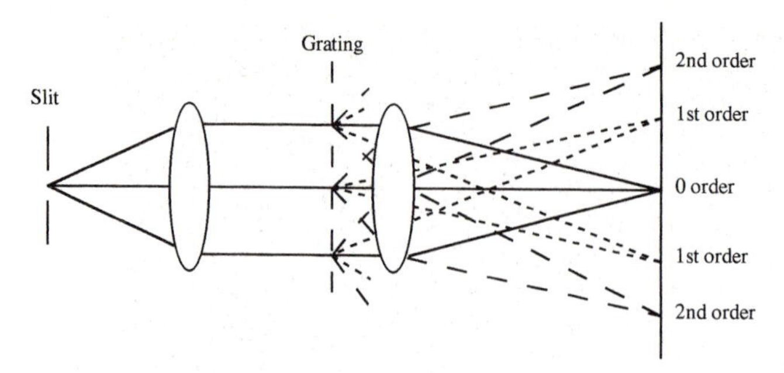

Fig. 11.4. Transmission grating as a disperser.

One may see immediately that the grating made of spaces and opaque strips is not at all efficient. Half of the incident light is blocked by the opaque strips. Optical theory indicates that half of what is left goes into the central image or zero order. That does not leave much to be distributed among the progressively fainter orders on the two sides. It was found many years ago, however, that if the grating consisted of a series of non-symmetrical grooves, great improvements became possible. The non-symmetrical shape is referred to as a 'blaze'. When a blazed grating is used, as much as 70% of the light can be concentrated into one order on one side of the zero order. Fig. 11.5 shows this result quite vividly. This photograph was made with a transmission grating in front of the lens of a camera with a focal length of only about twelve inches. The short lines are the zero order spectra for several stars. To the right of the zero order for the brightest star shown here one may see an overexposed first order and a portion of the second order. At the extreme left can be seen the first order on the other side. Clearly, a large fraction of the light has gone into the first order on the left. In the discussion which follows, we shall assume that all gratings are blazed.

Gratings are made by pulling a carefully shaped diamond tool across a metal surface. The diamond cuts one groove, is moved slightly to the side and cuts the next one. It is slow work and must be performed with great care to insure uniformity in the shape and spacing of the grooves. As a result, original gratings are very expensive. Fortunately, it is possible to make plastic replicas of expensive original gratings. A liquified plastic is poured over the grooves, and after the plastic has hardened it is lifted off. The grooves are efficiently transferred to the plastic. The plastic may then be rigidly mounted to produce a replica grating. The master grating is unchanged and may be used to produce many replicas at a moderate cost. The replica may be mounted on a glass plate to produce a trans-

Fig. 11.5. Spectra photographed by means of a transmission grating. The spectra were widened by periodically moving the telescope in declination. Note that the brightest first-order spectrum is much brighter than the second-order spectrum just to the right of it. A faint first-order spectrum to the left of the zero order can be seen in the original. (Photograph by the author.)

mission grating, or it can be mounted and coated with a reflective surface to produce a reflection grating.

Explanation of the complete theory of the diffraction grating requires a chapter or more in a typical textbook on optics, so we shall not attempt that in this book. For the present purposes it is sufficient to proceed directly to the practical usage of gratings.

In order to define some necessary terms, let us refer to Fig. 11.6 which shows cross-sectional views of a transmission and a reflection grating. In these diagrams θ is the blaze angle. We shall discuss criteria for the choice of this angle below. α is the angle of incidence and is measured with respect to the normal to the surface of the grating. β is the angle of diffraction, that is, the angle in which a ray of wavelength, λ, will be directed. These angles are related to each other by the grating equation which may be stated as follows

$$m\lambda = a(\sin\alpha \pm \sin\beta)$$

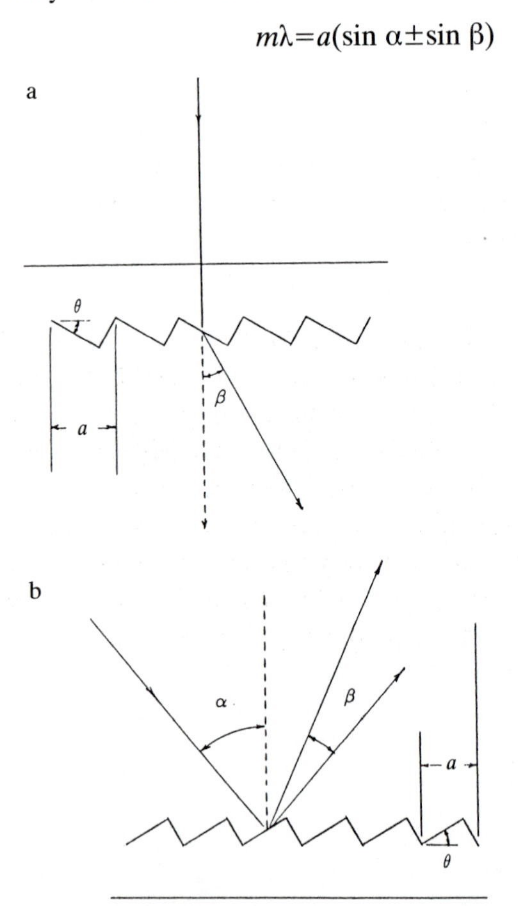

Fig. 11.6. Specification of angles in (*a*) a transmission grating and (*b*) a reflection grating.

where m is the order number and a is the grating spacing, that is, the distance between adjacent grooves. The positive/negative sign shows that a different spectrum is formed on each side of the zero order.

The choice of the blaze angle, θ, is ultimately governed by the portion of the spectrum in which the user intends to work, because the direction of a ray reflected directly from the face of one of the grooves should be the same as the direction of the diffracted ray of the desired wavelength. Let us illustrate this with an example and see what the blaze angle should be for a grating of 600 lines per millimeter which is to be used in the second order at 5000 ångströms (5000 Å=0.0005 mm). We shall assume also that the angle of incidence is 5 degrees. Then, substituting in the grating equation we have

$$2\times 0.0005\ \text{mm}=1/600\times(\sin 5+\sin\beta);\ \beta=30.853$$

Theta is now found from

$$\theta=(\alpha+\beta)/2$$

$$\theta=35.853/2=17.927$$

Further consideration of the grating equation reveals that a potentially serious problem exists because of the manner in which the orders overlap. Clearly, as m increases we see that

$$1\lambda_1=2\lambda_2=3\lambda_3\ \text{etc.}$$

Thus, if α remains constant, then β will be the same for 8000 ångströms in the first order, 4000 ångströms in the second order and 2666.7 ångströms in the third order. Fig. 11.7 indicates the overlap between orders for wavelengths from 4000 to 8000 ångströms. One may note that the overlap increases in the higher orders. In practical work the problem is alleviated somewhat by the limited spectral sensitivity of most detectors and by the inclusion of filters to eliminate the unwanted light from adjacent orders. For example, a grating might be intended for use in the second order with blue light on both sides of 4000 ångströms. If the spectrum was recorded

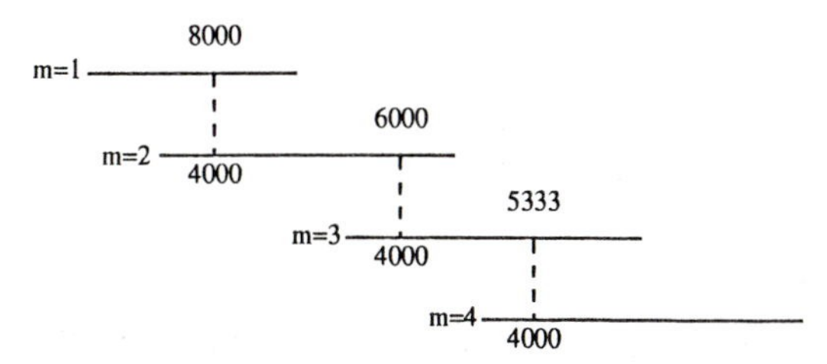

Fig. 11.7. Overlapping orders in grating spectra.

on a photographic emulsion sensitive to blue light, then the overlapping red from the first order would not be recorded.

Angular and linear dispersion

In reference to spectra we use the term 'dispersion' to indicate the degree to which light has been spread out. We speak of angular dispersion to indicate the manner in which the angle β varies with wavelength, and we speak of linear dispersion to describe the variation of wavelength with distance in the final image formed by the camera. It is the linear dispersion that we are ultimately interested in, but we must first understand how the angular dispersion is found. Angular dispersion may be thought of as the rate of change of β with respect to λ. Therefore, let us assume that α is constant and differentiate the grating equation with respect to λ. The result is

$$d\beta/d\lambda = m/(a \cos \beta)$$

Several important points are immediately seen from this. As mentioned above the dispersion is directly proportional to the order. It should also be clear that dispersion is inversely proportional to a, the spacing of the lines, so higher dispersion may be achieved by increasing the number of lines per millimeter. Finally, since β is usually small, $\cos \beta$ is usually close to unity. This means that the dispersion is very nearly linear, and this can be a great advantage in identifying lines and measuring wavelengths.

One could also find the dispersion simply by calculating β for a number of wavelengths in one order. Differences in β for uniform intervals of wavelength would then show just how β changes with wavelength.

Linear dispersion, the more useful quantity, is the rate of change of l with respect to wavelength where l is distance in the focal plane of the camera (plane ff′ in Fig. 11.1). Thus the linear dispersion may be calculated simply by multiplying the last equation by f, the focal length of the camera lens. If β is measured in radians, this gives

$$dl/d\lambda = mf/(a \cos \beta)$$

This illustrates quite clearly the obvious point that the image in plane ff′ is made larger if the focal length of the camera is increased.

The most common practice among astronomers is to make reference to the reciprocal dispersion for which the units are ångströms per millimeter. Low dispersion might mean something on the order of 80 Å/mm while high dispersion suggests 2 or 3 Å/mm. Low dispersions are suitable for the classification of stellar spectra, but high dispersions are needed when precise wavelengths are sought. More will be said about applications of low and high dispersion in Chapter 12 and in Chapter 13.

In the case of the prism the angular dispersion may be found through a series of steps such as those used to calculate Table 11.2. From that table we may calculate the values listed next in Table 11.3. Differences in l and values of the reciprocal dispersion are based on a focal length of 120 mm for the camera. Here the angular dispersion is simply the difference in ϕ' (column 2) divided by the difference in wavelength (column 1). Difference in l is found by multiplying the values in column 2 by the plate scale which is in this example 2.0946 mm/degree, so the units are millimeters. (See p. 78 for a discussion of plate scale.) Reciprocal dispersion is found by dividing values in column 1 by values in column 4, so the units here are ångströms/mm. From this example it is quite clear that the reciprocal dispersion for a prism is not constant but is greatest at the long wavelength end of the spectrum.

Objective prism and objective grating

Referring again to Fig. 11.1, we recall that the function of the collimator is to ensure that parallel rays should enter the disperser. When we consider the use of the spectrograph in astronomy, we note that the stars are so far away that rays from the star are in fact parallel. Therefore, if we simply place a prism or transmission grating in front of the objective, we should be able to record spectra in the focal plane. This arrangement has the great advantage that in one exposure a spectrum may be recorded for every star in the field of view down to some magnitude limit which depends upon the size of the telescope. A prism used in this manner is referred to as an objective prism. The vertex angle in such cases is usually only a few degress, and a prism of this type should

Table 11.3. *Dispersion by a prism*

Difference in λ(ångströms)	Difference in ϕ' (degrees)	Angular dispersion (degrees/ångströms)	Difference in l	Reciprocal dispersion
730	1.17782	0.00163	2.467	295.897
520	0.53493	0.00103	1.120	464.092
1030	0.61759	0.00060	1.294	796.222
670	0.26262	0.00039	0.550	1217.979

be used with a telescope with a low-numbered f-ratio, that is, a fast lens system.

A transmission grating may be used in the same way, and the spectra shown in Fig. 11.5 were made with an objective grating.

Practical design of a spectrograph

In designing a spectrograph for use on an astronomical telescope one must begin by deciding first what the required dispersion will be and in which range of wavelengths the spectra are to be obtained. These two questions govern the choice of the grating and the camera lens. Many different gratings and camera lenses are commercially available, so the designer must consult catalogs of each to find those which most nearly match the ideal. The dimensions of the grating must be large enough to insure that the diffracted beam is nearly as large as the aperture of the camera lens, and the size of the grating in turn dictates the diameter of the collimating lens. The geometry of these elements may be seen in Fig. 11.8. If the grating is too small, light will be wasted. If the camera lens is

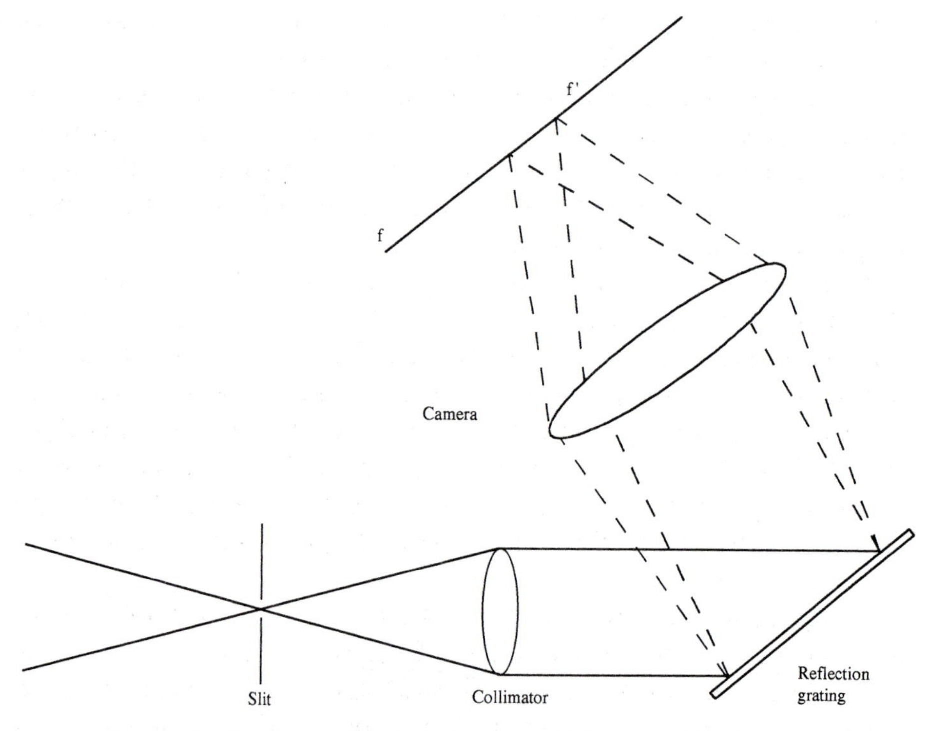

Fig. 11.8. The dimensions of the grating will determine the sizes of the camera lens and the collimator.

too large it will be unnecessarily expensive. If the focal length of the camera lens is too long, the lens will be optically slow and exposure times will be long. The choice of these major elements requires careful analysis and a number of compromises.

The focal length of the collimator should be chosen so that the f-ratio of the collimator closely matches that of the objective lens of the telescope.

One must now provide a structure in which the slit, collimator, disperser, camera lens and photographic plate may be rigidly held in their proper positions relative to each other. This structure is often in the form of a box with one side removable to provide access to the components. Provision must also be made for focusing the camera lens and for rotating the grating through a small angle. The latter is needed in order to bring the desired portion of the spectrum to a point near the optical axis of the camera lens. For example, the grating could be rotated slightly when the observer changes from observations in the first order red to observations in the second order blue.

The structure described above is to be mounted on the telescope, so it must be reasonably compact and light in weight. When in position, the slit should be right in the focal plane of the objective, and the axis of the collimator should coincide with the axis of the objective.

The image of a star is to be brought into focus on the slit, so the observer must be able to see the slit. For this purpose a small telescope is usually mounted in the side of the box and focused on the slit. During an exposure the observer will have to watch carefully to see that the image of the star remains on the slit at all times.

Astronomers have found that it is necessary to widen their spectra if the important features in the spectra are to be seen. This is due to the fact that the image of the star is really quite small, and its spectrum will normally appear simply as a straight line. In order to widen a spectrum the astronomer usually causes the image of the star to move slowly along the slit. This may be accomplished by using the slow-motion controls of the telescope either in declination or right ascension. The choice depends, of course, on the orientation of the slit in the north–south or east–west direction.

A comparison spectrum is often added to a photograph of a stellar spectrum in order to provide reference points for the precise determination of wavelengths. For this purpose two more additions are made to the spectrograph. The first of these is a light source. This may be a metallic arc or a lamp filled with a gas or combination of gases which are excited and caused to glow when a current passes through the lamp. Argon and neon are good choices since the spectra of these gases contain large numbers of lines of known wavelength. The second is a mask which can be used to cover and uncover portions of the slit. Such a mask is sometimes called a 'decker', and a sketch of a typical mask is included in Fig. 11.9. The mask pictured here may be moved left or right to either of two positions. In the first position the central opening is in front of the slit, so the ends of the slit are covered. When the spectrum is being photographed, the star's image is moved back and forth along that part of the slit which is not covered by the mask. After a suitable time, the mask is moved to its second position so that the ends of the slit are open and the central part is covered. The comparison lamp is now turned on for some predetermined period. The final developed photograph will then show the spectrum of the star in the center and the spectrum of the comparison source on either side of it. This is shown in Fig. 11.10. In Chapter 13 we shall show how wavelengths in a stellar spectrum may be determined when wavelengths in the comparison spectrum are known.

It was mentioned earlier that the overlap of one order onto another could sometimes be eliminated through the combination of the spectral sensitivity of the photographic emulsion and a suitable filter. Referring again to Fig. 11.7 we note that a grating blazed for the second order red could also be used for the third order blue. For the second case a film sensitive only to blue light could be used so that the red light would not be recorded. A filter which does not transmit blue light could be inserted into the beam if the red light from the second order was to be recorded. For this reason the spectrograph is usually equipped with a slide which can serve both as a shutter and a carrier for one or more filters.

Finally, astronomers must be able to identify the star for which a spectrum is desired. This is usually facilitated by the provision of a wide-

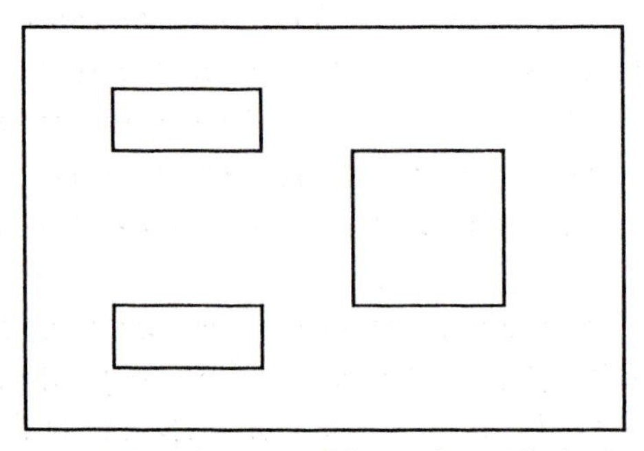

Fig. 11.9. The moveable mask or 'decker'.

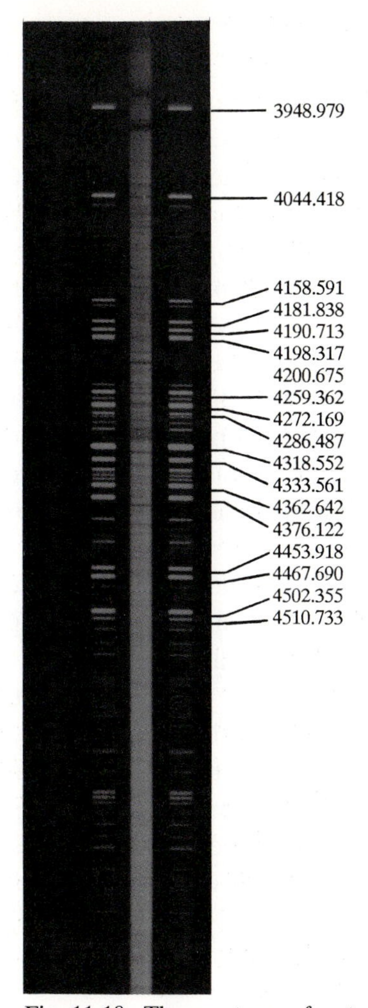

Fig. 11.10. The spectrum of a star with the spectrum of argon/krypton photographed on either side of it. (Wellesley College photograph.)

field eyepiece on the side of the instrument. Just as in the case of the photometer, a mirror may be moved in and out of the light path to reflect light to this second eyepiece. The star to be studied is identified and centered in the wide–field eyepiece. The mirror is then moved out of the way, and the observer looks through the small, guiding telescope to bring the star's image into position on the slit.

We have tried here to describe the common features which govern the design of an astronomical spectrograph. There will be many differences in detail from one instrument to the next, so the user should make sure that all features are well understood.

Resolving power

Stellar spectra contain a tremendous amount of information, so we seek to record the spectra in ways which will maximize the amount of detail which can be seen or measured. As in the case of the images of stars we speak of resolving power or resolution, and here we say that two lines are resolved if the image of one can be distinguished from the image of another one very close to it. If we assume for the moment that the spectra are recorded photographically, we may list the factors which affect the resolution in the final result. First, there are optical effects due to diffraction. These are the same as those which produced the disc and rings which were discussed in Chapter 7 for the optical images of stars. In the case of the grating spectrograph this effect is substantially reduced when the total number of lines in the grating is large. It also helps if the aperture of the camera is made as large as possible. Second, we must consider the actual width of the entrance slit since this is projected into the focal plane of the camera. If the slit is made narrower, its image will, of course, be made narrower and resolution will be improved. A third factor is the graininess of the film or plate on which the spectrum is recorded, and this will often impose the most serious limitations.

In order to illustrate these principles let us consider the resolution to be expected in a small spectrograph which might be used at a dispersion of about 80 Å/mm. Spectra made with an instrument of this sort would be suitable for a program of classification. Let us assume that the major elements are defined as follows

collimator	focal length 39.4 cm
	aperture 2.92 cm
grating	ruled area 5 cm by 5 cm
	600 lines per mm
	blaze angle 17 degrees, 27 minutes
camera	focal length 10.16 cm
	aperture 3.63 cm
slit width	0.040 mm

We calculate the reciprocal dispersion by substituting into the grating equation. Let us assume that the angle of incidence is 5 degrees and calculate β for wavelengths of 3800 Å and 4000 Å. The two values of β are 21.644 degrees and 23.132 degrees, so $\delta\beta$ is 1.488 degrees. The plate scale in the focal plane of the camera is 1.75 mm/degree. Therefore, the distance between lines at 3800 Å and 4000 Å will be 1.488×1.75 or 2.603 mm on the plate. The reciprocal dispersion is 200 Å/2.603 mm or 76.8 Å/mm.

In order to calculate the projected width of the slit in the focal plane of the camera, we must find the demagnification factor which results from the fact that the focal length of the camera is shorter than that of the collimator. This factor is simply the ratio of the focal lengths of the two lenses, so in this example it is 101.6/394 or 0.258. Now if the original width of the slit is 0.040 mm, the projected width will be 0.010 mm. Combining this with the reciprocal dispersion, we note that the projected width of the slit is 0.77 Å.

Taking a look at the graininess of the developed photographic emulsion we find that for the materials often used by astronomers the resolution is only about 0.02 mm. In the light of the previous paragraph, we see that in this example the slit could be widened with no loss of detail. A wider slit means, of course, that more of the light of the star can pass through the slit and that the exposure time can be shorter. The resolution in ångström units is easily seen to be 1.54 (0.02 mm×76.8 Å/mm).

Microdensitometer

In most astronomical applications the spectra which are recorded photographically are quite small and must be measured or studied under a microscope. For example, the low-dispersion spectra made with the spectrograph described in the previous section are only about 25 mm long and about 5 mm wide. Experience quickly shows that when such small spectra are being studied visually, best results are found when a minimal magnification is used. If one tries to see more detail by increasing the magnification, one soon begins to see only the graininess of the emulsion. Often one sees just faint suggestions of spectral features and cannot be certain that they are real. Under these circumstances some more objective means of looking for detail is desirable. One auxiliary instrument for doing this is the microdensitometer – a device for measuring the photographic density along the length of a spectrum.

The arrangement of the major parts of the microdensitometer is shown in Fig. 11.11. The spectrum to be measured is placed in a carriage which can move the spectrum through a beam of light. A photomultiplier measures the amount of light transmitted by the plate, and the output of the photomultiplier is fed to a chart recorder or to a computer. As the plate moves through the beam, the variations in density are recorded on the chart. A portion of a tracing made in this way is seen in Fig. 13.10. An adjustable slit is also placed in the light path, and a lens projects the

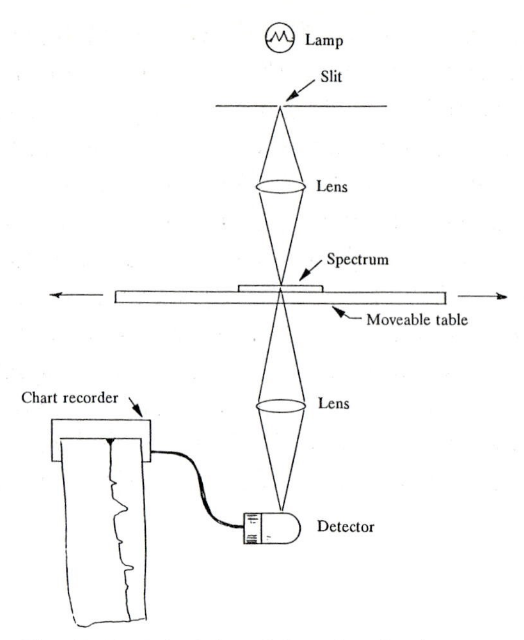

Fig. 11.11. Principles of the microdensitometer.

image of this slit onto the plate. This slit should be made as narrow as is practical, since its width determines the resolution which can be achieved in the tracing. The length of the slit should be less than the width of the spectrum, and the image of the slit should be parallel to the spectrum lines on the plate. When attention has been paid to all necessary details, the tracing can reveal faint spectral features which cannot be detected at all by the eye.

Other recording methods

In an attempt to take advantage of the sensitivity and efficiency of the photomultiplier, astronomers have modified the conventional spectrograph to develop the spectrum scanner. An exit slit is mounted in the focal plane of the camera and made parallel to the entrance slit. A photomultiplier is then mounted behind the exit slit and connected to a chart recorder or to a computer. In the design shown in Fig. 11.12 a concave grating eliminates the need for the lenses of the collimator and camera. The grating is now mounted in such a way that it may be rotated on an axis parallel to the slits. Now when the entrance slit is illuminated by the light of a source such as a star, the grating is rotated through a small angle. This causes the spectrum to be swept past the second slit. The photomultiplier measures and records the variation in intensity with wavelength along the spectrum. In some other designs the grating is fixed, and the exit slit and detector are moved. Over the years many designs have been tested and used with considerable success.

The photomultiplier can detect only one small range of wavelengths at a time, and astronomers have sought other electronic devices which could record or measure the intensity of starlight at many wavelengths simultaneously. Detectors for this application took several forms. One was essentially a television camera which could detect, amplify and record an entire spectrum all at once. Another was the image intensifier which is an evacuated tube with a photocathode at one end and a phosphor screen at the other. A strong magnetic field is provided by magnets

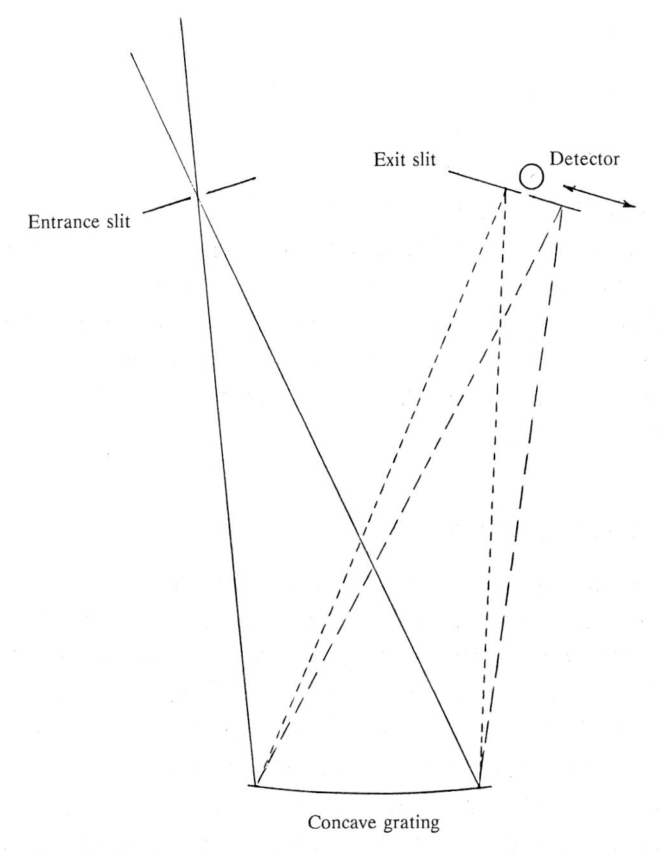

Fig. 11.12. One form of a spectrum scanner using a concave grating. The spectrum is formed in the plane of the exit slit and is scanned as the slit and detector are moved along the spectrum.

surrounding the tube, and a high voltage is applied between the two ends. Light striking the cathode causes the emission of electrons and these are accelerated toward the opposite end by the voltage difference. Ordinarily, the electrons would move in undefined paths and cause a large glowing spot to appear on the screen. In the image intensifier, however, the magnetic field causes the electrons to move along well-defined spiral paths. At the screen the positions of the electrons relative to each other are the same as they were at the cathode, and a picture is produced. The picture on the screen can be brighter than the original image on the cathode. Furthermore, there can be what amounts to a change in wavelength between the two ends of the tube. The cathode may be a surface which is sensitive to infrared radiation, and the phosphor on the screen at the other end may be one that glows at some shorter wavelength in the visible region. Finally, a second cathode may be put inside the tube to provide a two-stage image intensifier.

Another type of detector is the linear array of photodiodes. The photodiode is a crystal of silicon with electrical properties which change when light falls on it. A number of these can be arranged in a row to provide the linear array. In several designs there are more than one thousand individual detectors in a device only about twenty-five millimeters long. If a spectrum is projected onto a detector of this sort, the intensity in the spectrum may be measured at each of these many individual points. The resolution obviously depends upon the size of the elements in the array, but the detector described here should give resolution as good as that which can be achieved photographically. The rapid measurement of the voltages on such a large number of elements requires the use of a small computer as the measuring and recording unit, and the computer must be controlled by quite an extensive program. Nevertheless, systems based on these principles have a great many advantages, and they are likely to be widely used in the future. Fig. 11.13 shows a spectrum of the sun recorded by means of a linear array.

A slightly different but very successful array is the charge-coupled device (CCD) so named from the manner in which the charges on the individual detectors are read into a computer. Two-dimensional CCDs are finding broad use in observational astronomy, and their applications in photometry were discussed in Chapter 9. CCDs are described more fully in Appendix 4.

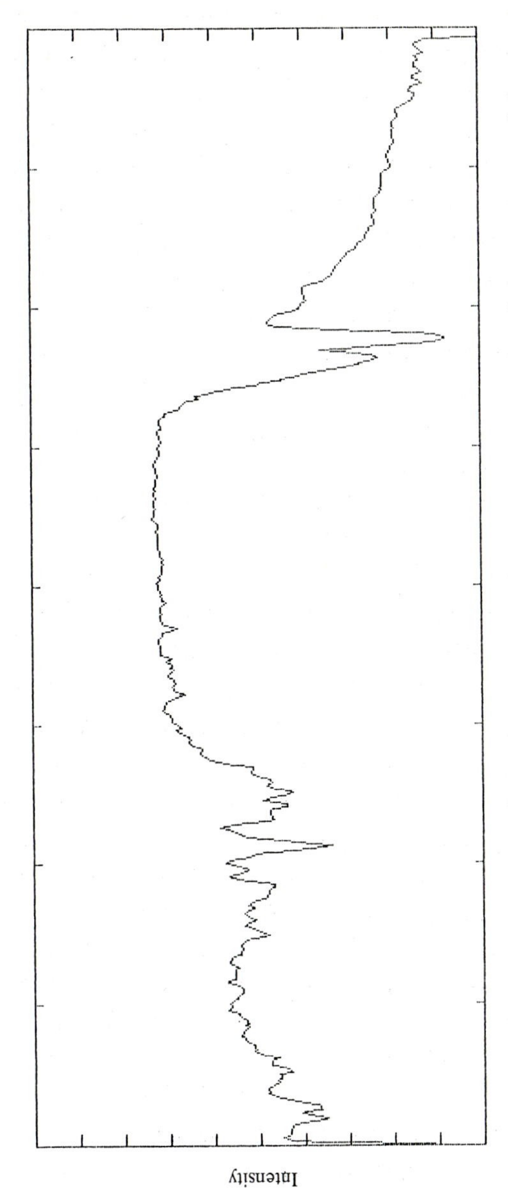

Fig. 11.13. Spectrum of the sun obtained with a linear array. Wavelength increases toward the left. The two conspicuous lines toward the right are the H and K lines of CaII. The pixels are numbered from 0 at the left to 1024 at the right. (Courtesy of Richard G. French, Wellesley College.)

The coudé spectrograph

Some of the earliest high-dispersion spectrographs were designed to operate at the coudé focus of the telescope. We remind the reader that the coudé focus is a point at the lower end of the polar axis as shown in Fig. 11.14. This axis must be made in the form of a hollow tube, and a system of mirrors reflects light down this tube. The curvature of the secondary mirror is chosen so that the effective focal length of the primary–secondary combination is long enough to reach a convenient place below the telescope. The coudé focus is a stationary point, so large instruments may be placed adjacent to it. Since the focal length of the coudé system is necessarily long, the reciprocal dispersion can be high, 1 or 2 Å/mm, for example, and the focal length of the camera can be short. This means that the camera will be optically fast.

The echelle spectrograph

In order to provide for high-dispersion spectra without the disadvantages of the coudé spectrograph, instruments using the echelle as the dispersing element have been designed and built. The echelle is simply a very coarse grating having on the order of only 70 lines per millimeter and a blaze angle near 60 degrees. These are used, however,

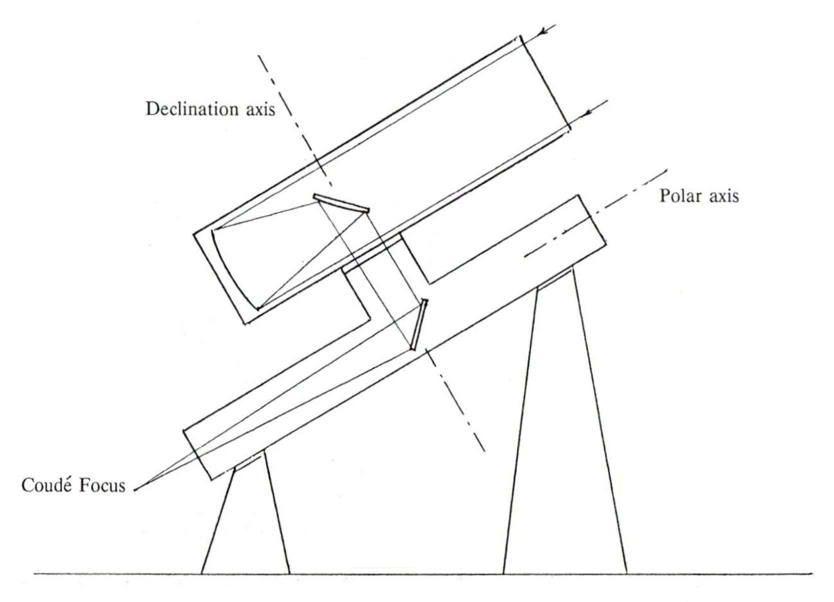

Fig. 11.14. The light path to the coudé focus of a reflecting telescope.

in very high orders where they are capable of producing high-dispersion. If we recall again the grating equation

$$m\lambda = a(\sin\alpha + \sin\beta)$$

then we can easily see how these principles are applied. Large values of β may be obtained simply by working in the higher orders i.e. by using higher values of m. A major problem arises because of the serious overlap of the spectra in the higher orders, but this may be overcome by means of a cross-disperser placed in the beam between the grating and the camera. An arrangement of this sort is shown in Fig. 11.15. The cross-disperser is a standard grating with its direction of dispersion at right angles to that of the echelle, and the grating equation again explains how it functions. β will be larger in successively higher orders, and β will be larger at the longer wavelengths than at the shorter ones in each order. The result is that each order appears to be tilted (and perhaps curved) and successive orders are arranged beside each other. An echelle spectrogram is reproduced in Fig. 11.16.

The advantages of the echelle spectrograph arise chiefly from the fact that high dispersion may be achieved in an instrument that is compact and optically fast.

The applications of spectroscopy in astronomy will be described in the next two chapters. Chapter 12 will be devoted to low-dispersion spectra

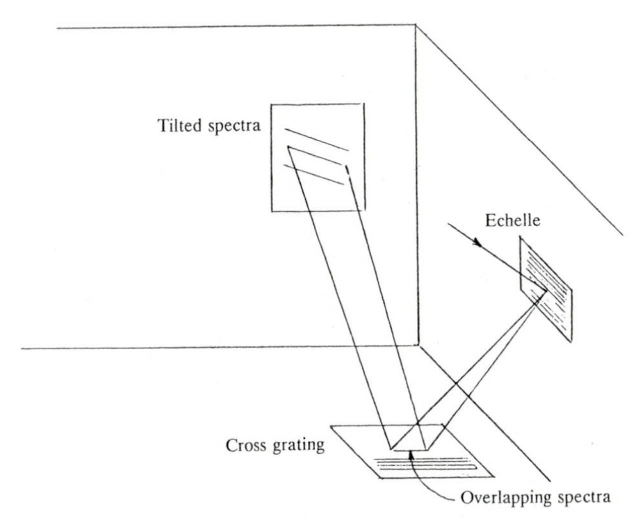

Fig. 11.15. The principal components and the light path in an echelle spectrograph.

Fig. 11.16. An echelle spectrogram of the star T Microscopium. Orders from the red are at the top. The broad feature in the seventh row from the bottom is the absorption line of neutral calcium at 4226 Å. The small white dots are the comparison spectra. (Courtesy of Irene Little-Marenin, Wellesley College.)

which means that it will cover the classification of stellar spectra. Chapter 13 will cover high-dispersion spectra, so it will present the methods of determining radial velocities. Other features which require high dispersion will also be described in Chapter 13.

QUESTIONS FOR REVIEW

1. Describe the function of each of the principal parts of a grating spectrograph.
2. What are the advantages of the grating over the prism as the dispersing element in a spectrograph?
3. What is meant by the term 'blaze' when used in reference to a grating?
4. Imagine that a transmission grating of 600 lines/mm is used as an objective grating with a small, fast camera designed for use from aircraft. The focal length is twelve inches (305 mm), and the aperture is four inches (102 mm). Calculate the linear dispersion for wavelengths near 4500 Å.
5. Why is it often desireable to provide a means for photographing a comparison spectrum? How is this done?

6. In what way does the width of the entrance slit affect the resolution obtainable in a spectrum? Name another factor which might limit the resolution even more than the width of the slit.

7. Describe a method of scanning (i.e. directly recording) a spectrum with a detector such as a photomultiplier or a photodiode.

8. What is the role of the CCD in the design of a modern spectrograph?

9. Describe the way in which the echelle grating is used in the design of a spectrograph. How is the overlap of orders eliminated in the echelle spectrograph?

10. Why is it sometimes useful to mount a spectrograph at the coudé focus of a large telescope?

11. Specify the components needed to design a spectrograph for use with a twenty-four-inch telescope of twenty-six foot focal length. The desired dispersion is 120 Å/mm. Price and availability dictate that the grating will be three inches square.

Further reading

Hearnshaw, J. B. (1986). *The Analysis of Starlight*. Cambridge: Cambridge University Press. This is a complete and detailed history of astronomical spectroscopy from its beginnings to the 1980s.

Kitchin, C. R. (1984). *Astrophysical Techniques*. Adam Hilger Ltd. Chapter 4 presents a good treatment of the basics. The emphasis is on prisms as the dispersing element.

Miczaika, G. R. and Sinton, W. M. (1961). *Tools of the Astronomer*. Harvard University Press. Read Chapter 6 for a good, basic treatment of the spectrograph.

Pannekoek, A. (1961). *A History of Astronomy*. Interscience Publishers. The early applications of spectroscopy in astronomy are described very well in Chapter 37 and in parts of some later chapters.

Walker, G. (1987). *Astronomical Observations*. Cambridge University Press. In Chapter 5 Walker gives a treatment of the spectrograph which is more technical than that in the present chapter.

12

Classification of stellar spectra

When we examine the spectra of a large number of stars, certain differences are easily recognizable. We need only to look at Fig. 12.1 to understand that this is so. The spectra shown here were recorded by the Schmidt telescope at Warner and Swasey Observatory with a prism mounted in front of the corrector plate. The use of an objective prism for spectra of this type was described in the previous chapter. This photograph shows only part of the entire field recorded on the original plate, but even here one finds considerable variety among the spectra. The complete interpretation of these differences was one of the greatest achievements of astronomers in the first half of the twentieth century, and it could be accomplished only after the full range in types of spectra had been established. The first step was to photograph the spectra of a large number of stars, and the second step was to classify those spectra according to some logical scheme. In this chapter we shall describe first the origins of the present system of classification and then show how details in the spectra distinguish giants from stars of the main sequence. At the end of the chapter we shall summarize the early history of the classification of stellar spectra.

Fraunhofer lines in the solar spectrum

Joseph von Fraunhofer, the distinguished nineteenth century optical worker, made significant studies of the solar spectrum, and he published some of his results in 1817. He mapped more than five hundred lines and assigned the letters A through K to some of the most conspicuous ones. We still find it useful to use some of these designations in discussing lines in the spectra of stars as well as that of the sun, so we include the Fraunhofer lines, their wavelengths and their identifications in Table 12.1. Their positions in the solar spectrum are shown in Fig.

Fig. 12.1. Differences in the appearance of spectra are apparent in low-dispersion spectra made with an objective prism. One may note emission lines in two of these spectra. (Stellar spectra obtained with the Burrell Schmidt-type telescope of the Warner and Swasey Observatory of Case Western Reserve University.)

12.2. Today, Fraunhofer's designations are in common use only for the D lines, the G band and the H and K lines.

Careful examination of a large number of spectra shows that the absorption lines of hydrogren are common if not quite universal. These lines, the Balmer Series, are the principal lines visible in the white stars, and they are readily recognized. They become progressively less con-

Table 12.1. *The Fraunhofer lines*

A	7594	Molecular oxygen in earth's atmosphere
B	6867	Molecular oxygen in earth's atmosphere
C	6563	Hα
D1	5896	Sodium
D2	5890	Sodium
E1	5270	Iron
Eb	5183–5168	Magnesium
F	4861	Hβ
G	4308	Blend of band of methane and iron
H	3968	Ionized calcium
K	3933	Ionized calcium

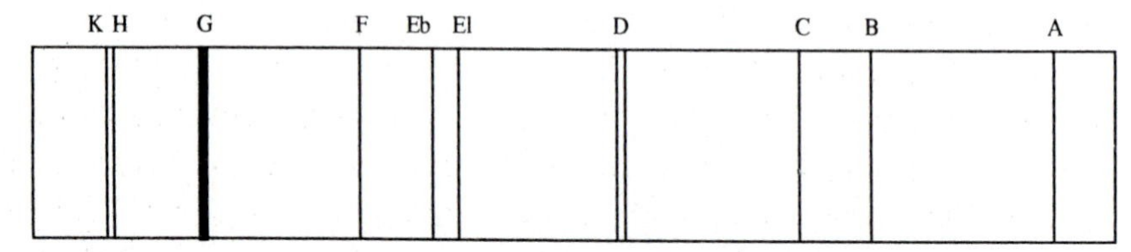

Fig. 12.2. The locations of the Fraunhofer lines in the solar spectrum.

spicuous in the yellow, orange and red stars at the same time that other lines become more prominent. It is hardly surprising, then, that in one of the first large-scale programs of spectral classification the strength of the hydrogen lines was chosen as the principal criterion. This seemed to be quite a reasonable approach at a time when the causes for differences in spectra were not understood.

The Henry Draper Catalog – 1918

Recognizing that a comprehensive catalog of spectral types would be valuable in a variety of ways, astronomers at Harvard Observatory began a major program of spectral classification in 1886. This program continued for nearly forty years. Previous small-scale programs had established classes based on the most obvious features, but the Harvard astronomers were able to recognize more subtle gradations. Thus, in the early stages of the new program they were able to recognize sixteen classes on the basis of the relative strength of the hydrogen lines and of some other readily recognizable features. The classes were labelled with the letters of the alphabet, A, B, C, D, etc. with Class A being used for the stars with the most conspicuous lines of hydrogen in their spectra.

It was soon apparent, however, that for a very clear reason some re-ordering was needed. Color could be expected to be an indicator of temperature, and the blue and white stars could be expected to be hotter than the yellow and red ones. In the alphabetical sequence based on the hydrogen lines, the classes O and B were definitely out of place among the yellow, orange, and red stars. The lines of helium were also a problem, for in the alphabetical sequence they appeared abruptly only in classes B and O. Eventually, the entire sequence was rearranged and many of the classes were dropped. The new sequence of spectral classes was simply OBAFGKM. In this order the lines of hydrogen became progressively stronger in classes from O through B until they reached a maximum in class A. Then they became progressively weaker in the classes F, G, K and M. A similar increase and decrease in line strengths held true for the lines of helium, calcium and some other elements, but the classes in which the lines became strongest were different for each element. Those individuals who actually made the classifications learned to perceive small differences in appearance of the spectra, and from these differences they were able to specify subdivisions within each class. The subdivisions are specified by the numbers from 0 through 9. Thus, for example, the spectra of Sirius, Aldebaran and the sun are classified respectively as A1, K5 and G2.

The magnitude of the Harvard work can best be appreciated when one realizes that in its final form it contained 359 082 spectral classifications. The guiding force behind the project was Edward C. Pickering, the director of the observatory, and major contributions were made by Annie J. Cannon and Antonia C. Maury. The results were published beginning in 1918 as the *Henry Draper Catalog*. Henry Draper had been an American doctor who was keenly interested in astronomy and particularly in stellar spectroscopy. He had been one of the first persons to record the spectrum of a star photographically. His widow donated money to Harvard so that Draper's interest could be continued.

The value of the *Henry Draper Catalog* to astronomers is inestimable. It has served as the data base for statistical studies of spectral type, for studies of galactic structure, as the source material for high-dispersion spectral studies and for much more. It is reliable and internally consistent, and upon it much later work has been based. It stands as a true monument to the persistence, dedication and high standards of all who worked on it.

The *Henry Draper Catalog* was published as Volumes 100 to 105 in the *Annals Of The Harvard College Observatory*. Within the catalog the stars are consecutively numbered in order of increasing right ascension. As an example of the numbering system, the star, α Cygni, Deneb, is specified as HD197345 in the *Henry Draper Catalog*.

Even before the publication of the catalog, the Harvard sequence of classes was recognized as a temperature sequence principally on the basis of color. The red stars of class M were assumed to be the least hot, and the blue stars of Class O were assumed to be the hottest, but the relationship between temperature and appearance of the spectrum was not immediately apparent. Then in 1920 M. N. Saha, a physicist from India, brilliantly solved the problem. He showed that going from cool stars to hotter ones, the lines should become stronger as the level of excitation of the atoms in a star's atmosphere increased. Maximum strength should occur just before the temperature at which ionization begins. Then, as more atoms become ionized, there are fewer and fewer atoms which are able to absorb, so the lines become weaker. Since some atoms are ionized at lower temperatures than others, the maximum line strengths occur in different classes for different chemical elements. Saha's analysis was a major turning point in the development of astrophysics.

Classification features

In outlining the means by which the student can learn how to assign a spectrum to a particular class, let us begin by indicating and identifying some of the most conspicuous features that are the keys to the classification scheme. In Fig. 12.3 we have reproduced photographs of some typical spectra, and we have identifed some of the key features. The second spectrum shows clearly the absorption lines of hydrogen, and one should note here that the series of lines begins with the line Hα at the red (right) end of the spectrum. The Fraunhofer B line due to oxygen molecules in the earth's atmosphere is also conspicuous.

In the next spectrum, that of an F0 star, two lines of equal intensity are seen. These are the H and K lines of the Fraunhofer spectrum and they come from ionized calcium (CaII). Notice here that the H line is at almost exactly the same place as the Hε line.

The next spectrum, G0, shows even stronger H and K lines, but it also shows a broad feature, Fraunhofer's G band. In spectra made with higher dispersion a line of neutral calcium at a wavelength of 4227 Å is also important in classification.

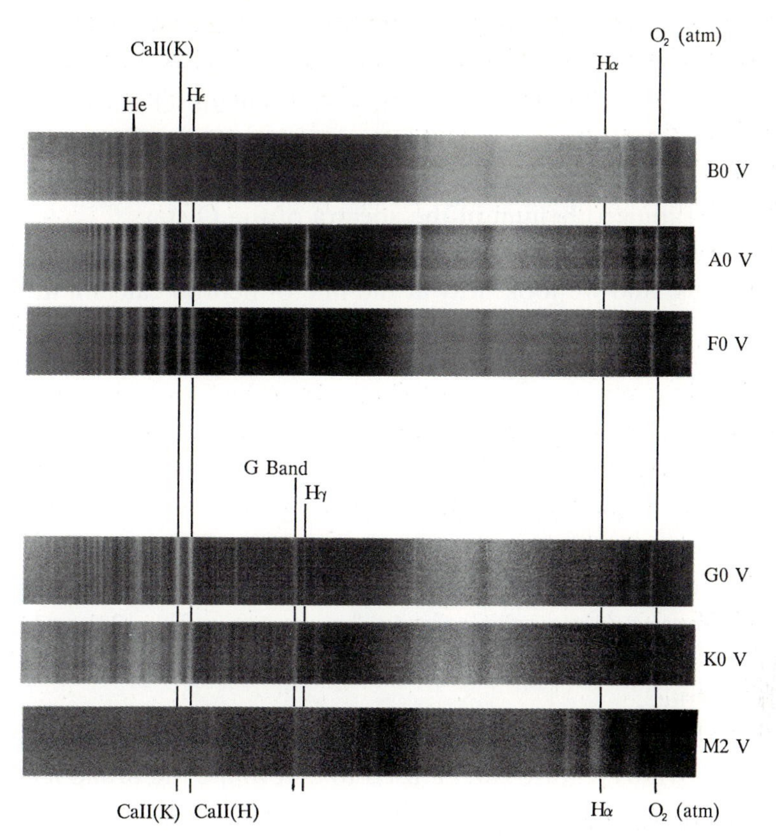

Fig. 12.3. Comparison of spectra and identification of some lines which are useful as criteria for classification. (From Seitter, W. C., *Atlas for Objective Prism Spectra*, 1970, Ferd. Dummlers Verlag, Bonn).

Finally, the last spectrum, M2, shows several broad features in the red which are due to molecules of titanium oxide.

The reader should study Fig. 12.3 carefully in order to be able to recognize the few features listed here. One may also look back at Fig. 12.1 to see how readily some of these can be picked out in low-dispersion objective prism spectra. We may now proceed to see how these lines and bands can be used to make an assignment to one of the seven classes.

We make an initial division of spectra into two groups just on the basis of the hydrogen lines. In classes B, A and F the hydrogen lines are stronger than any other lines. In classes G, K and M the hydrogen lines are not the strongest.

In the first group we may distinguish stars of type B simply by the presence of helium lines. That means that if the helium lines are not visible, the class is either A or F. In the A stars the G band cannot be seen, and the line of calcium at 4227 Å is very faint or not visible either. In addition, the combination of Hε and the Fraunhofer H line is stronger than the K line. Here we have not separated the O stars, since their spectra are similar to those of the B stars. These two classes, O and B, may be distinguished from each other by the weaker hydrogen lines and the lines of ionized helium in the spectra of the O stars.

In the second group, classes G, K and M, the H and K lines are always conspicuous, and we look next at the line of neutral calcium at 4227 Å. Now we assign a star to type G if the 4227 Å line is comparable in strength to the Hγ line which is at 4340 Å. In the K and M stars the line of calcium is noticeably stronger than the lines of hydrogen. The choice between K and M may be made from the presence or absence of the bands of titanium oxide. The criteria described above have been summarized in a convenient form in Table 12.2.

The reader should again review the spectra in Fig. 12.3 and then look carefully to see how the above criteria are applied in the classifications. As an exercise one may also refer to the spectra in Fig. 12.4 and attempt to classify those. The correct classifications are listed on p. 209 at the end of this chapter.

In the material presented here we have only attempted to show how the most obvious feature can be used to determine the broad letter class of a spectrum. Specification of the numbered sub-class requires more detailed examination of the features listed in Table 12.2 and other more subtle features as well. In addition, practice and experience are necessary for the development of real expertise in this art. Several important and useful references on spectral classification are listed at the end of this chapter. Finally, we should point out that less detail can be seen in spectra made with an objective prism or grating than in spectra made with a slit.

Table 12.2. *Criteria for spectral classification*

H lines strongest	He lines present	B
	He lines absent	A or F
	Hε+H>K, no G band, 4226 very faint and 4481 present	A
	H and K about equal, 4226 visible, G band present	F
H not strongest	4226=Hγ and Hβ	G
	4226>Hγ and Hβ	K or M
	no TiO bands	K
	TiO bands clearly seen	M

Fig. 12.4. Test spectra. The reader should review the criteria described in the text and try to classify these spectra. The correct classifications are to be found on p. 209. (From Seitter, W. C., *Atlas for Objective Prism Spectra*, 1970, Ferd. Dummlers Verlag, Bonn).

Luminosity classification

Detailed examination of stellar spectra can also reveal the relative luminosities or absolute magnitudes of the stars, and the modern designation of spectral type includes a Roman numeral to specify this. The present scheme was introduced by W. W. Morgan, P. C. Keenan and E. Kellman in *An Atlas Of Stella Spectra And An Outline Of Spectral Classification* which was published in 1943. The following six classes are described:

Ia	Most luminous supergiants
Ib	Less luminous supergiants
II	Bright giants
III	Normal giants
IV	Sub-giants
V	Main-sequence stars

The positions of these classes in the familiar H–R diagram may be readily seen in Fig. 12.5. As examples of this two-dimensional classification the solar spectrum is type G2V and that of Rigel is B8Ia.

The choice of a luminosity class is based on differences noted in two criteria, namely, the breadth of the lines and the strengths of the lines of certain ionized atoms. The lines in the spectra of the most luminous stars are narrower and more sharply defined than those in the spectra of the stars on the main sequence. This is quite clearly seen in Fig. 12.6 taken from the *MKK Atlas*. In addition, the lines of ionized atoms are stronger in the spectra of the giants. Here again, experience, patience and careful study of reference materials such as the *MKK Atlas* are needed to develop skill in recognizing luminosity class. We list in Table 12.3 just three inter-line comparisons which can provide a crude key to the distinctions between supergiants, giants and main-sequence stars. The lines used in these comparisons are helium lines at 4009 and 4144, a line of triply ionized silicon (SiIV) at 4089, a line of SrII at 4072, and a blend of Fe and Co at 4119.

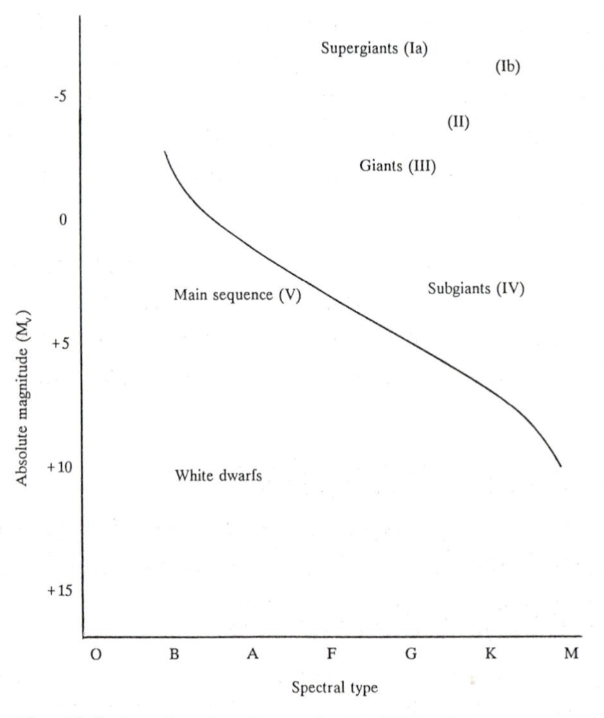

Fig. 12.5. Luminosity classes in the H–R diagram.

Fig. 12.6. Comparison of spectra of giants and main-sequence stars (Yerkes Observatory.)

Table 12.3. *Criteria for distinguishing giants and main-sequence stars*

Supergiants	Giants	Main Sequence
4009<4089	4009=4089	4009>4089
4072<4089	4072=4089	4072>4089
4144<4119	4144=4119	4144>4119

Other spectral types

The seven spectral classed described in the preceding section are sufficient to take in a very large percentage of all of the stars. There are, however, small numbers of stars with spectra which do not fit into any of these classes. Some of these do fit into additional classes which have small populations, but others can only be described as 'peculiar'. Representative spectra are shown in Fig. 12.7. Stars of these classes may be described as follows:

W, the Wolf–Rayet stars. These are hot, luminous stars with spectra resembling those of the O stars. Broad emission features are superimposed, however. These emission features are due to the presence of multiply-ionized helium, carbon, nitrogen and oxygen.

Fig. 12.7. Spectra of some of the less-common types of stars. (*a*) a Wolf–Rayet star, HD193793. (*b*) a carbon star, HD52432. (*c*) a long period variable, o Ceti. (*d*) a metallic-line star, 63 Tauri. (Yerkes Observatory.)

Subgroups are designated WC and WN to indicate spectra in which the emission features are due to carbon or nitrogen.

C, the carbon stars. Spectra of carbon stars show molecular bands which indicate temperature less than 4600 K. The hotter ones were formerly designated as spectral type R and they show bands of molecular carbon (C_2) and of cyanogen (CN). The less hot ones comprise the old class N and show bands of methane (CH) in addition to those of C_2 and CN.

S, zirconium stars. These spectra are distinguished by bands of zirconium oxide. The temperatures of these stars are comparable to those of the M stars. Hydrogen lines are sometimes seen as emission lines.

Other spectral features

Certain prefixes and suffixes are often included in the older designations of spectral types in order to indicate some other unusual features. Those which are most frequently encountered are listed below.

Prefix	c	very narrow absorption lines which are characteristic of supergiants
	g	recognized as a giant on the basis of lines of ionized elements such as SrII; also show bands of cyanogen (CN)
	d	dwarf stars recognized by stronger lines of neutral elements such as Ca and Sr.

(NOTE: The g and d characteristics can only be identified in stars of classes F to M.)

Suffix	n	broad, shallow lines indicating rapid rotation
	e	emission lines in the spectrum
	p	'peculiar' in that the spectrum does not match any normal standard

Examples of spectra containing these features are found in Fig. 12.8.

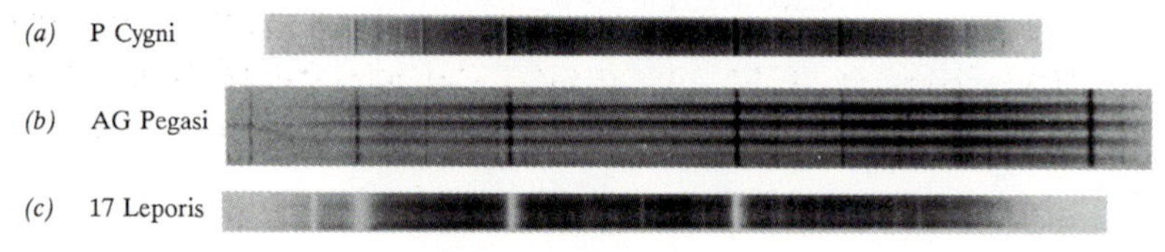

Fig. 12.8. Samples of spectra showing special characteristics. (*a*) P Cygni showing emission lines. (*b*) the emission line star, AG Pegasi. (*c*) the peculiar A star, 17 Leporis. (Yerkes Observatory.)

Some historical notes

It is very likely that Fraunhofer was the first person to examine the spectrum of a star, for we know that before 1823 he had seen the spectra of both Sirius and Castor. He reported that these spectra showed patterns of dark lines which did not match those seen in the solar spectrum. Unfortunately, he died at the age of only thirty-nine and had not been able to carry this part of his work any further.

The next thirty years were ones in which spectroscopy was considerably advanced. Several people were able to match Fraunhofer's D lines with the laboratory spectrum of sodium, and in about 1860 Gustav Kirchoff described the manner in which gases can be expected to absorb and emit producing dark and bright lines. After this it became possible to identify the dark lines in the spectra of the sun and the stars. In 1868 A. J. Ångström determined the wavelengths of spectral lines and adopted as a convenient unit one ten-millionth of a millimeter. This is the unit that now bears Ångström's name. In recent years the nanometer has begun to replace the ångström unit in the measurement of wavelengths. The nanometer is one billionth of a meter, so 10 Å=1 nanometer. 6000 ångströms is equivalent to 600 nm.

It was in this same period that two others undertook serious and productive studies of stellar spectra. The first of these was William Huggins who maintained his own observatory at his home in England. Huggins examined the spectra of a number of the brighter stars and was able to identify lines of sodium, calcium, iron and magnesium. He became convinced that the chemical elements which were known on the earth were present also on the sun and would someday be identified there. He was the first to measure the radial velocity of a star from the Doppler shift in its spectrum, and he was the first to examine spectra of gaseous nebulae.

Father Angelo Secchi of the Vatican Observatory near Rome was the second of the important early spectroscopists. He was interested in the variety to be found in the compositions of the stars, so he began an extensive program of visual observations of their spectra. In a five-year period from 1863 to 1868 he studied the spectra of more than four thousand stars. From this body of data he was able to define four types as follows:

Type I — four conspicuous lines later identified as Hα, Hβ, Hγ and Hδ. These stars were all white or bluish.

Type II	many narrow lines in spectra that were quite similar to that of the sun. These stars were yellow.
Type III	spectra containing broad bands which were well-defined on the violet edge and faded out toward the red. These stars were all red.
Type IV	broad bands which were not the same as those found in Type III. These were also red stars.

It is easy now to see that Secchi's Type I includes the earlier classes O, B, A and F while his Type II includes classes G and K. Type III represents the M-type spectra, and Secchi's Type IV are spectra of the carbon stars. It is interesting to see how much he was able to accomplish through careful visual observations and great patience. Examples of Secchi's four types are included here in Fig. 12.9.

Antonia C. Maury was mentioned earlier in this chapter as one of the people who made major contributions to the *Henry Draper Catalog*. It was Maury's job to study spectra of some of the brighter stars which had been photographed with a somewhat higher dispersion than was normally used in the program. In the course of this she noticed that within a given class the lines could often be described as very broad, broad, or narrow, and she used the letters a, b and c to specify these three cases. It is from this that the prefix c has come to indicate that a spectrum is that of a supergiant.

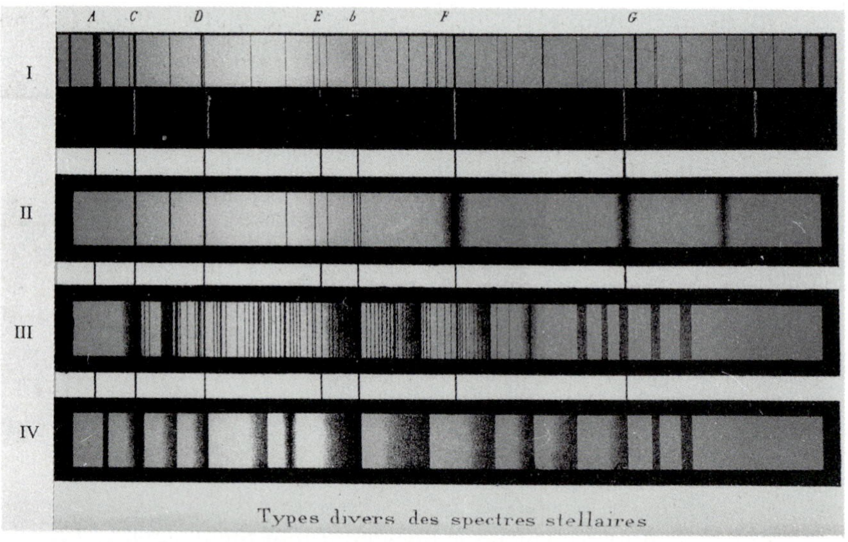

Fig. 12.9. Fr. Angelo Secchi's four types of spectra. Wavelength increases toward the left.

Almost as soon as the first classifications of spectra became widely known, astronomers began to speculate that the appearance of the spectrum was somehow an indication of the evolutionary stage of a star. Even those who were most directly involved with this realized that such speculations were premature since no one really understood the full message of the spectra. Logic suggested that the stars should be cool and therefore red when they were either contracting at a very young stage or cooling off in their most advanced stages. That left the blue and white stars as the ones in their most energetic phases. At this time contraction was still thought to be the principal source of energy, and the relationship between temperature and appearance of the spectrum was not yet recognized. Thus, the giants were thought to be the youngest stars. They seemed to be in the process of contracting to become bright blue stars which would then slowly contract and grow cooler and dimmer. This implied that on the H–R diagram stars would move down the main sequence as they became older. This led to the introduction of the terms 'early' and 'late' to describe respectively stars of Classes O and B in the first case and Classes G, K, and M in the second.

QUESTIONS FOR REVIEW

1. One pattern of absorption lines is seen in the spectra of nearly all stars. Which chemical element causes those lines?
2. Going from cool stars to hot stars, explain the processes that cause the lines of hydrogen to progress from weak to strong to weak again.
3. In what spectral class are the Fraunhofer H and K lines strongest? In which spectral class are the bands of TiO (titanium oxide) most conspicuous?
4. How can the spectra of giants be distinguished from the spectra of main-sequence stars?
5. Test yourself with the spectra in Fig. 12.4.
6. Describe the classification scheme used in the *Henry Draper Catalog*.
7. Why is the classification scheme of the MK system described as a two-dimensional system?
8. Describe the contributions to stellar spectroscopy of Fraunhofer, Secchi and Huggins.

Identification of spectra in Fig. 12.4:

1. G2V HR483
2. A7V θ Cass
3. K5V 61 Cyg A
4. A5V β Ari
5. F5V ι Peg

Further reading

Jaschek, C. and Jaschek, M. (1987). *The Classification of Stars*, Cambridge: Cambridge University Press. This is a useful and interesting book in which Chapters 2 and 3 deal with the classification of spectra.

Morgan, W. W. and Keenan, P. (1973). Spectral classification. *Annual Review of Astronomy and Astrophysics*, **11**, 29. Here Morgan and Keenan review the changes that have become necessary through thirty years of use of the MK system.

Morgan, W. W., Keenan, P. and Kellman, E. (1943). *An Atlas of Stellar Spectra*. University of Chicago Press. The Introductory Remarks published as part of this atlas contain the original definition of the MK system. The atlas itself contains photographic prints of the actual spectra which were used in defining the system.

Pannekoek, A. (1961). *A History of Astronomy*. Interscience Publishers. On the pages from 450 to 462 Pannekoek describes the steps leading up to the Harvard system.

Seitter, W. C. (1970). *Atlas For Objective Prism Spectra*. Ferd. Dummlers Verlag. The spectra in this atlas cover the range from the UV all the way to the Hα line and beyond. Examples of all of the classes are included and show the progress from one class to the next. All of the lines are clearly identified.

Stromgren, B. (1966). Spectral classification through photoelectric narrow-band photometry. *Annual Review of Astronomy and Astrophysics*, **4**, 433. This is the paper in which the work leading to the Hβ system is summarized.

Struve, O. and Zebergs, V. (1962). *Astronomy of the Twentieth Century*. Macmillan Company. Struve's review of spectroscopy from Fraunhofer to Stromgren is found in Chapter X.

13

Analysis of high-dispersion spectra

In our earlier discussion of spectra and spectrographs we referred to the fact that when we speak of high-dispersion spectra, we are indicating that the dispersion is approximately 10 ångströms per millimeter or less. Review of the material in Chapter 11 will show that dispersion of this sort may be obtained in several ways. One may use a grating with a very large number of lines per millimeter, a long focal-length camera, or a spectrograph designed around an echelle grating. In order to make the exposure times as short as possible and still not be limited to only the brightest stars, a very large telescope may be necessary. In practice it is likely that most systems designed to operate at high dispersions will make use of several techniques for increasing the overall efficiency.

One of the earliest and most widespread uses of high-dispersion spectra has been in the determination of radial velocities of the stars, and in this chapter we shall describe the fundamentals of the method by which the radial velocity of a star may be determined. There are, however, other interesting characteristics which may also be recognized in high-dispersion spectra. These include the determination of chemical abundances, rotational velocities and magnetic fields among other things.

Radial velocities

The method of determining the radial or line-of-sight velocity is, of course, based on the Doppler Effect applied to light waves. Christian Doppler had orginally discussed the effect of the velocity of a source of sound waves on the detected pitch of a note in 1842, and Armand Fizeau suggested that in the spectrum of a star a change in the wavelength of the absorption lines should be an indication of the radial velocity of the star. It was not until spectra could be routinely recorded photographically that astronomers were able to measure radial velocities with any reasonable degree of confidence. The early workers showed that effects of changes in temperature and flexure of the spectrograph would have to be eliminated before precision could be achieved. By 1890 all of the needed equipment and skill had been brought together by J. E. Keeler at Lick Observatory, and since that time radial velocities of stars have been determined in large numbers. The precision of radial velocities is very likely better than that of any other measurable astronomical quantity.

The fundamental idea behind the determination of radial velocities is really quite simple, and may be expressed as

$$\Delta\lambda/\lambda = \text{velocity/velocity of light}$$

Here λ is the normal wavelength of a spectrum line, and $\Delta\lambda$ is the amount by which that line has been displaced – the Doppler shift. This formula does not hold in cases in which the velocity is a significant fraction of the velocity of light. In addition, by velocity here we mean velocity of the source relative to the observer, and we are dealing only with that component of velocity which lies along the line of sight between

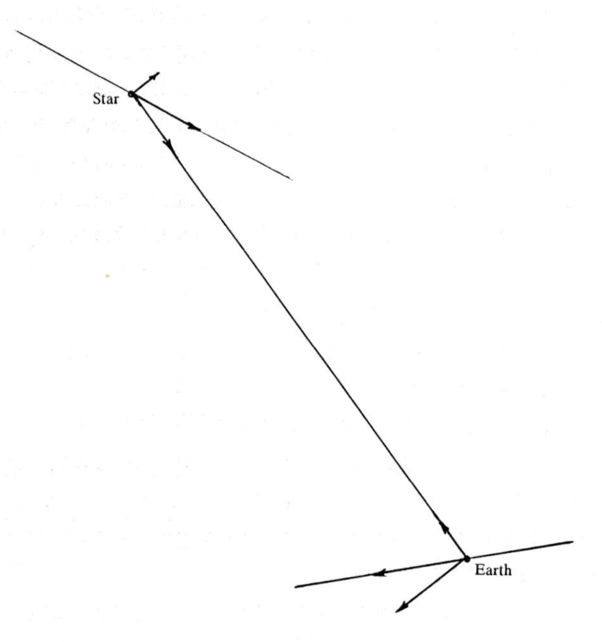

Fig. 13.1. The observed radial velocity results from both the motion of the star and that of the earth.

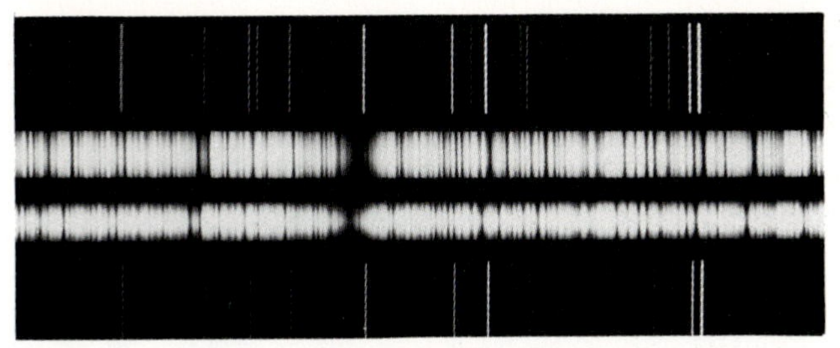

Fig. 13.2. Portions of two high-dispersion spectrograms of Arcturus. The two spectra were made six months apart, and the change in wavelength over that period is clearly seen. (The Observatories of the Carnegie Institution of Washington.)

source and observer. As indicated in Fig. 13.1 the radial velocity of a star results from components of the motion of both the earth and the star.

When we look at the spectrum of a star, we can often identify a line simply from its position relative to other lines. For example, the lines of hydrogen and of ionized calcium are quite readily identified as we pointed out in Chapter 12. The 'rest' or laboratory wavelengths of the important stellar lines are, of course, well known, so λ in the above equation is easily known. It is, however, somewhat more difficult to know what the displaced wavelength of the stellar line is so that $\Delta\lambda$ may be found. Some sort of reference scale is needed, and as we mentioned in Chapter 11, this is usually provided by means of the comparison spectrum photographed along with the spectrum of the star. Again the 'decker' or mask is moved in front of the slit so that an electric arc or a gas discharge lamp within the spectrograph can illuminate the ends of the slit before or after the star illuminates the central portion. The result in a typical case has the appearance of the spectra shown in Fig. 13.2. Two spectra of the star are in the center, and the spectrum of the comparison source is on either side of it.

Let us assume for a moment that there is a linear relationship between position in the spectrum and wavelength such as might be the case in a spectrum recorded with a grating spectrograph. We have shown this schematically in Fig. 13.3(*a*). A well-defined line near one end of the comparison spectrum may be chosen as a zero point, and distances from this line to a number of the other comparison lines may be measured with the greatest possible precision. Measured distances to lines in the stellar spectrum should be recorded at the same time.

Measurements of this sort are usually done with a microscope mounted above a moveable stage. The spectrogram is placed on the stage, and the stage is moved until the crosswire of the microscope appears to bisect a

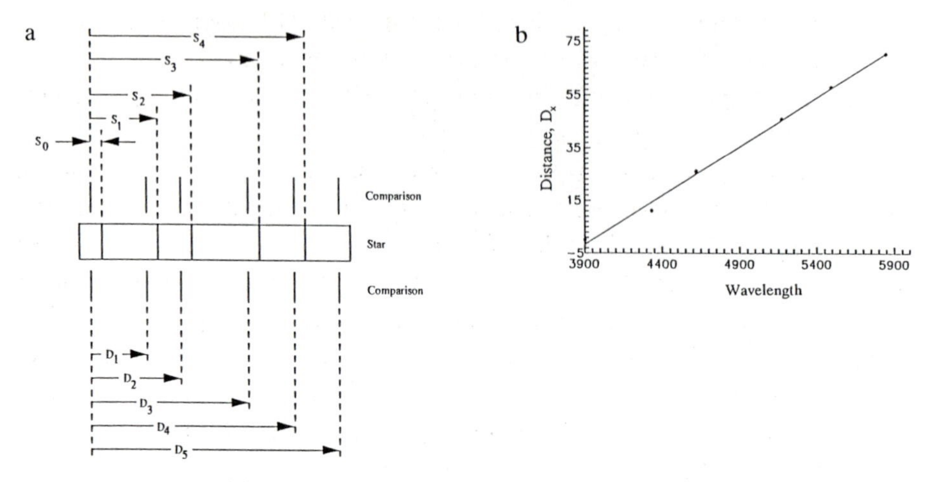

Fig. 13.3. (*a*) Hypothetical spectra of comparison source and star with positive radial velocity. (*b*) Calibration curve in which distances on the comparison spectrum have been plotted against assumed wavelengths.

line in the spectrum. The position of the stage is then recorded and the stage is moved until the next line is centered. The stage is attached to a precision screw which is in turn attached to a graduated drum, and the operator moves the stage by rotating the drum. The position of a line is read from the drum. In the more modern machines the operator does not have to write down the setting of the drum as there are provisions for automatic readout of the data. Some machines are equipped with scanning systems which help the operator know when a line is properly centered in the microscope.

To give an idea of the subsequent steps in this procedure, we have shown in Fig. 13.3(*b*) how wavelength might be plotted versus distance measured on the comparison spectrum. This diagram might be referred to as a calibration curve. The wavelengths of the displaced stellar lines could now be determined graphically by entering the scale on the left with each measured value, S, and reading the wavelengths from the lower scale. In practice it is usually simpler and more precise to do this numerically. The method of least squares may be used to find the equation of the best line through the points on the graph, and this equation may be used with the measured positions of the stellar lines to find the Doppler-shifted wavelengths. The method of least squares is described in detail in Appendix 3. Earlier in this discussion we suggested that since the dispersion by a grating is linear, the calibration curve should be a straight line. In fact, the calibration curve may be a curve. One should examine carefully the residuals from the numerical straight-line solution in order to see whether or not a second-order curve might actually give a better fit. It should be obvious that all steps in this entire procedure deserve the greatest possible care.

In summary, the comparison lines are used to produce a calibration curve, and this curve or the equation of the curve may be used to find the wavelengths of the Doppler-shifted stellar lines. Each of these wavelengths may then be combined with the rest wavelength of the same line in order to find values of $\Delta\lambda$. By using $\Delta\lambda$ in the simple relationship stated at the beginning of this section, we may calculate a value of the radial velocity for each of the measured stellar lines. In theory, all such values should be the same. In practice the values will not all be the same because of inevitable errors in making the measurements.

In Table 13.1 we have listed the hypothetical data on which Fig. 13.3(*b*) is based. The numerical solution for the equation of the best line through the five points gives

$$Y = -142.0010 + 0.03642864X$$

The standard deviation of a wavelength determined in this problem is ± 0.239 Å.

At this point the student may have the idea that the determination of radial velocities is really very simple, so let us consider several numerical examples to put this in perspective. In the first case let us assume that a star has a radial velocity of just one kilometer per second, and let us choose 4200 Å as the rest wavelength of some hypothetical line. We quickly calculate

$$\Delta\lambda/4200\ \text{Å} = 1\ \text{km per s}/300\,000\ \text{km per s}$$
$$\Delta\lambda = 0.014\ \text{Å}$$

Table 13.1. *Measurements of spectrum lines*

Wavelength	Measured Distance, D_m	Computed Distance, D_c	Residual $D_m - D_c$
4332.2	16.000	15.815	0.185
4619.7	26.000	26.288	−0.288
5174.5	46.500	46.499	0.001
5498.2	58.500	58.291	0.209
5850.0	71.000	71.107	−0.107

If we make use of a spectrograph in which the dispersion is 10 Å/mm then the linear shift on the spectrogram is only 0.0014 mm. This is a very small distance to try to measure with precision, but nevertheless it can be done. Second, if the radial velocity was 100 km/s and if the dispersion was 1 Å/mm, then the linear displacement would be 1.4 mm. One might feel a bit more confident in measuring a displacement of this size. Finally, for the same radial velocity of 100 km/s, if one attempted to use a spectrograph with a dispersion of only 80 Å/mm, the linear displacement would be only 0.0175 mm. This should be convincing evidence that high dispersion is desirable in the determination of radial velocities.

As we showed in Fig. 13.1, the motion of the earth can contribute to the observed radial velocity of a star, so we are usually interested in removing the effects of the earth's motion. Only then can we know what the true radial velocity of the star really is. Two motions of the earth must be considered. The first of these is daily rotation and the second is annual motion. Components of both may be directed toward or away from a particular star, and both of these components will be changing continuously.

Correction for daily rotation

To see how corrections for these two motions may be calculated, let us first consider the earth as it might appear to an observer looking at it from a point above the north pole. This view is sketched in Fig. 13.4. Here an observer is presumed to be at point Y in the middle latitudes. A

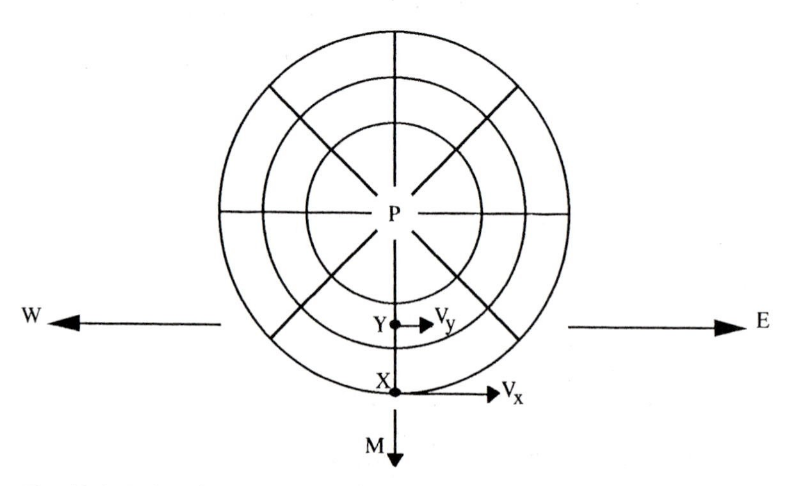

Fig. 13.4. Polar view of the rotating earth.

point at the same longitude but situated on the equator is at X. The eastward velocity of X is easily found from

$$V_x = \text{circumference/seconds per sidereal day}$$

$$V_x = 2\ \pi\ 6378\ \text{km}/86\ 164$$

$$V_x = 0.47\ \text{km/s}$$

This velocity is reduced at higher latitudes by the factor, cos ϕ, where ϕ is the observer's latitude. Thus,

$$V_x = 0.47 \cos \phi\ \text{km/s}$$

Again looking at Fig. 13.4 one can see that if a star is in the western sky for the observer at Y and is on the celestial equator, it will be seen in the direction, W. The observed radial velocity of the star will then be increased by 0.47 cos ϕ km/s. A second star on the celestial equator in the direction, E, will have its observed radial velocity diminished by 0.47 cos ϕ km/s. Finally, a star in the direction, M, will be on the meridian for observers at X and Y, so the earth's rotation will have no effect at all on that star's observed radial velocity.

We must now see how the earth's rotation will affect the radial velocity of a star at some declination, δ, and hour angle, τ. The geometry of this situation is sketched in Fig. 13.5. A star is at S, and the points D and F represent the directions east and west as discussed in connection with Fig. 13.4. M is a point on the observer's meridian. In this figure MD=90 degrees, and GD=90+τ. We wish to find the angle DES, so we must solve for the side DS in the spherical triangle GSD.

$$\cos DS = \cos \delta \cos GD + \sin \delta \sin GD \cos SGD$$

Since angle SGD is a right angle, the second term is zero, and we see after substituting

$$\cos DS = \cos \delta \cos(90° + \tau) = -\cos \delta \sin \tau$$

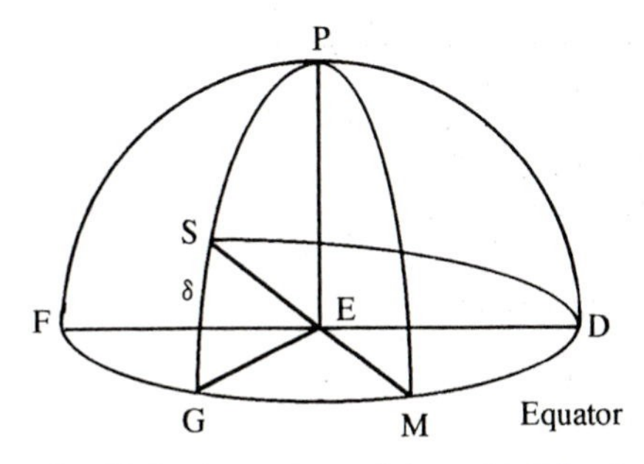

Fig. 13.5. A portion of the celestial sphere with a star at S.

Now consider just the plane, SED, as drawn in Fig. 13.6. V' is the eastward velocity of the observer at latitude ϕ; V is the velocity of a star in the west resulting from V'; and V_d is the component of V which is in the direction of the star. We see readily that

$$V_d/V = \cos(180° - DS)$$

$$V_d = V\cos(180° - DS) = -V\cos DS = V\cos\delta\sin\tau$$

Thus, $V_d = 0.47\cos\phi\cos\delta\sin\tau$ km/s.

The hour angle, τ, increases toward the west from zero when a star is on the meridian. Therefore, $\sin\tau$ is positive for stars in the western sky and negative for stars in the east. Cos δ is always positive since δ can range only from $-90°$ to 0 to $+90°$.

It is very important to recall the convention for the sign of the radial velocity. When the distance between source and observer is increasing, the radial velocity is taken as positive. Consideration of the formula given above for V_d will show that the signs will follow the convention.

In summary, we correct for the effect of the earth's rotation by subtracting V_d from the measured radial velocity of a star. Even without looking at a typical numerical example, one may see that this correction will usually be quite small. It can be ignored when the precision of the measured radial velocity is so poor that this correction will not substantially improve the final value of the radial velocity.

Correction for orbital motion of the earth

As it moves in its orbit around the sun, the earth will always be moving toward some stars and away from others. Quite clearly this orbital motion of the earth will affect the observed radial velocity and must also be removed in order to find the true radial velocity of a star. For simplicity and in order to have an idea of the approximate size of this correction, let us begin with an analysis in which we consider that the orbit of the earth is circular rather than elliptical.

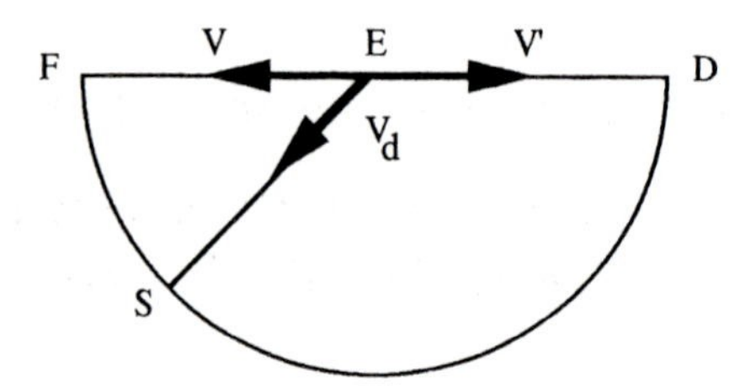

Fig. 13.6. Velocity component toward a star at S.

The orbital velocity, V_c, is found from

$$V_c=2\,\pi\,R/t$$

where R is the earth–sun distance, i.e. the radius of the earth's orbit, and t is the number of seconds in one year. Using $R=149.5\times10^6$ km and $t=31.5\times10^6$ s/yr, we find for the circular velocity

$$V_c=29.82 \text{ km/s}$$

This is considerably more significant than the correction for the earth's rotation. We must now see how we can find the magnitude of a component of this velocity directed toward a particular star. We shall first derive an expression in terms of the star's ecliptic coordinates, celestial latitude and longitude. We can then modify that to provide a more convenient expression in terms of the star's equatorial coordinates, declination and right ascension.

At any instant the earth's orbital velocity will be directed toward a point indicated as F in Fig. 13.7(*a*) and (*b*). The angle between F and the direction to the sun will always be ninety degrees. The longitude of the sun is defined as the angle at the earth between the direction of the sun and the direction of the vernal equinox, and this angle is tabulated in the *Astronomical Almanac* for every day of the year. It is customary to use the symbol, $\odot$, to represent the longitude of the sun. Clearly, the longitude of F is readily known also since $F=\odot-90°$.

Now consider Fig. 13.7(*b*). Here β and λ are the ecliptic coordinates of a star at S. The sun, $\odot$, and F are shown on the ecliptic and are ninety degrees apart. We may represent the earth's orbital velocity, V_c, with a vector (not shown) directed along EF. Our problem is simply to project V_c onto the line ES. To do this we must first find angle SEF. We consider the spherical triangle SFG and note that we must solve for the side SF. In this triangle β is known and angle SGF is a right angle. Side FG is found from

$$<GEF=\lambda-(\odot-90°)=\lambda-\odot+90°$$

Then in triangle SFG

$$\cos SF=\cos\beta\cos(\lambda-\odot+90°)+\sin\beta\sin(\lambda-\odot+90°)\cos SGF$$

Because angle SGF is 90°, the second term is equal to zero. Further, we may rewrite the factor involving λ to read $\cos(90°-(\odot-\lambda))$. As a result we may write

$$\cos SF=\cos\beta\sin(\odot-\lambda)$$

a

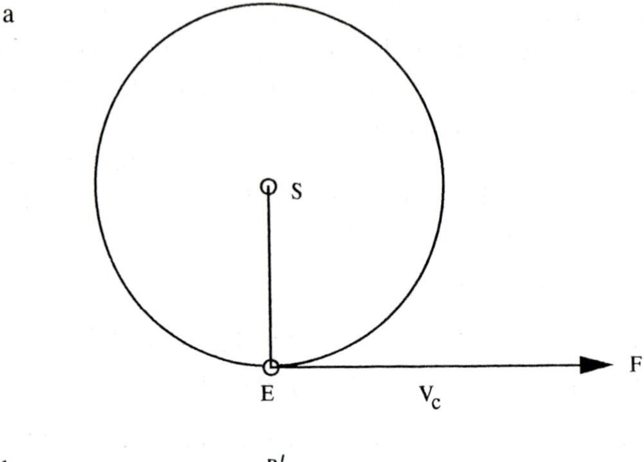

b

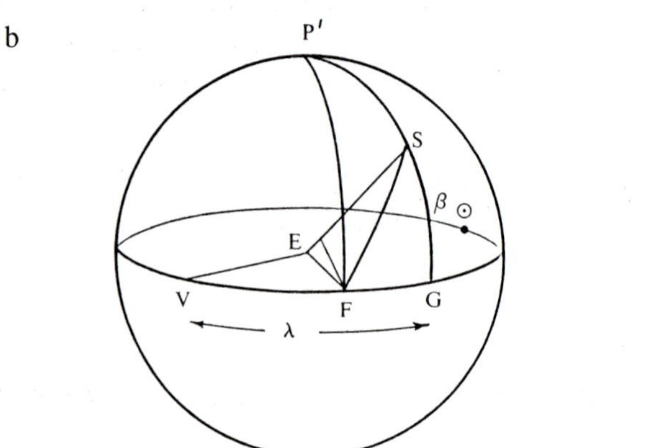

Fig. 13.7. (*a*) The earth's orbital velocity directed toward F. (*b*) The position of a star at S in the ecliptic coordinate system.

In the plane SEF the projection of V_c in the direction of S is V_c cos SF. If we designate this as V_y, velocity due to the earth's annual motion, we have

$$V_y = -29.82 \cos \beta \sin(\odot - \lambda)$$

In this equation it has been necessary to add the negative sign in order to maintain the convention that a radial velocity is positive when the distance between source and observer is increasing.

Corrections in right ascension and declination

In the previous section we showed how the orbital motion of the earth affects the radial velocity. The necessary correction, V_y, was expressed in terms of the star's ecliptic coordinates, β and λ and the longitude of the sun, $\odot$. The practical application of this correction would be much simplier if the star's equatorial coordinates, α and δ, could be used, since these coordinates are the most readily available. We may derive an appropriate formula as follows.

We begin with the final term in the above equation for V_y, i.e. $\sin(\odot-\lambda)$. Using the rules for the expansion of a function which is the difference between two quantities, we write

$$\sin(\odot-\lambda)=\sin\odot\cos\lambda-\cos\odot\sin\lambda$$

After multiplying both sides by $\cos\beta$ and rearranging a bit we have

$$\cos\beta\sin(\odot-\lambda)=\cos\beta\cos\lambda\sin\odot-\cos\beta\sin\lambda\cos\odot$$

We must now try to find expressions for $(\cos\beta\cos\lambda)$ and $(\cos\beta\sin\lambda)$ in terms of α and δ. This can be done if we first take note of the spherical triangle SPP′ in Fig. 13.8. In this triangle $A=90°-\lambda$ and $B=90°-\alpha$. Now we use the sine formula to see that

$$\frac{\sin(90°-\delta)}{\sin(90°-\lambda)}=\frac{\sin(90°-\beta)}{\sin(90°+\alpha)}$$

If we simplify this we may write

$$\cos\delta\cos\alpha=\cos\beta\cos\lambda$$

Looking again at Fig. 13.8 and using the Formula C from p. 50 we may write

$$\sin(90°-\beta)\cos(90°-\lambda)=\cos(90°-\delta)\sin\varepsilon-\sin(90-\delta)\cos\varepsilon\cos(90°+\alpha)$$

$$\cos\beta\sin\lambda=\sin\delta\sin\varepsilon+\cos\delta\cos\varepsilon\sin\alpha$$

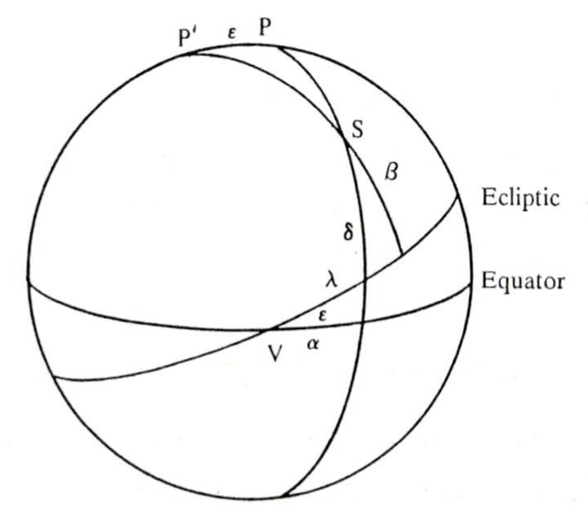

Fig. 13.8. Transferring from ecliptic to equatorial coordinates.

We are now in a position to eliminate β and λ from the earlier expression for V_y, and we write

$$V_y = -29.82\{\cos\alpha\cos\delta\sin\odot - \cos\odot(\sin\delta\ \sin\varepsilon + \cos\delta\cos\varepsilon\sin\alpha)\}$$
$$= -29.82\{\cos\delta\cos\alpha\sin\odot - \cos\odot\sin\delta\sin\varepsilon - \cos\odot\cos\delta\cos\varepsilon\sin\alpha\}$$

If we now define V_M as the measured radial velocity before any corrections have been applied, and if we call V_R the true radial velocity, then we note that

$$V_R = V_M + V_d + V_y$$

In order to see that this is the correct way in which to apply the corrections, we may examine Fig. 13.9 which shows six positions of the

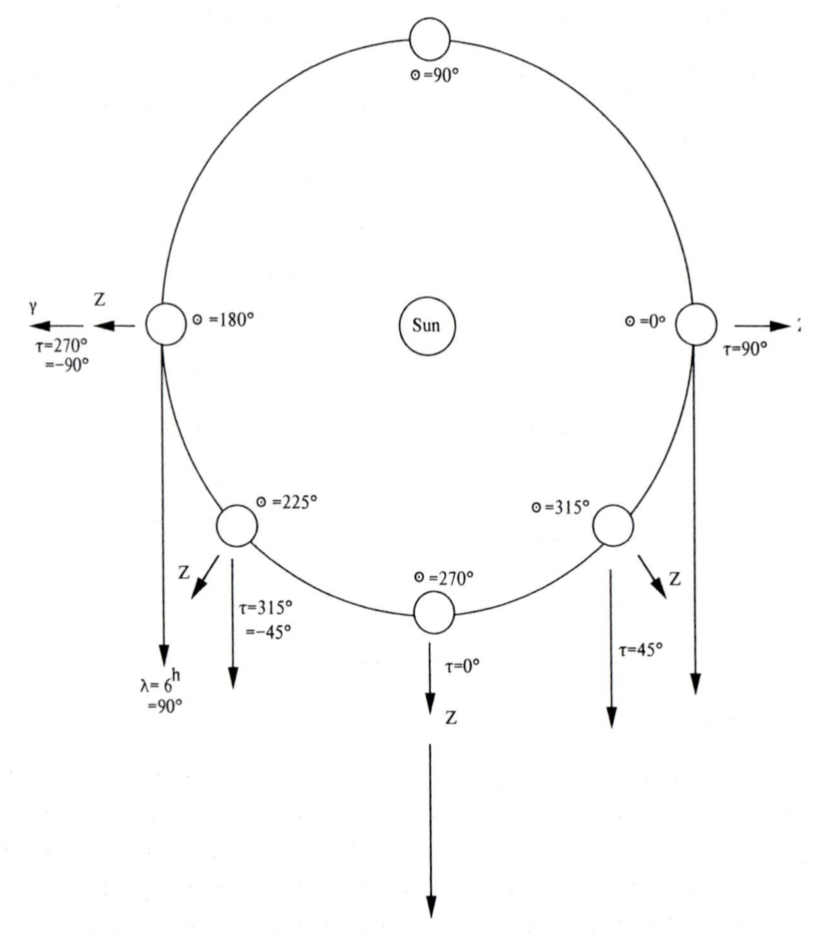

Fig. 13.9. Geometry for the computation of corrections due to rotation and orbital motion of the earth. The local time for the observer is midnight. The earth is rotating in the counterclockwise direction and revolving around the sun in the counterclockwise direction.

earth and the direction to a star at $\lambda=90°$. Let us first take the case in which the longitude of the sun is 225°. Then the star is in the east, and τ is $-45°$. Assume also that the observer is on the earth's equator and that the star is on the celestial equator. When we calculate the two corrections, we find

$$V_d=-0.332 \text{ km/s}$$

$$V_y=-21.043 \text{ km/s}$$

Inspection of the figure shows us that both of the motions of the earth (rotational and orbital) are taking the observer toward the star, so both of the corrections should be negative. In a similar way, when the earth is in such a position that $\odot=315°$ both rotation and revolution are taking the observer away from the star, so both corrections should be positive.

Example: To demonstrate more completely the computations, consider the bright star, Regulus, as it might have been seen on the evening of March 1, 1988. We show the pertinent data below:

$$\alpha=10 \text{ h } 07 \text{ m } 28 \text{ s}$$

$$\delta=+12° \; 03' \; 03''$$

$$\text{longitude of the sun, } \odot=340° \; 50' \; 32.78''$$

$$\text{eastern standard time}=20 \text{ h}$$

$$\text{local sidereal time}=6 \text{ h } 52 \text{ m } 47 \text{ s}$$

latitude of observatory=42°.29363887[1]

$$\text{longitude of observatory}=71°.305 \text{ W}$$

When we substitute into the two equations to compute V_d and V_y, we have

$$V_d=+0.255 \text{ km/s}$$

$$V_y=+5.805 \text{ km/s}$$

Line profiles

As astronomers continued their study of high-dispersion spectra in the early years of this century, they began to ask questions about the differences in the appearance of lines in some stars compared to others. We have already mentioned that sharp, narrow lines separated the spectra of giants from those of main-sequence stars. In addition, there was a noticeable gradation in the width of lines within spectral types. Many astronomers joined the effort to understand these differences, and by

[1] Whitin Observatory, Wellesley College.

combining their work with that of a number of physicists, the astronomers were successful. Some of the quantitative data needed in the analysis was obtained from line profiles.

A 'line profile' may be defined as a record of the variation in intensity of light from one side of a spectrum line to the other. An example is shown in Fig. 13.10 along with the spectrum from which it was made. In Chapter 11 we described the use of the microdensitometer to make a record of variations in density along a spectrogram. We pointed out there that on such tracings we can sometimes see spectral features that were not detectable in a visual examination of the spectrogram. The line profiles with which we are now concerned may be derived from similar tracings made on high-dispersion spectra, and they offer insight into several important astrophysical problems.

We must remember that the tracings produced by the microdensitometer are density tracings. These can be useful, but we are usually more interested in variations in the intensity along the spectrum rather than photographic density. Therefore, we must convert the density tracing to an intensity tracing. This conversion is not a trivial procedure, and in order to effect it, we must know how variations in intensity in the original light source caused variations in density in the developed negative. Studies of the relationship between intensity and density are often referred to as sensitometry.

The customary practice in achieving the goal outlined above begins with the exposure of part of a plate to a series of lights varying in intensity by known amounts. The techniques for doing this have been described in Appendix 1 on photography, but the procedure results in a series of spots of graded densities on the plate. If the photographic densities of these spots are measured, then the densities can be plotted against the original intensities. The resulting graph is the characteristic curve of the emulsion. As noted in Appendix 1 the slope of the nearly straight central portion of this curve is a measure of the speed of the emulsion. If the curve is steep, then a small increase in the amount of incident light will result in a large increase in density.

If the characteristic curve has been plotted for a particular emulsion, then densities on the plate can be converted to relative intensities in the incident light. Keep in mind that the characteristic curve for any photographic emulsion depends on the wavelength. If all of this is to be done by hand, it will involve a considerable amount of work. Once this conversion to intensity has been done, however, a true intensity profile for a complete spectrum or for a single line can be displayed. It is more

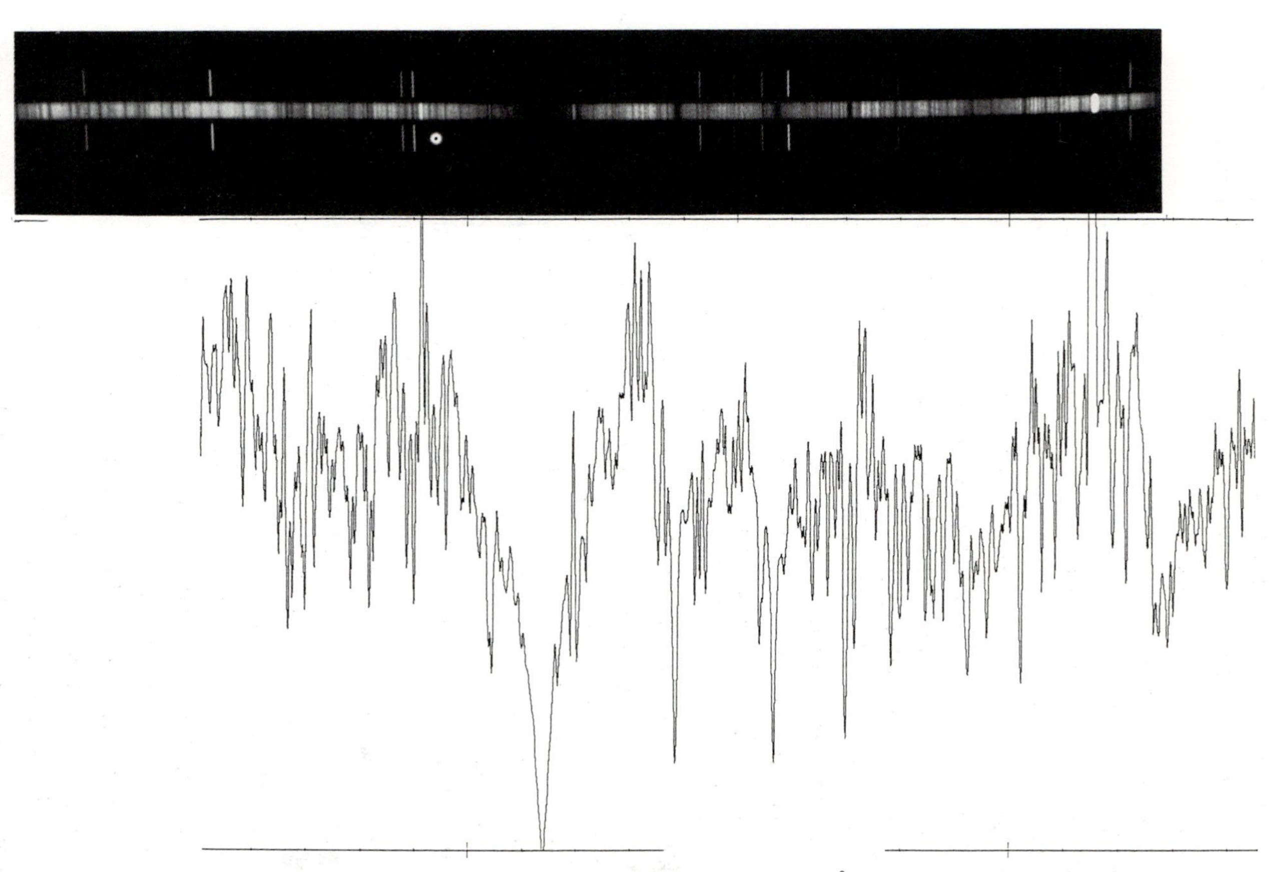

Fig. 13.10. A portion of the spectrum of R Cancri showing the 4226 Å line of calcium just to the left of the center. Below the spectrum is the densitometer tracing in which a remarkable amount of detail may be seen. At the far right there is a strong emission line of iron. (Courtesy of Irene Little-Marenin, Wellesley College.)

common to see an intensity profile for a limited range of wavelengths than it is to see one for a spectrum spanning one thousand or more ångströms.

Thus, the term 'line profile' has come to be used to describe an intensity profile that has been centered on the wavelength of some known absorption line.

By making use of the electronic detectors described on p. 190 the production of line profiles can be somewhat simplified. The responses of photodiodes to changes in intensity of light is linear over broad ranges in intensity, and one can concentrate on variations in response to differences in wavelength. Again, over a short range in wavelength centered on some absorption line the response of the detector may give an accurate line profile.

Widths of spectrum lines

The detailed study of line profiles has opened up several important areas of astrophysics, and we shall describe some of these in the following paragraphs. The discussion will proceed only in a qualitative way because the analyses are very specialized. Several factors are involved, and they all act together, in one degree or another, to affect the appearance of an absorption line. Throughout all of the discussion which follows the reader should keep in mind the roles of temperature and pressure in stellar spectra. As temperature increases from about 3000 K in the atmospheres of the cooler stars to about 30 000 K for stars of spectral class O, absorption lines become progressively stronger and then weaker. The strengthening is due to the increase in excitation as the temperature increases. The hydrogen atom, for example, cannot absorb at the wavelength of Hα until it has been excited from level one to level two. The weakening of the lines results from ionization. After an atom reaches some maximum level of excitation, a small increase in absorbed energy causes the atom to lose an electron (become ionized). That atom is effectively removed from the gas which causes the absorption line, since the ionized atom absorbs and emits as if it were another atom altogether. Four of the factors which can affect the shape of a line profile may be described as follows.

First, each line has a certain 'natural width' which results from the fact that the actual energy levels are not exact. A transition may begin and end slightly above or below the nominal energy levels which are associated with a particular line, so photons with a small range of wavelengths will be absorbed.

Second, the number of absorbing atoms in a column of gas will affect the appearance of the line. It has been found experimentally and confirmed in stellar spectra that the width of a line increases as the number of atoms increases.

Third, the wavelength which we detect as an individual atom absorbs or emits can be affected by the motion of the atom along the line of sight. This is referred to as Doppler broadening. It was not unexpected to find that Doppler broadening is more noticeable in the spectra of hot stars than it is in the spectra of cool stars.

Fourth, a second effect due to line-of-sight motions in an atmosphere is referred to as turbulent broadening. The difference between this and Doppler broadening is that here we are dealing with large masses of gas which may be rising or falling as they absorb or emit.

There are several other effects that can alter the shape of a line profile, but only one more will be discussed here. That one is axial rotation of the star, and it will be treated in the next section. More detail on the causes of line broadening are to be found in the books by Aller and by Struve and Zebergs listed at the end of this chapter.

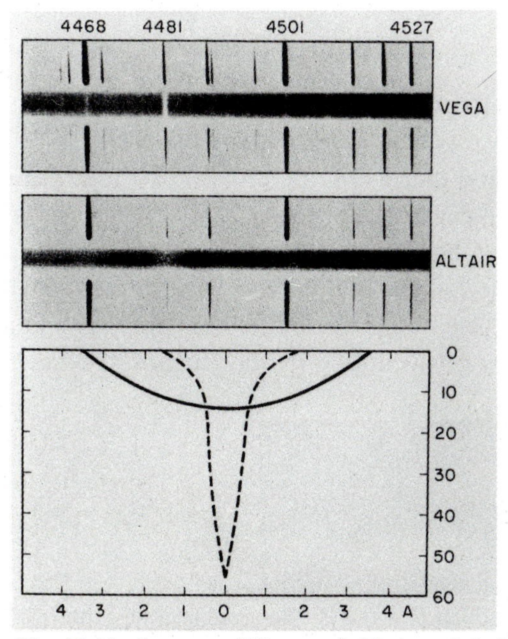

Fig. 13.11. Spectra of Vega and Altair and their line profiles. Altair is rotating rapidly and its line profile has been broadened considerably. (Yerkes Observatory.)

Axial rotation

Let us imagine for a moment that we are by chance looking into the equatorial plane of a star and that the star is rotating rapidly. Then we should expect that one edge of the star is approaching us and the other edge is receding from us as the star rotates. Surface speed and therefore radial velocity will decrease toward the poles. In the observed spectrum of such a star we should expect some recognizable effects. Fig. 13.11 shows how the line profile might be altered by rotation. The line appears to be shallower and broader in the intensity tracing. It appears to be 'washed out' in the actual spectrogram. By measuring the width and shape of the line profile, one can estimate the value of the surface velocity, $v \sin i$, where i is the inclination of the axis of the star to the line of sight. i is equal to 90° when the rotational axis of the star lies in the plane of the sky.

Analysis of rotational velocities by this means showed that most rotational velocities fell in the range from 0 to 250 km/s. Further, the most rapidly rotating stars are the A stars. None of the M stars have been found to be rapidly rotating. In some rare cases $v \sin i$ might be as large as 400 km/s.

Interstellar lines

Direct evidence for clouds of interstellar gas is sometimes found in high-dispersion spectra in the form of extra absorption lines. These lines are usually narrow and sharp when compared to the normal lines which form in the atmospheres of stars. Also, they are often at wavelengths slightly longer or shorter than the stellar lines. Fig. 13.12 shows a portion of the spectrum of HD 47240 with interstellar lines on either side of the K line. The stellar line is the conspicuous feature, and interstellar lines appear as narrow lines on either side of it. Along the line

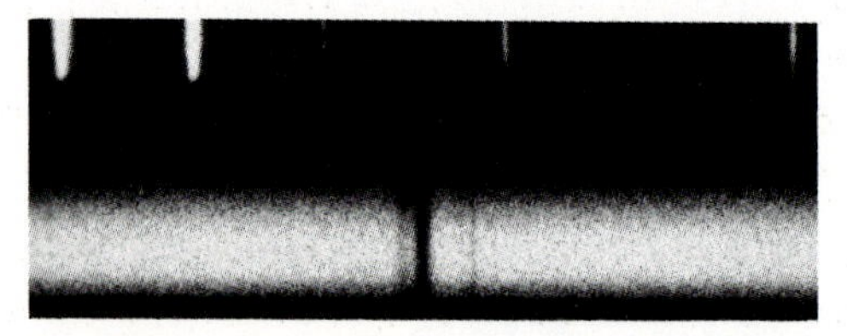

Fig. 13.12. The spectrum of HD 47240. The prominent absorption line is the K line. One interstellar line may be seen to the left of the stellar line, and two more interstellar lines to the right of it. (The Observatories of the Carnegie Institution of Washington.)

of sight to this star there are apparently two or three separate clouds of gas. The displacement of these lines tells us that each of the clouds has its own radial velocity, and that these are slightly different from that of the star.

Students who go beyond the level of this book will find that there are other phenomena which have been recognized in high-dispersion spectra. These include magnetic fields, gaseous shells and absorption lines with emission features in their centers. We hope that all readers will have an appreciation of the deep relationship between theory and observation that has advanced our interpretations of the details of stellar spectra.

QUESTIONS FOR REVIEW

1. When we speak of 'high-dispersion' spectra, what do we mean with regard to numbers of ångströms per millimeter?
2. Describe two ways of achieving high dispersion in the design of a spectrograph.
3. Outline the steps required in the determination of the radial velocity of a star.
4. Upon measuring a spectrogram made with a grating spectrograph, an observer finds that the Hγ line has been shifted toward longer wavelengths by 1.6 mm. The dispersion in the spectrograph is 2.4 Å/mm. What is the radial velocity of the star?
5. Under what circumstances would you feel comfortable in omitting a correction for the contribution of the earth's daily rotation when computing the radial velocity of a star?
6. Redo the problem on p. 223 using Procyon rather than Regulus.
7. How does a density profile differ from an intensity profile? How can one go from the first to the second?
8. Explain the effect of rotation on the line profile of a star.

Further reading

Aller, L. H. (1971). *Atoms, Stars and Nebulae*. Harvard University Press. This book is clearly written and well illustrated. It is a good place to begin the detailed study of stellar spectra and the understanding of stellar atmospheres.

Binnendijk, L. (1960). *Properties of Double Stars*. University of Pennsylvania Press. Beginning on page 108 Binnendijk discusses the measurement of spectra and the determination of corrections for motions of the earth.

Struve, O. and Zebergs, V. (1962). *Astronomy of the Twentieth Century*. Macmillan Company. See Chapter XI for Struve's discussion of spectroscopy applied to stellar atmospheres, abundances, rotation and luminosity. An appendix entitled 'Stellar spectroscopy' contains a rare summary of the mechanisms of line broadening.

14

Variable stars

Spread out on the celestial sphere there are thousands of variable stars, and an example of one is seen in Fig. 14.1. Their periods range from a few hours to hundreds of days. Some are visible to the naked eye, and others can only be detected with large telescopes. Some behave in an erratic fashion, but others are as predictable as the sunrise. Within this large group there is something for the interests and equipment of every observer, and serious contributions to the overall body of astronomical data can be made by anyone who is willing to exercise care in all phases of the collection of data. Thousands of astronomers all over the earth, both amateurs and professionals, find the study of variable stars to be both pleasant and rewarding. We shall begin this chapter with some

Fig. 14.1. The variable star WW Cygni at maximum and minimum brightness. (The Observatories of the Carnegie Institution of Washington.)

information on nomenclature and on reference materials. We shall then proceed to principles that apply to the observation of all types of variable stars, and we shall then discuss the advantages of observing some specific classes of stars.

Naming variable stars

Variable stars are named in accordance with a scheme that was introduced in the middle of the nineteenth century when variability was first being recognized as a common phenomenon in the stars. The originator of the current practice was the German astronomer, Friedrich Argelander, who was mentioned in Chapter 3 as the force behind the BD charts and catalog. In each constellation the first variable to be discovered was identified with the letter 'R' followed by the possessive form of the Latin name. The second variable to be discovered was given the letter 'S' and so forth through the letter 'Z'. The next one was RR, then RS, RT *etc*. T Cygni would have been the third variable star found in the constellation, Cygnus. RR Lyrae was the tenth variable found in Lyra. This scheme continues to ZZ, then goes to AA, and finally ends at QZ. The next one found after that was called V335, and subsequent variables were designated in this way with numbers increasing beyond 335. The awkwardness of the original scheme is obvious to us now, but in the early days no one had any idea of how many variable stars might someday be discovered. Whatever its faults may be, this is the scheme used by all astronomers. Variables which already had some other designation when variability was discovered were not renamed in Argelander's scheme. δ Cephei is an example.

General Catalog of Variable Stars

The ultimate source of information on variable stars is the General Catalog of Variable Stars published by the Academy of Sciences of the USSR. The fourth edition, published between 1985 and 1987, contains data on more than twenty-eight thousand stars arranged by constellation. The catalog was published in Russian, but those who do not read Russian have no trouble using it. The preface has been printed in both Russian and English, and the column headings are not hard to decipher. A sample page has been reproduced as Fig. 14.2. In addition to the table reproduced in this figure, there are other tables to provide cross references to many other identification systems.

ЗВЕЗДА		КООРДИНАТЫ 1950,0		ПРЕЦЕССИЯ 1950,0		L	B	ЛИТЕРАТУРА		ТИП	ЗВЕЗДА
LACERTA				LAC				ЯЩЕРИЦА			
R	*	22 41 02	+42 06,4	2,67	0,314	99,05	-14,48	0001	0002	M	R
S	*	22 26 49	+40 03,6	2,63	0,307	95,59	-14,87	0001	0002	M	S
T		22 20 09	+34 09,7	2,69	0,303	91,03	-19,06	0458	BD	E:	T
U	*	22 45 40	+54 53,7	2,47	0,317	105,82	-3,55	0882	0098	SRC	U
V	*	22 46 35	+56 03,4	2,45	0,317	106,47	-2,57	9288	0884	DCEP	V
W	*	22 05 21	+37 29,4	2,58	0,293	90,52	-14,61	0001	0100	M	W
X	*	22 47 00	+56 09,8	2,45	0,317	106,57	-2,50	8632	0884	DCEPS	X
Y		22 07 08	+50 48,0	2,30	0,294	98,72	-4,03	8300	0102	DCEP	Y
Z	*	22 38 53	+56 34,1	2,37	0,313	105,76	-1,62	9029	0884	DCEP	Z
RR	*	22 39 27	+56 10,3	2,39	0,314	105,64	-2,01	8632	0884	DCEP	RR
RS		22 10 47	+43 30,1	2,49	0,297	94,99	-10,35	0001	0002	SRD	RS

ЗВЕЗДА		ЗВЕЗДНАЯ ВЕЛИЧИНА MAX	MIN		ЭПОХА J.D,2400000+	ПЕРИОД	M-m ИЛИ D	СПЕКТР	ЗВЕЗДА
LACERTA					LAC			ЯЩЕРИЦА	
R	*	8,5	14,8	V	41526	229,86	41	M5E-M8,5E	R
S	*	7,6	13,9	V	43804	241,50	46	M4E-M8,2E	S
T		11,0		P				K0	T
U	*	9,4	12,1	P				M4EPIAB+B	U
V	*	8,38	9,42	V	28901,285	4,983458	25	F5-G0	V
W	*	10,3	(15,0	P	39439	328,5		M7E-M8E	W
X	*	8,20	8,64	V	42738,132	5,444990	38	F6-G0	X
Y		8,76	9,50	V	41746,745	4,323776	34	F5-G0	Y
Z	*	7,88	8,93	V	42827,123	10,885613	43	F6IB-G6IB	Z
RR	*	8,38	9,30	V	42776,686	6,416243	30	F6-G2	RR
RS		9,6	12,5	V	40884	237,26	53	K0	RS

Fig. 14.2. Top portions of facing pages from the *General Catalog of Variable Stars*. The left page is at the top.

Two of the columns on the left-hand page give numbers which lead the user to two important references for each star. These references are listed in the front of the catalog. The first reference is to the most recent definitive article listing period and epoch, and the second reference is to a publication in which a 'finding chart' has been included. A finding chart is a map of the region surrounding the variable, and it is a necessity when the user is trying to identify the variable on the photograph or at the telescope. A finding chart from a publication of 1931 is seen in Fig. 14.3.

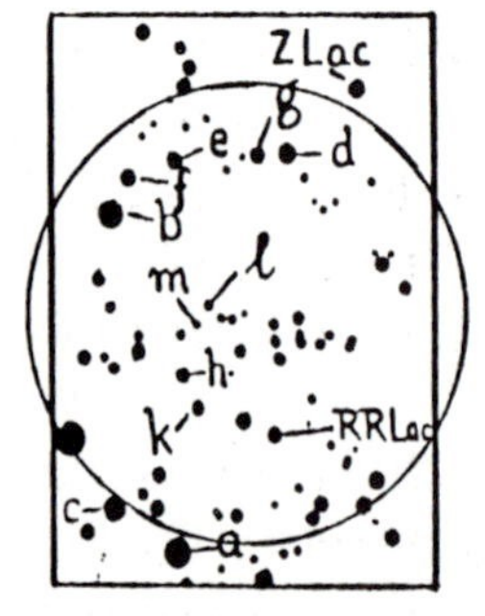

Fig. 14.3. A finding chart for the cepheid variables, Z Lacertae and RR Lacertae. The angular diameter of the circle is one-half degree. (Harvard College Observatory.)

The user must try to match the finding chart to the actual sky, and this is not always as easy as it may sound. The observer should first determine the angular size of the field of the eyepiece and then try to relate that to the scale of the finding chart. One may find it helpful to draw a circle representing the field of the eyepiece on the finding chart in order to know what to look for.

Copies of the *General Catalog of Variable Stars* may be difficult to locate for observers who do not have ready access to an astronomical library. Even those who do consult the catalog may have problems when it comes to finding the references which hold the finding charts. Some of those references are in obscure publications which are by now quite old. Fortunately, there is a world-wide organization which provides essential services to all who participate in the study of variable stars and can supply copies of the *General Catalog of Variable Stars*.

American Association of Variable Star Observers

The American Association of Variable Star Observers (AAVSO) was founded in 1911 to encourage and facilitate the observation of variable stars. The offices are at 25 Birch Street, Cambridge, Massachusetts, 02138, USA, and inquiries regarding membership may be sent there. Most of the members of the AAVSO are amateur astronomers, and in terms of expertise they range from beginners to skillful, well-equipped experts. Each member has a list of stars which he or she tries to monitor as time and weather permit. The members record magnitude and time for each of their observations, and they submit the data monthly to the headquarters. Here the data are processed and added to the computerized data base.

As the years progress the data accumulate, and light curves begin to take shape. Every few years light curves are published and distributed to the astronomical community, and very often such data have been of great value in theoretical work on questions such as stellar evolution. A light curve from AAVSO Report 38 is reproduced in Fig. 14.4. It is always impressive to think of the great cooperative effort that produced a light curve such as this one.

One of the most valuable and essential services of the AAVSO is the publication of its charts. A typical example is shown in Fig. 14.5. Each chart is centered on a variable star and identifies a group of comparison stars for that variable. Magnitudes are indicated to the nearest tenth of a magnitude without the decimal point. The variable R Cass is indicated

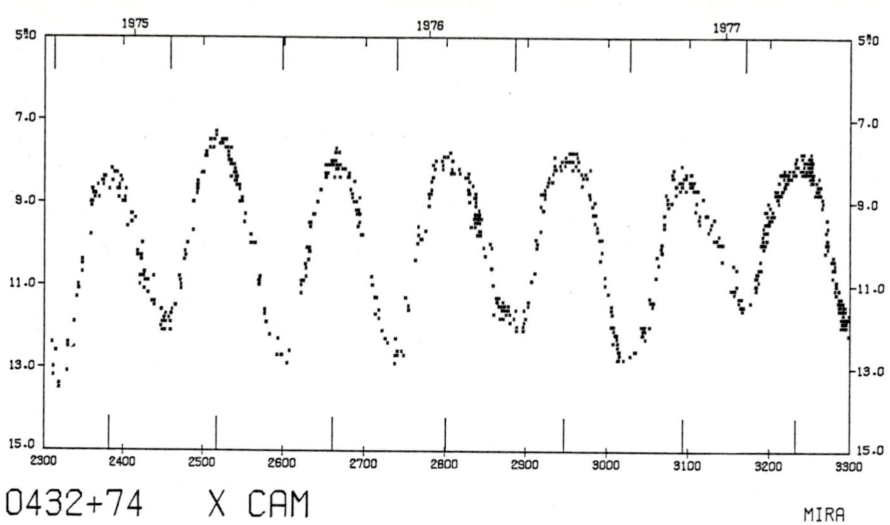

Fig. 14.4. Light curve for the Mira-type variable, X Camelopardalis as published in AAVSO Report 38. (Courtesy of AAVSO.)

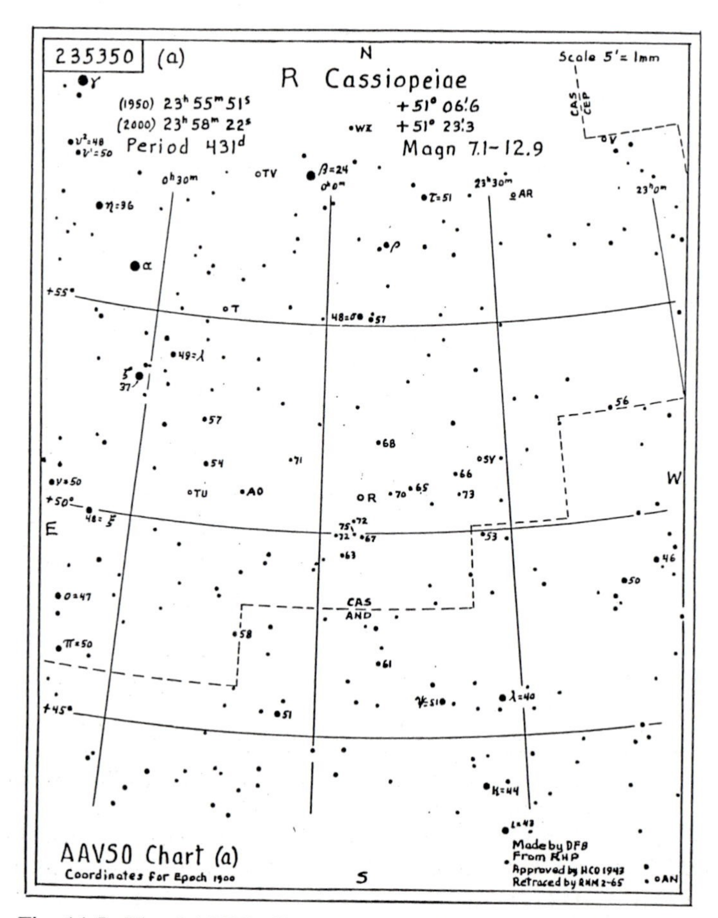

Fig. 14.5. The AAVSO Chart for R Cassiopeiae. This chart has been reduced from the original, so the scale shown here is not correct. (Courtesy of AAVSO.)

with a small, open circle, and the star just to the right of it is marked '70'. This means that the visual magnitude of that star is 7.0. The chart shown here is one of the charts in the 'a' series as indicated by the letter to the right of the number in the upper left corner. The AAVSO publishes the 'b', 'd' and 'e' series of charts as well in order to show the variable and its surrounding stars on different scales. The choice of the appropriate series of charts depends on the brightness of the star at the time of observation and the size of the telescope being used. An example of a chart from the 'd' series, also for R Cass, is shown in Fig. 14.6. Notice that the first chart

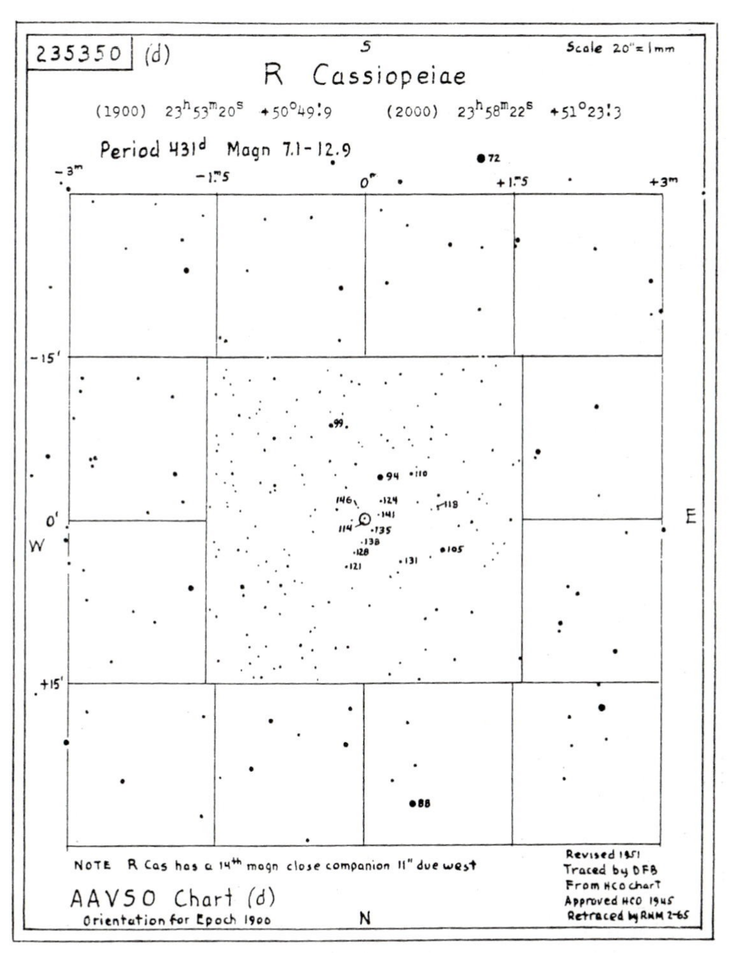

Fig. 14.6. The AAVSO d series chart for R Cass. Note the difference in scale between this chart and the previous one. Also note that north is at the bottom on this chart and that east is toward the right. This chart has been reduced from the original, so the scale shown here is not correct. (Courtesy of AAVSO.)

is drawn with north at the top to show the sky as it would appear to the unaided eye. On the other chart north is at the bottom, and the stars appear as they would in a telescope. The chart in Fig. 14.7 is from a fifth series of charts, the photoelectric series.

IBVS publications

Another useful publication comes at irregular intervals from the Konkoly Observatory in Budapest, Hungary, and is known as the *Information Bulletin on Variable Stars* (IBVS). The IBVS is sponsored by Commission 27 (variable stars) of the International Astronomical Union. Observers around the world send their finding charts, light curves and analyses of data to Budapest for inclusion and distribution to interested

STAR	COORDINATES- EPOCH 1900		TYPE	MAX-MIN	SPECTRUM	S.A.O. #
	h m s	° '		m m		
RU Cam 071069	07 10 54	69 51	C	9.3-10.4B	K0-R2	014157
Comparison	07 12 10	70 12		9.088V		006167
Check	07 15 00	69 39		8.909V	F8	014183

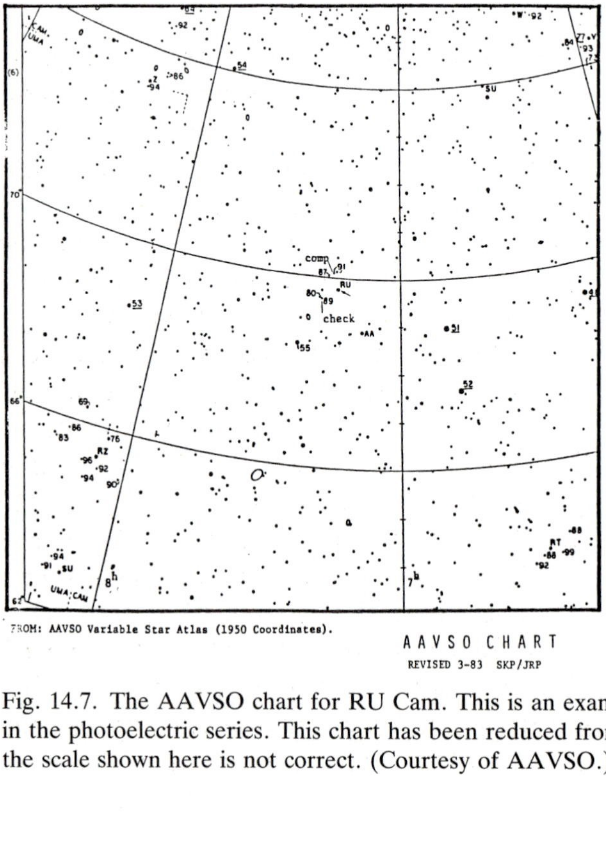

FROM: AAVSO Variable Star Atlas (1950 Coordinates).

AAVSO CHART
REVISED 3-83 SKP/JRP

Fig. 14.7. The AAVSO chart for RU Cam. This is an example of the charts in the photoelectric series. This chart has been reduced from the original, so the scale shown here is not correct. (Courtesy of AAVSO.)

observers. The IBVS is especially useful to astronomers who wish to distribute their results as quickly as possible.

Observational data – magnitudes

Since these stars as a group are ones that change in brightness, it is obvious that the two important quantities that define each observation are apparent magnitude and time. In Chapters 7, 8, and 9 we dealt with the determination of apparent magnitudes and colors, so we need not repeat that material. It is expected that an observer will always strive for the greatest possible precision in determining magnitudes, but realistically we know that the methods, equipment and, above all, the experience of the observer will determine the precision of the recorded data. Those who are starting out with photographic methods should strive for magnitudes which are accurate to within ±0.01 magnitude, but they should not give up if they can only achieve results good to within 0.05 or 0.10 mag. Observers with good photoelectric equipment and a good site should work toward results good to ±0.001 mag. Fortunately, there are stars for which observations of lower precision are still helpful, and much useful data is routinely recorded by visual observations at the telescope. There are even a few variables for which not even a telescope is needed. At the other extreme are stars for which the total variation is only a few thousandths of a magnitude, so the most precise methods must be applied. By choosing the right stars to observe with the equipment which one has, every observer can produce satisfying results.

On p. 146 the concept of differential magnitudes was introduced, and differential magnitudes were said to be useful in the study of variables. Since we are looking for changes in the magnitude of a star, we may make a series of observations of the difference in magnitude between the variable star and a star assumed to be constant in magnitude. The latter is described as the 'comparison star'. In photoelectric observations of variables it is customary to make comparisons of this sort and to compute differences in magnitude between variable and comparison stars. Then, when the magnitude of the comparison star has been found either from reference catalogs or from original observations, the magnitudes for the variable star can be determined. It is also good practice to include a 'check star' in the program so that the constancy of the comparison star can be established. A number of new variables have been discovered because comparison stars were found to have slow or very small variations. In photographic observations the situation is not the same because of the need to establish both scale and zero point on each photograph, and several comparison stars must always be measured.

Observational data – time

The second quantity that must be part of every observation is, of course, the time. Again, one should always work toward the best possible data, but one should be rational about it. For example, in making a visual estimate of a star's magnitude, one must look first at the star and then at the clock. Time recorded to the nearest minute or half minute is probably as good as one can hope for. In making a photograph one should be able to record the times of beginning and ending of the exposure to within a few seconds provided that care has been taken in setting a clock at the telescope. When using a photoelectric photometer, the precision of the time will depend upon the method used for recording the data. A paper chart can easily be read to a tenth of a minute if the chart is moving at one inch per minute. Somewhat less accuracy would be expected if one had to read and record a digital display on a counter and then note the time. If data is recorded by means of a computer which receives and records counts from the photometer, then the time of an observation can easily be known to the nearest second.

It has become standard practice in variable star work to convert the time to the Julian Day and fraction thereof as this makes the measurement of intervals of time more convenient. This conversion should be the first step after the observer leaves the telescope. Procedures for doing this were described in Chapter 2, and the formidable-looking expression can easily be programmed into a computer.

At the end of Chapter 2 we defined heliocentric time as the time as it would be recorded by a hypothetical observer located at the sun, and we noted that if observations over a period of months are to be combined in a meaningful way, they should be corrected from geocentric to heliocentric time. Light travels from the sun to the earth in about eight minutes, so the correction for light-time can be significant. Therefore, this additional correction should be applied to the recorded time whenever precise times are needed. The details of a method for computing the correction are included at the end of this chapter.

Preliminary light curve

When the magnitude and the time have been determined and corrected in whatever ways are appropriate, then one may plot magnitude against time in order to produce the light curve. If the period of a star was ten days, for example, and a fortunate observer was able to measure the star's brightness on every night for thirty days, then three cycles of the variation should be seen on a preliminary plot. A careful examination of the plot would reveal a value for the period, the time between successive maxima on the light curve. If the star's period is longer than the interval covered by the observation, then only a trend will appear. And if the star is one of those described as 'irregular', then continual but non-periodic changes may be seen.

In the real world it is not likely that one could obtain successful observations on thirty consecutive nights as in the example above, so the determination of a good value for the period is not so straightforward. This process may require observations over many cycles. For the moment, however, let us continue with the example which we began above. If the preliminary light curve did permit us to establish an approximate period, we could take the next steps to plot a mean light curve in which successive cycles will be plotted on top of each other.

In Fig. 14.8 we have plotted some hypothetical data which show slightly more than four cycles in the variation of magnitude. We note the time of the first maximum, and we define this as the 'epoch'. By measuring the intervals between maxima we find the period (or at least a preliminary period). Keep in mind that the period is the length of one cycle from maximum to minimum and back to maximum. Now in order to

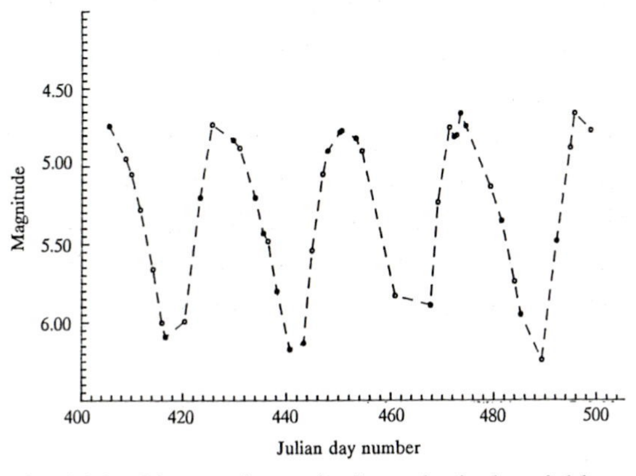

Fig. 14.8. Observations of a hypothetical variable star.

superimpose the four cycles in this example, we compute the 'phase' for each observation. The phase is simply the decimal fraction of the cycle in which an observation lies and is measured from the time of the last maximum preceding the observation. The phases of all points have been listed in Table 14.1 for two values of the period, 23 days and 18 days. Phase may be calculated from:

$$\text{ph}=(t-E)/p-\text{Int}((t-E)/\,p)$$

Table 14.1. *Data for light curves*

Original data		Period=23 days		Period=18 days	
Magnitude	Julian Date	Phase	Magnitude	Phase	Magnitude
4.74	405.3480	0	4.74	0	4.74
4.95	408.6646	0.007978	4.74	0.0011512	5.2
5.05	409.8706	0.062826	4.83	0.0820397	5. 83
5.28	411.6332	0.085318	4.82	0.1102448	5.13
5.66	414.1612	0.116269	4.88	0.1105279	6.13
6.00	415.9470	0.134726	4.9	0.1274177	4.73
6.09	416.6428	0.144201	4.95	0.1842567	4.95
5.99	420.3536	0.196635	5.05	0.1968553	5.54
5.20	423.3687	0.216713	5.13	0.2512563	5.05
4.73	425.6415	0.247361	5.2	0.312817	5.05
4.83	429.7930	0.273270	5.28	0.3491787	5.28
4.88	431.0222	0.313913	5.43	0.358056	4.83
5.20	434.0373	0.351222	5.48	0.3643562	4.9
5.43	435.5680	0.383183	5.66	0.4263442	4.88
5.48	436.4261	0.412031	5.83	0.4634281	5.89
5.80	438.1655	0.426848	5.8	0.4896223	5.66
6.17	440.6703	0.460826	6	0.4932065	4.78
6.13	443.3375	0.491078	6.09	0.5138228	4.77
5.54	444.8914	0.535752	6.17	0.5420278	5.23
5.05	446.9787	0.651717	6.13	0.5888333	6
4.90	447.9064	0.652418	5.99	0.59385	5.2
4.78	450.2257	0.710509	5.89	0.6274889	6.09
4.77	450.5968	0.719278	5.54	0.6645728	4.82
4.82	453.3103	0.772022	5.23	0.6682943	4.75
4.90	454.4467	0.78351	5.2	0.678889	5.43
5.83	460.8247	0.810031	5.05	0.7172615	4.81
5.89	467.6897	0.850366	4.9	0.7265608	5.48
5.23	469.1045	0.870839	4.75	0.7277053	4.9
4.75	471.3773	0.882327	4.73	0.7470229	4.8
4.81	472.2587	0.909161	4.81	0.7881283	4.66
4.80	472.7944	0.932453	4.8	0.8231948	5.8
4.66	473.5343	0.951205	4.78	0.8336453	5.99
4.74	474.5315	0.964622	4.66	0.8435279	4.74
5.13	479.3324	0.96734	4.77	0.9623498	6.17

Here t is the time of an observation, E is the epoch (i.e., the time of an assumed starting point), and p is the period. The value $\mathrm{Int}((t-E)/p)$ is the integer value, so it is actually the number of the cycle since the beginning epoch. (ph) is again the fractional cycle or phase. E and t should be expressed in Julian Days.

The mean light curve for the data using p=23 days has been plotted in Fig. 14.9(*a*). Apparent magnitude has been plotted against phase which runs from zero to one. In this simple way a series of cycles in the variation can be superimposed. It is often helpful to show more than one full cycle so that the maxima at the beginning and end of the mean light curve are more apparent. This has been done in Fig. 14.9(*b*) in which the first and last parts of the data have been plotted twice.

In the simplified example described here we have implied that the correct period could be read from the plot of Fig. 14.8 showing data from several cycles. The more common situation is that the maxima are not defined in a reliable way by the data. Therefore, it is not realistic to imagine that anything more than a preliminary period can be read from the continuous data. If the assumed period is even a little bit inaccurate, then the mean light curve described in the previous paragraph will show considerable scatter. See, for example, Fig. 14.9(*b*).

Finding a precise period

We suggested above that the real world is not as kind to the observer as our example might suggest. A long string of observations is hard to produce, and even a preliminary period may be hard to recognize. Thus, the determination of a value for the period and the production of a mean light curve are not always easy. For many classes of variable stars the periods are essentially constant over many years. Examples are the Cepheids and the RR Lyrae variables. This constancy means that if good data have been accumulated over a time encompassing many cycles an extremely precise period may be found.

Let us imagine a new example in which a preliminary period is thirty days. Assume, also, that the data extends over a period of four years. Unless the star is circumpolar for the observer, there will be breaks in the data when the star is not visible. Now, if we select an initial epoch and the preliminary period of thirty days, we can make predictions of the times of maxima from

$$T_{\mathrm{max}}=E+xp$$

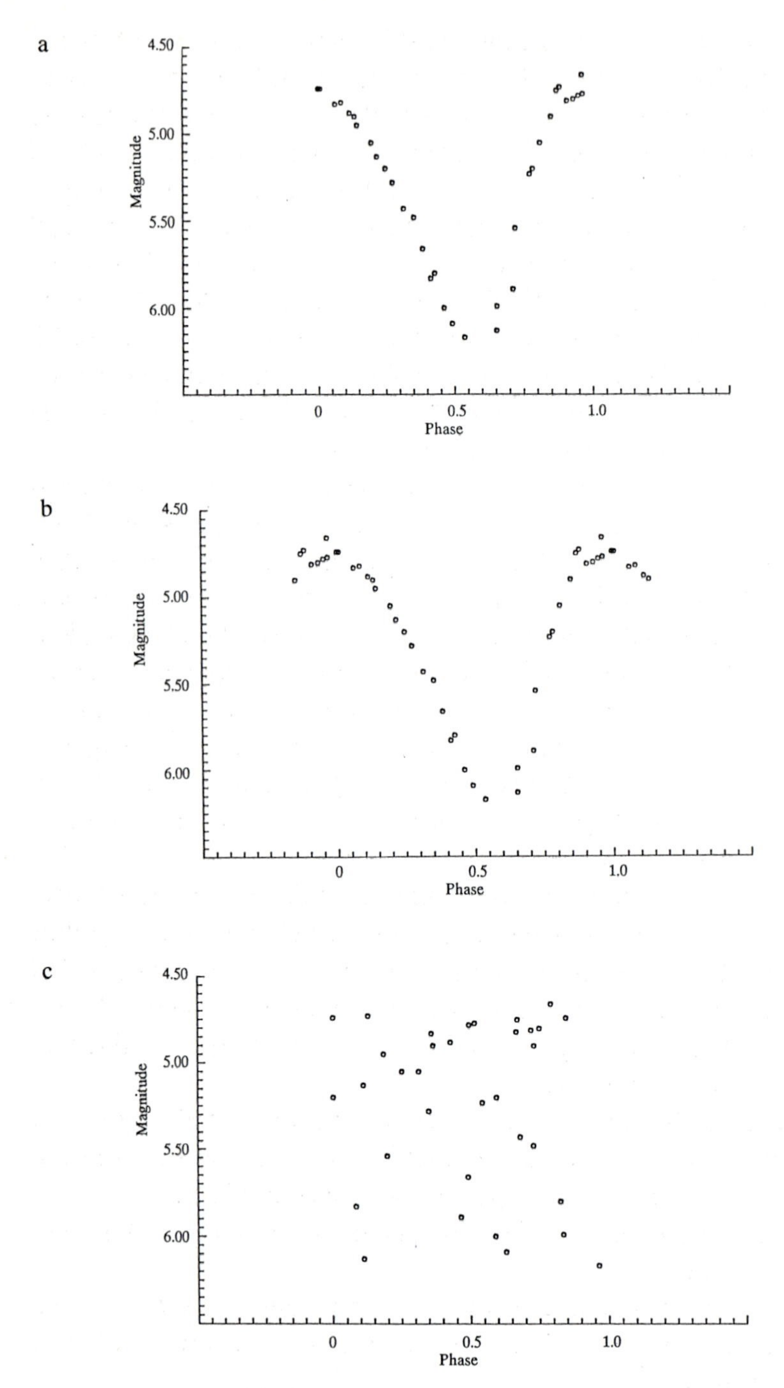

Fig. 14.9. (*a*) The mean light curve using the correct period. (*b*) The mean light curve from the previous figure with the first and last parts of the data plotted twice. (*c*) An attempt to plot a mean light curve when phases have been computed from an incorrect period.

T_{max} is the time of a future maximum, and x is a whole number of cycles. We may call this a computed time of maximum and designate it as C. Suppose that the data show many very well defined maxima. We refer to the observed times of maximum simply as O. Imagine that the last one of these observed maxima came at a time just about 24 hours later than $(E+47p)$, that is, 24 hours into the 48th cycle. Clearly, our original period was not quite long enough. A correction to the period might be found by trial and error or by dividing the 24-hour accumulated error by 47 cycles to get the error per cycle which in this example will be 30.638 minutes. As a fraction of a day this is 0.02127, so the corrected period should be 30.02127 days. Predictions based on this new period should exactly match the observed times of maxima.

The actual practice among workers in the field of variable stars is slightly different and makes use of a device referred to as the $O-C$ diagram. An extensive set of data covering many cycles will show many maxima, but the times of such maxima may be indefinite. Therefore, it is customary to make a table of observed times of maximum and couple it with predicted times of maximum calculated from an initial epoch and a preliminary period. For each cycle for which there is data we find the difference $(O-C)$, observed minus computed, and we plot these against time. The result should look like one of the graphs in Fig. 14.10, the $O-C$ diagrams. Several hypothetical diagrams are shown here so that we can understand the message conveyed by each.

In Fig. 14.10(*a*) the points representing maxima are scattered evenly above or below the horizontal axis, and none are very far above or below the line. We interpret this to mean that we have used the correct period in determining the values of C, the computed time of maximum. Scatter above and below the line suggests that errors were made when estimating the observed times of maximum.

The sloping line in Fig. 14.10(*b*) indicates that our initial estimate of the period was too short. Times predicted from that period were too early, so the difference, $O-C$, is positive and becomes larger with time. The scatter again means that times of maximum in many cycles were difficult to estimate.

Fig. 14.10(*c*) shows the case in which the preliminary period was too long. Computed times of maximum come later than the observed times, so $O-C$ is negative.

Finally, the curved line in Fig. 14.10(*d*) results from a case in which the period is not constant but has been changing slowly throughout the observations.

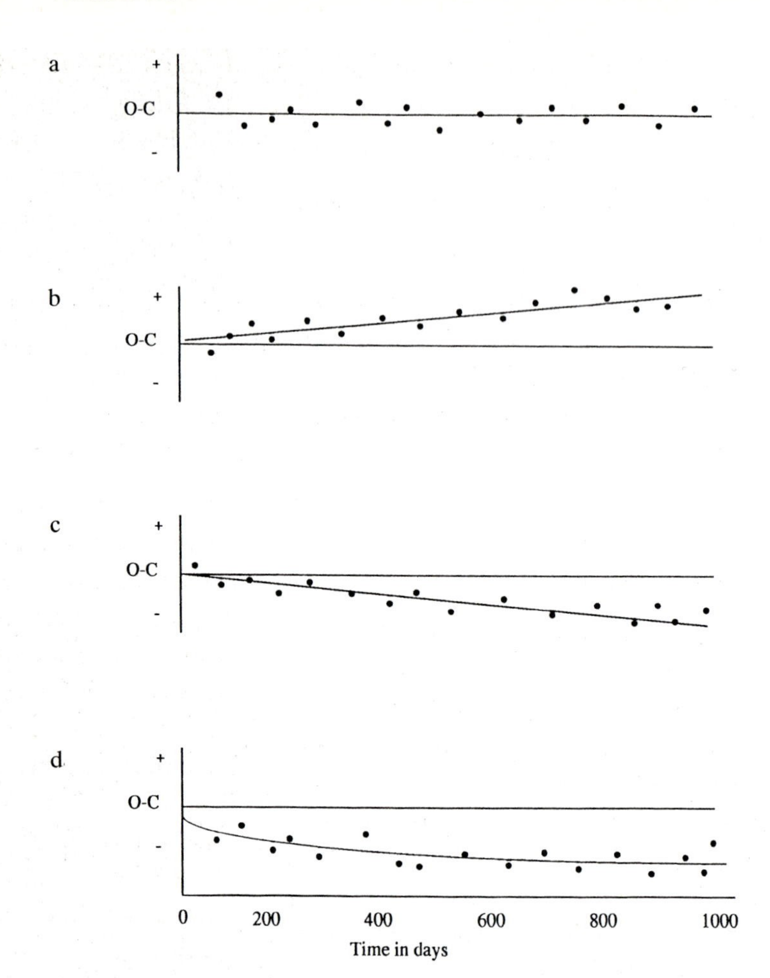

Fig. 14.10. O-C diagrams for (*a*) a period that fits the data; (*b*) an assumed period that is too short; (*c*) an assumed period that is too long; (d) a period that is changing.

If there had been an error in the determination of the time of the initial epoch, and the epoch was not at a time of maximum light, then the best line through the $O-C$ points would not pass through the origin but would cross the vertical axis above or below the origin.

Corrections to the preliminary period may be made by trial and error or by making a least squares analysis of the $O-C$ data. In the latter case we may begin by letting T_0 and P_0 be the preliminary values of the epoch and the period respectively. If x is some number of cycles since the beginning, then

$$T_0+xP_0=C_x$$

If dT and dP are corrections to be applied to the preliminary values of T and P, then we see that

$$(T_0+\mathrm{d}T)+x(P_0+\mathrm{d}P)= O_x$$

Taking the difference between observed and predicted

$$O-C=\mathrm{d}T+x\mathrm{d}P$$

An equation of this type may be written for each observation, and these may be solved by the method of least squares (see Appendix 3) to find the two corrections, dT and dP.

When an accurate value for the period has finally been determined, one can then plot a mean light curve such as the one shown in Fig. 14.9(*b*).

Spurious periods

There have been occasions in the past when periods which seemed to fit the data in quite a reasonable way later proved to be incorrect or spurious. It is quite likely that a star might be observed at just about the same hour on every clear night for a period of several months. The position of the star in the sky and the daily habits of the observer combine to insure this. Analysis of times and magnitudes might suggest a period of about thirty days. Then in a single evening the observer makes a large number of measurements of magnitude and sees that the period is actually very short, just under twenty-four hours, perhaps. What has happened is that the regularity of the first observations caused only a slight difference in true phase from one observation to the next. A series of slow changes would then suggest the longer period. The regularities in times of observation could be yearly in the case of stars with periods of years or daily for stars with much shorter periods.

In order to avoid such 'stroboscopic' effects the times of observation should be varied and extensive periods of frequent observations should be inserted on occasion.

Selection of variable stars for study

The important question for observers who wish to establish their own new program, is 'Which of the thousands of known variable stars should I begin to monitor?' The answer to this question depends partly upon the equipment available and the experience of the observer.

Obviously, one must select stars which are bright enough throughout their cycles to be readily seen in the telescope being used. Second, one should select stars for which the amplitude of the variations is significantly larger than the accuracy of the magnitudes that the observer is capable of producing. As stated earlier, the beginner should not attempt to monitor stars which vary by only a few hundredths of a magnitude. Third, one may wish to look for stars for which several cycles of the variation may occur during one observing season (i.e., the portion of the year during which the star can be seen). For example, several cycles could be followed in one season if the period was in the neighborhood of twenty-five days.

One should not underestimate the problem of identifying the variable being studied while at the telescope. This can be a time-consuming and, at times, frustrating process. If the star is a faint one, then an observer may see many similar stars in the field of the eyepiece. One must then look back and forth from the telescope to the finding chart to make the identification. This can still be confusing unless one has taken time to note the size of the field of view compared to the scale of the finding chart. When brighter stars are being observed, then the finding chart may cover an area larger than that of the eyepiece, and the observer will actually have to move the telescope from star to star in order to confirm the identification of variable and comparison stars. Once again, it is in the best interests of the beginner to monitor bright variables just because it is easier to be positive about identifications.

In Table 14.2 we have listed some of the many classes of variable stars and the characteristics of each class. Within these classes there are many stars to choose from and many of the stars are bright enough to be

Table 14.2. *Some classes of variable stars*

Type	Symbol	Period (days)	Amplitude (magnitudes)	Brightest example
Cepheids	C	1–50	0.2–2.0	δ Ceph
W Virginis	CW	1–50	0.2–2.0	W Vir
RR Lyrae	RR	0.2–1	0.5–1.5	RR Lyr
Long period	M	100–600	2.5–10	Mira, o Ceti
Semi-regular	SR	20–500	0.2–2.5	T Cen
RV Tau	RV	30–150	1–3	R Sct
U Geminorum	UG	20–600	2–6	SS Cyg
δ Scuti	δ Sc	0.05–0.2	<0.2	β Cas
Novae	N		7–>15	V603 Aql
Supernovae	SN		>20	1054 Tau (typ)

This expression can be used in practical situations since the right ascension and declination of the sun may be found for every day in the *Astronomical Almanac*. Where these calculations are to be repeated in a routine way, it is more convenient to express the position of the sun in terms of its celestial longitude and the obliquity of the ecliptic, ε. The longitude of the sun may be calculated for any date by means of the formula

$$\odot = L + (1^\circ.915 - 0^\circ.0048\mathrm{T}) \sin M + 0^\circ.020 \sin 2M$$

in which

$$T = (\mathrm{JD} - 2451545.0)/36525$$

$$L = 280^\circ.460 + 36000^\circ.772T$$

$$M = 357.528 + 35999^\circ.050T$$

taken from the *Almanac For Computers*.

Returning to Fig. 14.14 and the triangle VT$\odot$ we write the following identities:

$$\cos \odot = \cos D \cos A$$
$$\sin \odot \sin \varepsilon = \sin D$$

$$\sin \odot \cos \varepsilon = \cos D \sin A$$

Then

$$\cos u = \sin \delta \sin \odot \sin \varepsilon + \cos \delta \cos \alpha \cos \odot + \cos \delta \sin \alpha \sin \odot \sin \varepsilon$$

Finally, let us recall that the orbit of the earth is elliptical rather than circular and that the earth is closer to the sun in January than it is in June by some five thousand kilometers. It is possible to introduce further corrections to account for this variation, but this is not practical in the present context. The maximum correction to be added because of the shape of the earth's orbit is only 0.008 seconds.

QUESTIONS FOR REVIEW

1. From Table 14.2 select one class of variable stars for which meaningful observations could be made visually using a six-inch telescope. Explain your choice. Select a second class of stars which would have to be observed with precise photoelectric methods and a telescope of thirty-inch aperture or larger. Again, give reasons for your choice.
2. Refer to the *General Catalog of Variable Stars* and to Table 14.2 from this chapter. For each of the classes of Table 14.2 select one star that could be

observed in the fall and one that could be observed in the spring from your latitude.

3. Locate finding charts for at least half of the stars which were selected in the previous question.

4. Describe the procedure for finding the period of a variable star. Include the use of the $O-C$ diagram.

5. Using a period of 22.4 days, calculate the phase for each of the first five observations in Table 14.1.

6. Consider the bright star, Vega, observed at 23h on July 22, 1990. Calculate the helicocentric correction for Vega on that date.

7. Describe a situation in which you might need to consider the ellipticity of the Earth's orbit when finding the heliocentric time.

8. Explain why 'spurious' periods will sometimes be found from the analysis of a set of observations of a variable star.

9. The elements of the light variation for δ Cephei are 2 396 203.7454+5.36091x. Find the Julian Date at 10:00 am Eastern Standard Time on April 24, 1989. Now calculate the phase of δ Cephei at that time.

Further reading

Belserene, E. P. (1988). Rhythms of a variable star. *Sky and Telescope*, **76**, 3, 288. This article contains a computer program for determining the period of a variable star by means of the discrete Fourier transform.

Henden, A. A. and Kaitchuk, R. H. (1982). *Astronomical Photometry*. Van Nostrand Reinhold Company. See especially Chapter 10 for list of stars to observe.

Hoffleit, D. (1987). History of variable star nomenclature. *Journal Of The AAVSO*, **16**, 65. For more detail on nomenclature of variable stars see this paper.

Hoffmeister, C., Richter, G. and Wentzel, W. (1985). *Variable Stars*. Springer Verlag. The primary goal of this book is to describe the characteristics of the various classes of variables. However, there are chapters on discovery and techniques of observation.

Percy, J. (1986). *The Study of Variable Stars Using Small Telescopes*. Cambridge University Press. This is a collection of papers presented at a symposium held in Toronto in July, 1985.

Petit, M. (1987). *Variable Stars*. John Wiley and Sons. A translation of the book originally published in French in 1983.

Scovil, C. E. (1980). *The AAVSO Variable Star Atlas*. Sky Publishing Corporation. This atlas consists of 178 charts on which several thousand variable stars have been identified. All of the variables are brighter than visual magnitude 9.5 and all have amplitudes greater than 0.5 magnitude. The charts also include many clusters, galaxies, double stars and nebulae.

Strohmeier, W. (1972). *Variable Stars*. Pergamon Press. The emphasis here is on causes of variation of light rather than methods of observation.

15

Observing the sun

The importance of the sun as the most observable of all stars cannot be overstated. As shown in Fig. 15.1, no other star can be studied with the resolution that we achieve in even the simplest observations of this source of all of our light and energy. As a result, what we have learned from the sun we have applied in our study and analysis of the stars. Our knowledge of the sizes and distances of the stars is based upon our

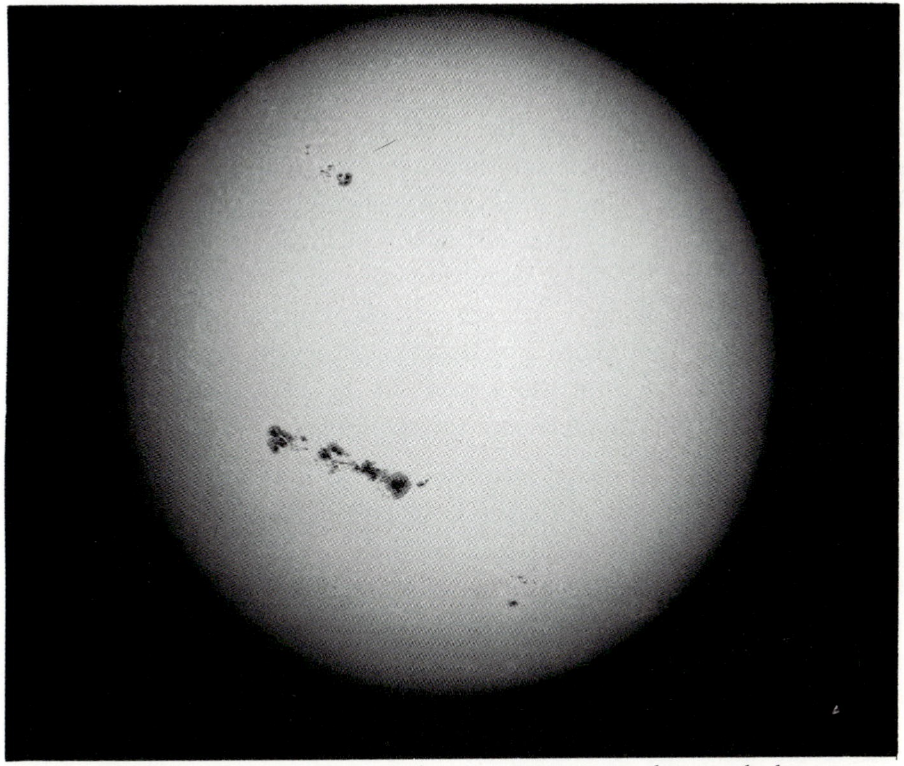

Fig. 15.1. The sun and a conspicuous sunspot group photographed on September 4, 1989. (Courtesy of A. W. Steinbrecher, Oswego, New York.)

knowledge of the sun. Also our notions of the output of energy by the stars is calibrated in terms of our measurements of the output of energy from the sun. In this chapter we shall first describe methods of observing the sun in simple ways that can be used by anyone with a telescope. Then, beginning with the discussion of the spectroheliograph, we shall move on to observational methods that are used at observatories dedicated mainly to solar research.

The sun with small telescopes

The sun is so bright that one should never try to make direct, naked eye or telescopic observations of it. This is an absolute rule, for the observer can be blinded by even a brief attempt. There are, however, safe ways to view the sun, and some of these require no complex equipment.

The most readily available method of seeing the sun's apparent surface or photosphere is by means of eyepiece projection. A screen may be mounted at some fixed distance behind the eyepiece, and the eyepiece can be moved back away from its normal visual position. When the proper position of the eyepiece has been found, a sharp image of the sun will be seen on the screen. A bright image of the sun will have been formed in the focal plane of the objective. The eyepiece in its new position serves as a simple lens rather than a magnifier. The bright image becomes the 'object' and the eyepiece forms an image of it at an appropriate distance, the image distance. The geometry of this is seen in Fig. 15.2, and the principles were discussed in Chapter 6. The object distance is now the distance from the eyepiece to the bright image, and the image distance is the distance between the eyepiece and the screen. The size of the image on the screen can be varied by changing the positions of both the eyepiece and the screen.

Example: A small telescope is to be used to look at sunspots, and a screen can conveniently be located 500 mm behind the eyepiece. The aperture of the telescope is 150 mm and the focal length is 1500 mm. The focal length of the eyepiece is 50 mm. How much should the eyepiece be moved back from its normal position for an image of the sun to be in focus on the screen? We begin with the lens formula

$$1/f = 1/I + 1/O$$

and we substitute the values of f and O. Solving for I we have

$$1/I = 1/50 - 1/500$$

$$I = 56 \text{ mm}$$

Thus, the eyepiece should be moved 6 mm back from its position for normal viewing.

We should also find out how large the image of the sun will be on the screen, and we may proceed as follows. The angular diameter of the sun is one half degree, so the radius of the sun's image in the focal plane will be given by

$$\tan 0°.25 = X/1500$$

$$1500 \times 0.00436 = 6.55 \text{ mm}$$

6.55 mm is the radius of the sun's image, so the diameter of the image is 13.1 mm. The diameter projected on the screen will be

$$D_p/500 = 13.1/56$$

$$D_p = 117 \text{ mm}$$

The brightness of the projected image will depend on the size at which it is projected and the aperture and focal length of the telescope. The latter, of course, determine the size and brightness of the original image of the sun. The brightness of the projected image may be reduced as necessary by spreading the light out into a larger image, or by partially covering the telescope's objective.

The sun's light is accompanied by quite a lot of heat, and this can be a serious problem even in a small telescope. Care must be taken to be sure that heat does not build up to the point that it can damage the eyepiece.

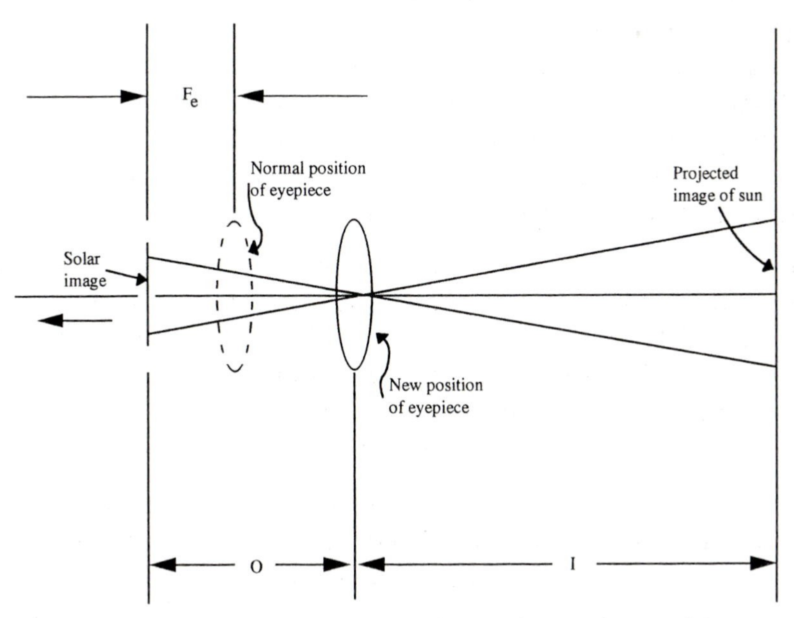

Fig. 15.2. The manner in which the eyepiece projects an image of the sun.

A vivid demonstration of the need for caution can be performed in the following way. With the sun's image focused on the screen behind the telescope, hold a piece of paper in front of the screen. Slowly move the paper toward the eyepiece, and note the way in which the sun's image becomes smaller and brighter. If the paper is held still when it is close to the eyepiece, the concentrated heat from the sun will soon set the paper on fire.

Surface features in white light

On most occasions the observer looking at the projected image of the sun will see sunspots, irregular dark areas surrounded by less dark borders. A photograph showing some especially large sunspots is included as Fig. 15.1. The dark, inner part of a spot is known as the umbra, and the surrounding part is known as the penumbra. Spots vary in size from those which are barely visible to spots similar to the ones in the figure. On rare occasions spots have been so large that they could be detected without a telescope (but with suitable precautions). Spots of some sort are almost always visible, but there have been periods when no spots were seen. In the seventeenth century almost no spots were recorded in a period lasting seventy years.

It is a worthwhile exercise to determine the sizes of typical sunspots, and the procedure is a simple one. The diameter of the sun is 1 396 282 km. The diameter of the sun's image in millimeters can be measured either on the screen or on a photograph such as the one included here. Then the largest dimension of the spot can be measured in millimeters on the photograph. By means of a simple proportion the diameter of the spot in kilometers is calculated. Students are often surprised to discover that even a small sunspot may be large enough to swallow the earth without even a splash. On the scale of this photograph the earth would appear as a small dot less than one millimeter in diameter.

The persistent observer who makes a series of sketches of the sun over a period of a few days will notice that the sunspots appear to move. This change of location is caused by the rotation of the sun on its axis, and the rate of rotation can be calculated from measurements of the change in position. A good series of observations may also show that the apparent path of a spot is not a straight line, and this is evidence that the equator of the sun does not lie in the plane of the earth's orbit. The tilt of the sun's axis is 7° 15′. As seen from the earth the sun's period of rotation is about twenty five days as measured by spots near the equator. The period becomes longer at higher latitudes.

It is typical for the size and shape of a sunspot to change from day to day. There is no rule concerning the time during which an individual spot may persist. Two weeks is not unusual, but as seen in Fig. 15.3 some groups of spots have been observed through more than one rotation of the sun.

The careful observer should also be able to notice three other phenomena in the white-light image. First, is the granulation, variations in brightness over the surface. This small-scale variation can be seen in Fig. 15.4 which is a closeup of the large spot group of Fig. 15.1. Second are the faculae, areas in which the photosphere appears to be especially bright. Faculae are usually seen near sunspots or groups of spots. Third is limb darkening. The edge of the apparent disk of the sun is referred to as the limb, and the brightness of the sun's surface appears to decrease as one looks from the center of the disk toward the limb. This, too, can be seen in Fig. 15.1. Limb darkening results from the fact that we are looking into hot gas when we look at the sun. Our view is blocked at some distance down into the gas, and the length of this path is nearly the same whether we are looking at the center of the disk or near the limb. The point at which the view is blocked is deeper inside the sun at the center of the disk than at the edges, and this is illustrated in Fig. 15.5. At this lower depth the gas is hotter and, therefore, brighter.

Galileo was the first person to make serious studies of sunspots in the course of his early applications of the telescope. This was in 1610. For a few years after Galileo's active years other observers also recorded the appearance and disappearance of spots. Then beginning in 1645 sunspots became rare indeed. From 1645 until 1710 almost no spots were seen. The few spots that did appear were so rare that they were carefully recorded. After 1715 spots were again seen on a more regular basis, and were no longer unusual sights. Even with modest equipment an observer today is far better equipped than Galileo and his immediate successors were. With practice and patience an observer can find pleasure and satisfaction in recording the development of sunpots.

Sunspot cycle

In the early years of the nineteenth century a German druggist named Heinrich Schwabe began to make regular observations of the sun and to record the locations of any sunspots that he saw. His observations

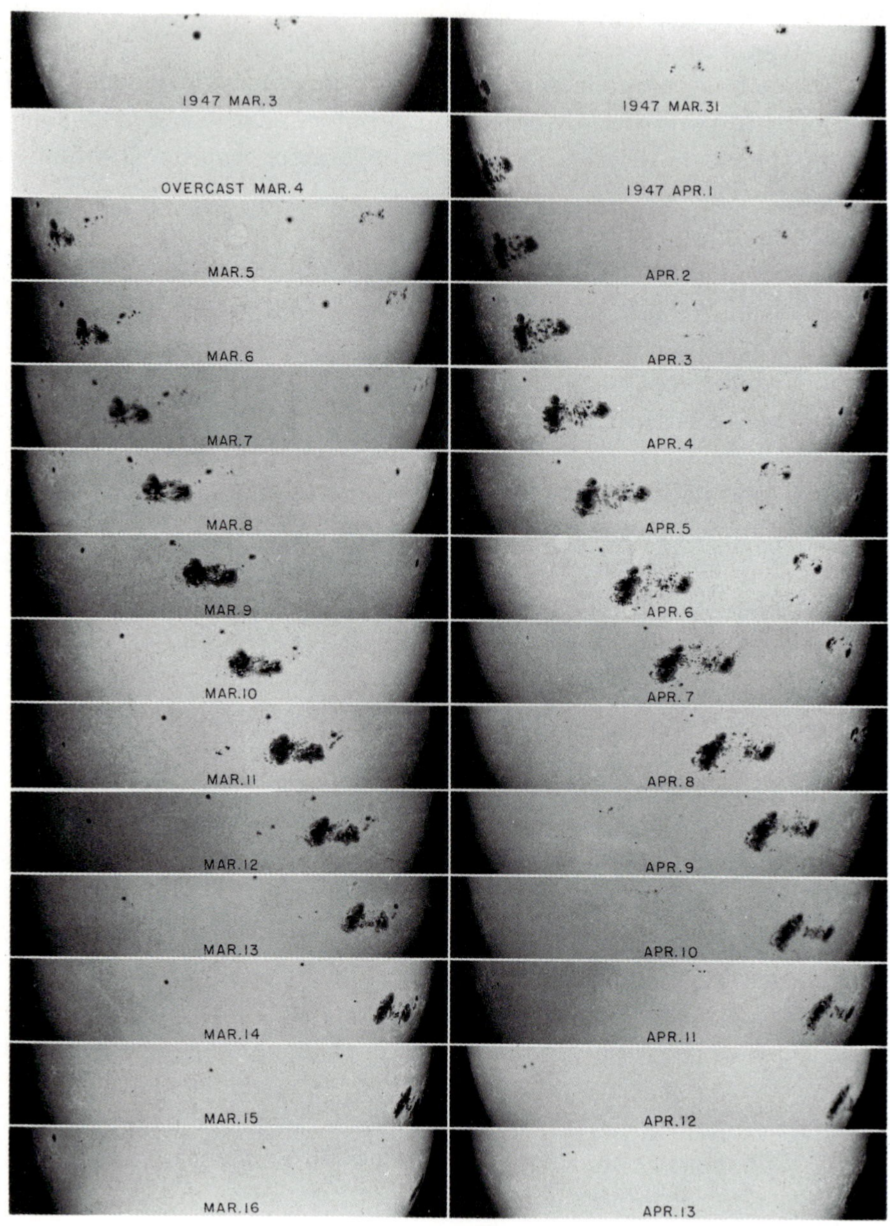

Fig. 15.3. A series of photographs in which the rotation of the sun is shown by the apparent motion of sunspots. (The Observatories of the Carnegie Institution of Washington.)

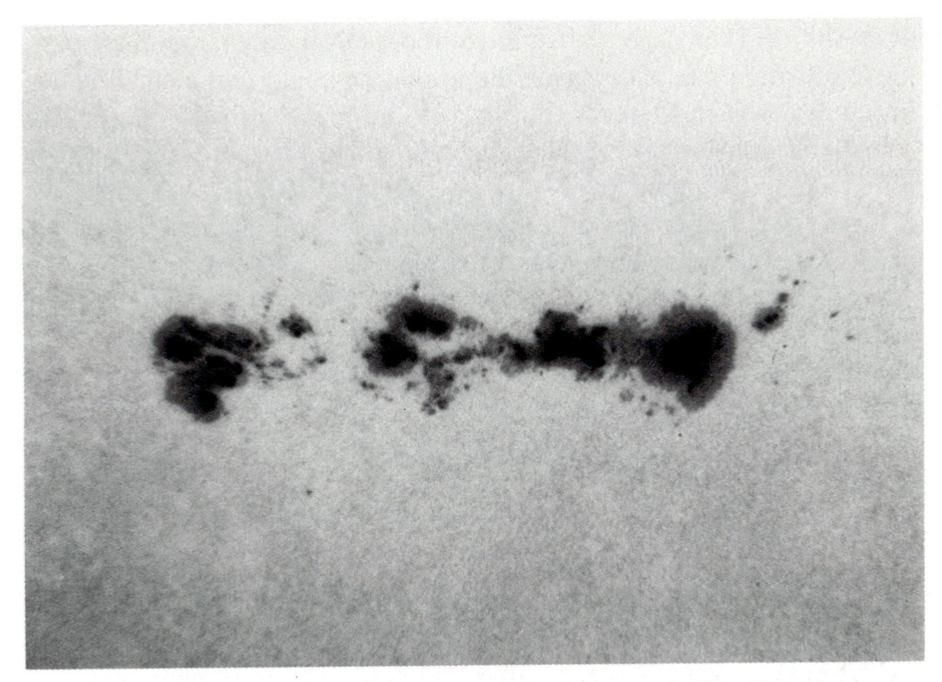

Fig. 15.4. An enlarged view of the group of sunspots in Fig. 15.1. Note especially the granulation in the areas outside of the group. (Courtesy of Gregory Terrance, Lima, New York.)

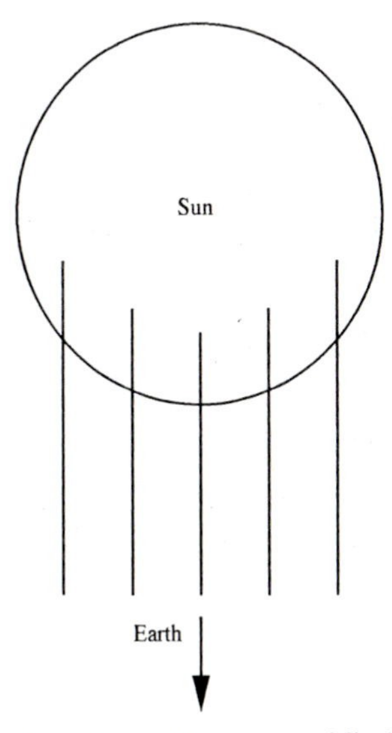

Fig. 15.5. The cause of limb darkening.

at his shop in Dessau continued for considerably more than twenty years. His motivation was actually the hope that he could find a small planet inside the orbit of Mercury, but by 1843 he realized that he had found something even more exciting. His data showed a nearly-regular variation in counted numbers of sunspots from year to year. The cycle that Schwabe discovered is now recognized to have a period of roughly eleven years, but cycles have been as short as nine years and as long as fourteen years. A contemporary of Schwabe, Rudolph Wolf in Switzerland, then went back through previous data and showed that the cycle could actually be traced all the way back to 1715. Wolf developed a uniform method for recording the number of spots seen each year, and his system has been in use ever since. The cycle continues much as it did when discovered by Schwabe.

Photographing the sun

Many observers will wish to accept the challenge of photographing the sun, and this can be done with fairly simple equipment. In principle this should be a straightforward procedure, but the very fact of the sun's extreme brightness makes it different from photographing a field of stars. There are, however, a few accessories that can simplify and improve the operation considerably. First, we must find some way of rejecting a large fraction of the solar energy reaching the telescope, and the best way to this is by means of a special solar filter placed over the objective lens or over the open end of the tube if the telescope is a reflector. Filters of this type are sometimes made by depositing a thin metallic film on a disc of high-quality glass or on a Mylar sheet. The filter should diminish all wavelengths equally, or, in other words, it should be a neutral-density filter. Solar filters can be purchased from suppliers of telescopes and accessories, and these have been carefully designed not just to dim the sun but to block harmful wavelengths in the ultraviolet and infrared as well. It is also possible to reduce the brightness of the solar image by means of a mask that covers a portion of the aperture. This makes it possible to use a smaller filter with a possible saving in cost. One should keep in mind, however, that in reducing the aperture with a mask, one is also reducing the theoretical resolving power of the telescope. In reality, resolution in the final photographic image will probably be limited by seeing conditions.

A small filter placed near the focal plane can be less expensive than one that covers the full aperture, but then one must face problems caused by heat inside the telescope. Especially in a reflector heat can cause turbulence inside the tube. As we have said earlier, heat concentrated by the objective can cause damage in any telescope.

Even with the sun's image greatly reduced in brightness the exposure time for a photograph is going to be very short, so a second important accessory is a reliable shutter. The most rational way of providing this is to mount a thirty-five millimeter camera in place of the eyepiece. The shutter in such a camera permits exposure times as short as 1/500 or 1/1000 second, and for two reasons short exposures will be necessary even with a very-high-density filter in front of the lens. First, it will minimize effects of vibration introduced by the operation of the shutter itself. Second, it will minimize blurring due to atmospheric and internal turbulence. A sturdy mounting and a cable release on the shutter should also be part of the equipment. One should keep in mind that if the focal length of the objective is too long, the image of the sun will overflow the twenty-four millimeter limits of the thirty-five millimeter format. Since the sun's angular diameter is just over thirty arcmin, the limit on focal length is 2750 mm ($\approx$108 in).

The choice of an emulsion type can in this case be made on the basis of resolution rather than speed. The flexibility offered through a selection of shutter speeds means that very slow films may be used. High contrast in the finished prints will show details of the photosphere to the best advantage, so this should govern the choice of film and processing. Kodak Technical Pan 2415 has been used with success by many observers.

Finally, the mid-morning is likely to be the best time of day for solar photographs. By then the sun is high in the sky, and turbulence due to air rising from the warm earth has not reached its maximum.

Spectroheliograph

By the end of the nineteenth century astronomers had begun to recognize that detailed photographs and high-dispersion spectra of the sun could be keys to the understanding of the physical properties of stars in general. Thus, in the early years of the twentieth century there was a movement to develop specialized telescopes devoted only to the study of the sun. One of the great figures in this effort was George Ellery Hale who began his work at the University of Chicago and later moved to California where he founded a solar observatory on Mount Wilson. It was this observatory that eventually developed into the Mt. Wilson Observatory.

One of Hale's early contributions was the development of the spectroheliograph, an instrument which could produce an image of the sun in the light of a single narrow spectral line. Lines in the spectrum appear dark in contrast to the adjacent regions, but in fact there is some light coming to us at the wavelengths of the apparently dark lines. Hale's instrument was essentially a filter which rejected all light except that at the wavelength of interest. Fig. 15.6 is an example of a photograph showing the sun just in the light of the Hα line. A schematic diagram of the spectroheliograph is shown in Fig. 15.7. At the top of the figure the sun is represented by the circle, and the entrance slit of a normal spectrograph is represented by the vertical line. In an ordinary application the spectrum would be formed from the light in just a narrow line across the image of the sun. Hale placed a second slit in the focal plane of his spectrograph and positioned it at the wavelength of some particular line, Hα, for example. A photographic film was placed behind the second slit. The entire assembly from entrance slit to exit slit was moveable in the direction of the dispersion, but the solar image and the film were stationary. When the entrance slit was moved across the image of the sun, an image of the sun

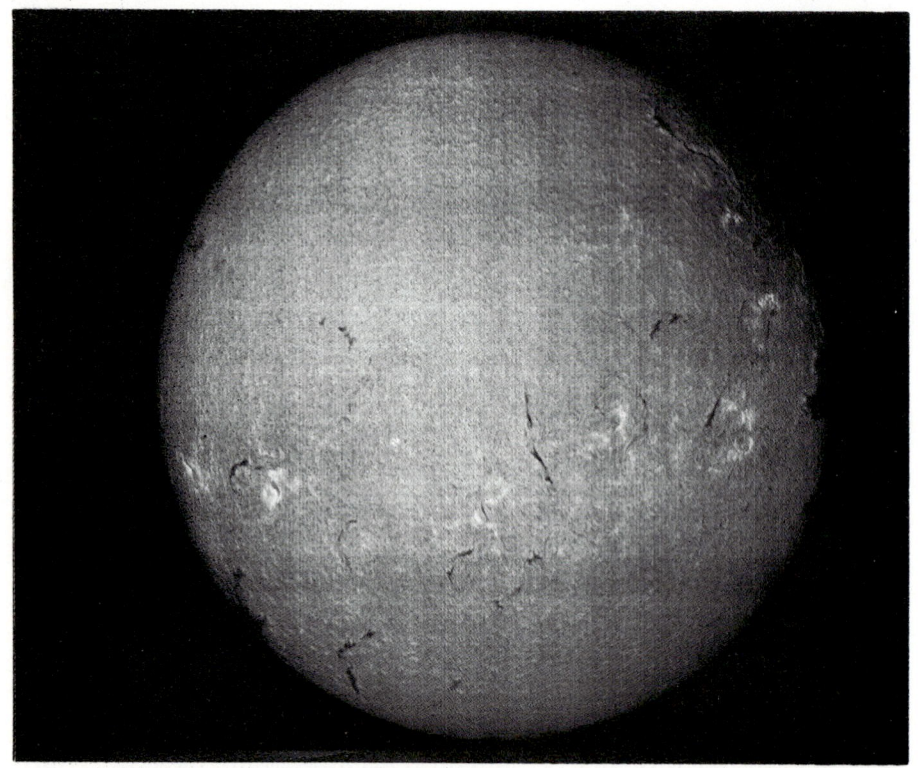

Fig. 15.6. The sun in the light of the Hα line at 6563 Å. (Manila Observatory photograph.)

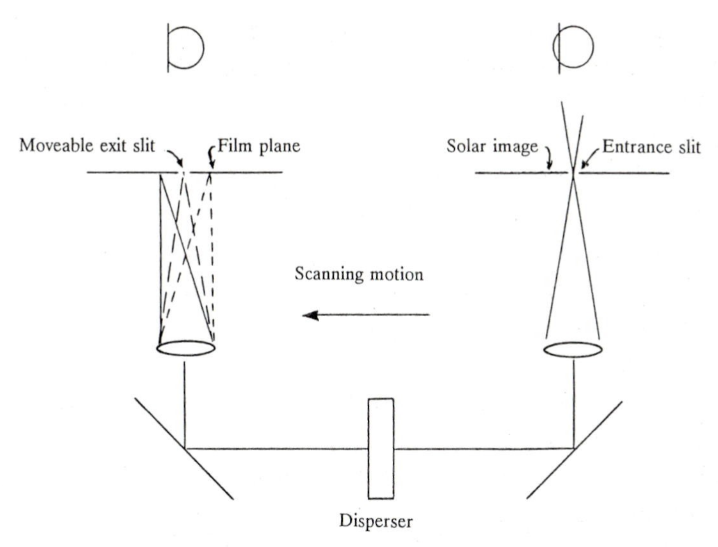

Fig. 15.7. A schematic diagram of the spectroheliograph.

could be built up, strip by strip, on the photograph. The position of the exit slit could be adjusted to produce a solar image at any wavelength, but the most commonly used wavelengths are 6563 Å (Hα) and 3933 Å (K line).

What the spectroheliograms really show is the distribution of temperatures in the chromosphere. In the Hα photograph of Fig. 15.6 the light regions are places where the hydrogen is hottest and are known as plages. The irregular, black lines are places where the hydrogen is coolest, and they are called filaments.

In more recent times the work of the spectroheliograph has been done with narrow-band filters which are essentially the same as those described in Chapter 10 for the Hβ photometric system. Fig. 15.8 shows an image of the sun produced with a filter which transmits only at the wavelength of the K line of ionized calcium. Large K-line filters can cover the entire objective of a telescope and make it possible to view the sun directly in this wavelength.

Solar telescopes

Hale and his contemporaries understood much of what we now know about astronomical 'seeing' as described in Chapter 6. They knew that the air was likely to be most turbulent close to the ground and that turbulence would increase during the day as the ground became warmer.

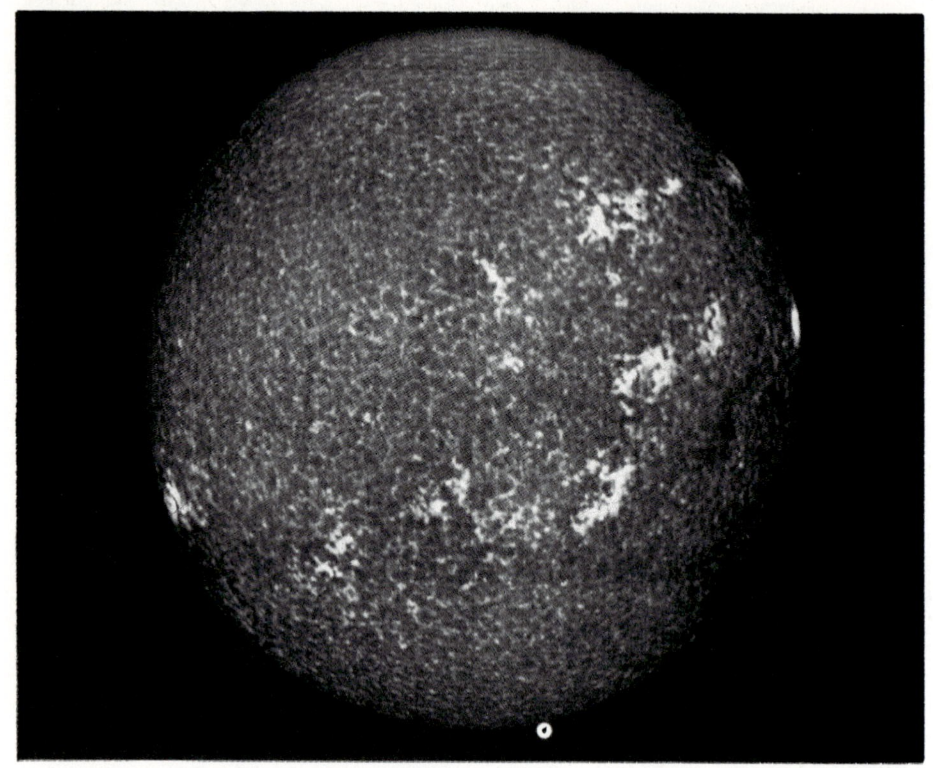

Fig. 15.8. The sun photographed in the light of the K line of ionized calcium. (Manila Observatory photograph.)

The air was also more stable if the area surrounding the telescopes was covered with trees. In working toward the most stable possible conditions and the best solar images, Hale developed the tower telescope. Two tower telescopes were built at Mt. Wilson. The first was sixty feet tall, and the larger had a height of one hundred and fifty feet. Mirrors at the top of the tower were arranged to track the sun and to reflect sunlight down through a tube which was insulated to reduce turbulence along the light path. In some configurations a lens near the top of the tower formed an image of the sun at ground level. Buried in the ground was a vertical tank which could be evacuated. This tank contained the spectrograph.

When the Kitt Peak National Observatory was being planned in the late 1950s, it was decided that an advanced and versatile solar telescope should be included. The resulting tower telescope went into operation in 1962, and was named in honor of Robert R. McMath, a solar astronomer from Michigan. The McMath telescope is pictured in Fig. 15.9. A flat, sun-tracking mirror at the top of the tower reflects sunlight downward to

Fig. 15.9. The McMath Solar Telescope at Kitt Peak station of the National Solar Observatory. The mirrors of the heliostat at the top of the tower reflect light down the long sloping section to instrument rooms below the level of the ground. (National Optical Astronomy Observatories.)

Fig. 15.10. An image of the sun at the prime focus of the McMath Solar Telescope. (National Optical Astronomy Observatories.)

image-forming optics. Here, however, the downward path of the light is through a sloping structure parallel to the earth's axis. This eliminates one reflective surface with its subsequent losses and cost. The sloping tube is 136 meters (450 feet) long, and at the bottom of it there is a concave mirror with a focal length of 82.5 meters (270 feet) which reflects the light back up towards the top of the tower. Another flat mirror intercepts the beam and reflects the light vertically downward to a large table on which an image of the sun is formed. This image is 73 centimeters in diameter and is so bright that the scientists studying it must wear very dense dark glasses.

Adjacent to the focal plane of the McMath telescope are a number of instruments including a spectrograph, a spectrohelioscope and equipment for studying the magnetic fields on the sun.

Eclipse observations

An eclipse of the sun can provide some of the most exciting and interesting of all possible astronomical observations, and the observer planning to observe eclipses can find predictions in the *Astronomical Almanac*. For some time after the moon first begins to come between the earth and the sun the observer is hardly aware of any changes at all, but as the time of totality comes closer, the sky begins to darken just as it does at twilight on a normal day. Birds begin to sing and chickens go to roost. Crescent-shaped images of the sun may be seen on the ground or on walls wherever a small opening can act as a pinhole camera. This can be anything from spaces between leaves on a tree to a hole punched in a card. As totality begins the sky becomes quite dark, and one can look skyward to see the corona extending beyond the edge of the moon's disc. This view lasts only about seven minutes in the most favorable cases, but it is one that will never be forgotten. The thrill of those moments of totality is well worth the effort to find a site on the path of totality and the risks of cloudy skies. Eclipses are such inspiring sights that they have been recorded in all parts of the world for thousands of years. Historians and archaeologists have used these records to relate calendars from different cultures to each other and to establish the dates of important events.

Astronomers still look for sites with the best prospects for clear skies, and they travel wherever necessary to make several types of eclipse observations. The first of these are direct photographs intended to show as much detail as possible in the image of the corona. Telescopes for this purpose can be conventional astronomical telescopes equipped with suitable means for holding films or plates. The combination of focal length and size of the plates must be such that the entire corona can be recorded, and the angular diameter of this may be as large as one and a half to two degrees. Even today in a world of very advanced scientific equipment the corona can best be seen and photographed only during a total eclipse. A spectacular view of the corona is shown here in Fig. 15.11.

The second type of observations that are often carried out are observations of the spectrum either just before totality or just after it. At these moments only a small arc of the sun's surface is still visible, and since this arc is narrow, it can serve as the light source for the spectrograph. A conventional entrance slit is not needed, but the 'lines' which appear in the spectrum are arcs just as the source was an arc. Fig. 15.12 is an example of this type of spectrogram, and it can provide some important details. A spectrum photographed as shown here is described as a 'flash

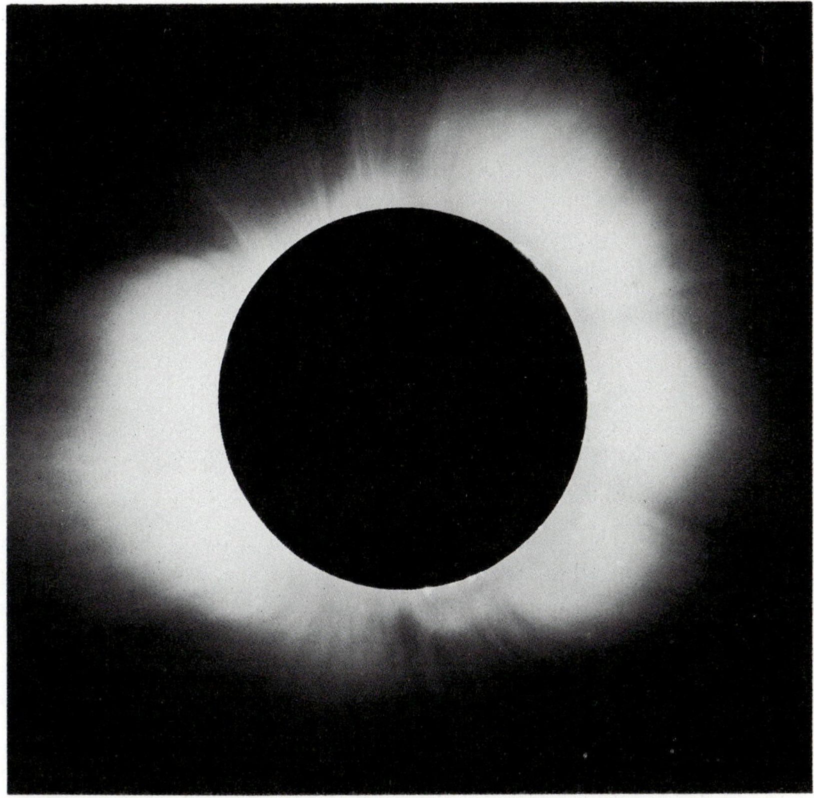

Fig. 15.11. Direct photograph of the solar corona photographed at the eclipse of June 8, 1918. (The Observatories of the Carnegie Institution of Washington.)

spectrum' because of the fact that the suitably narrow arc of the chromosphere persists for only a very short time. The flash spectrum is an emission line spectrum because the gases from which the light comes are hot.

It is interesting to take a close look at the flash spectrum, for some important points are easily recognized. If one recalls the discussion of spectra in Chapter 12, then one should be able to identify five of the curved spectrum lines in Fig. 15.12. The strong line near the right side of the photograph is Hβ; the next strong one toward the left is Hγ; then Hε. The two equally strong lines which show arcs longer than those of the hydrogen lines are the H and K lines of ionized calcium (CaII). To the right of H and K one can see a number of other weak lines. Some of these are hydrogen lines which continue the converging pattern of lines, the Balmer series, that is the dominant feature in spectra of type AO (see Fig. 12.3). The longer arcs in Fig. 15.12 indicate that light at those wavelengths comes to us from higher levels in the solar atmosphere. This is not an indication of a distribution of chemical elements with depth in the atmosphere. Rather, it is an indication of the changing conditions of excitation and ionization with depth.

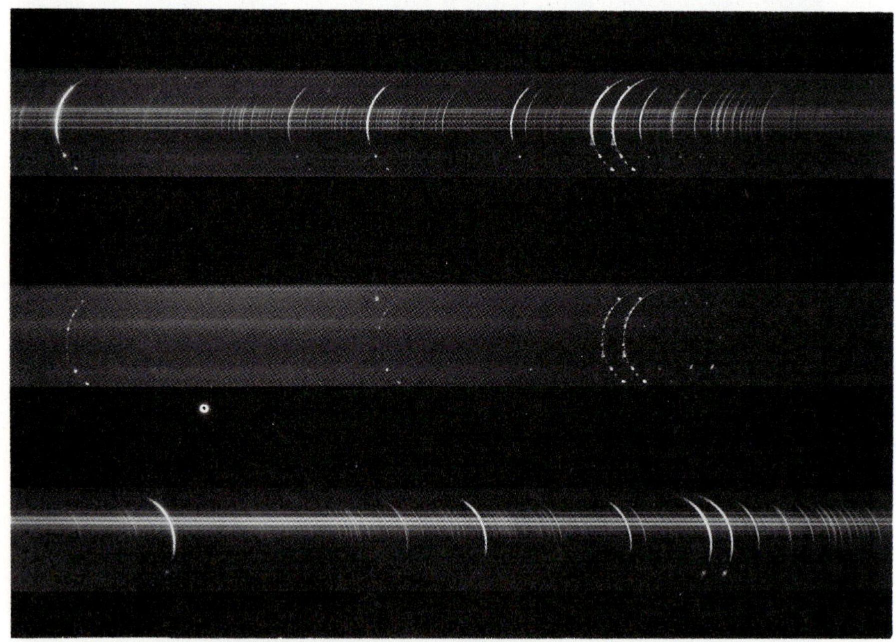

Fig. 15.12. The flash spectrum recorded at the eclipse of January 25, 1925. Wavelength increases toward the left. The pair of strong lines are the H and K lines of CaII, and lines of the Balmer series can easily be identified. (The Observatories of the Carnegie Institution of Washington.)

Coronagraph

With so many special features visible during solar eclipses, astronomers were understandably anxious to try to develop some sort of instrument which would permit study of the corona and chromosphere outside of eclipses. This was a difficult task, because the uneclipsed sun is about a million times brighter than the corona. The problem was at least partially solved in 1931 when Bernard Lyot in France began to use his version of the coronagraph. Lyot had learned a few lessons from Hale and others who had attempted to solve the same problem, and he went to great lengths to eliminate all sources of scattered light. To eliminate light scattered in the earth's atmosphere Lyot placed his instrument on a high peak in the Pyrenees mountains of southern France. To eliminate light scattered inside the coronagraph he was meticulous in his efforts to keep dust particles and scratches off of all optical surfaces. When atmospheric conditions were especially good, Lyot was able to study the inner, brighter parts of the corona. The outer corona could never be seen outside of eclipses until astronauts aboard Skylab in 1973 were able to make similar observations.

The principles of the coronagraph may be seen in Fig. 15.13. Light enters through the objective lens on the left, and an image is formed at the focal plane. A shiny cone with its base the same diameter as the image of the sun is located so that its base is in the focal plane. The image of the sun is then reflected to the inside surface of the tube where it is absorbed. The image of the chromosphere and corona are left undisturbed in the focal plane.

A second lens is situated just behind the original focal plane, and this lens forms an image of the objective at the plane of a diaphragm. The function of the diaphragm is to block light that is diffracted by the edges of the objective. The final element might be described as the camera lens, and it forms an image of the chromosphere and inner corona on a photo-

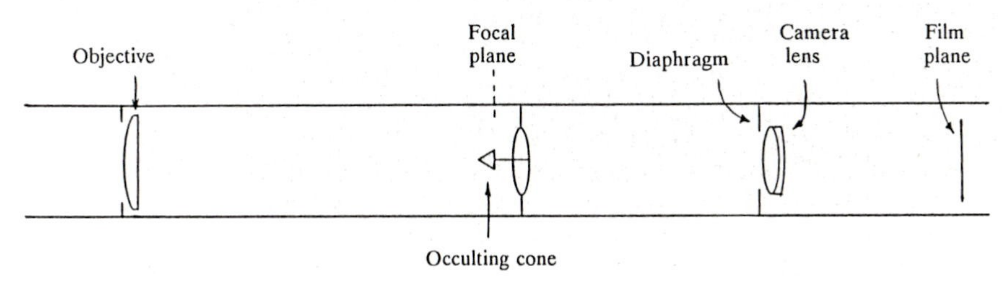

Fig. 15.13. The arrangement of parts in the Lyot coronagraph.

graphic plate. Only a small number of coronagraphs were ever built, but they were placed around the world so that a continuous watch on the sun could be maintained.

Some of the lenses in this system are not corrected for chromatic aberration, so a filter is used to transmit only the color that has been brought into focus at the base of the cone.

Even though the faint corona eluded him, Lyot was able to photograph, on a routine basis, the prominences, clouds of hot gas which are frequently seen above the sun's limb. Evidence of prominences may be seen as the bead-like spots of light in the H and K lines in the center panel of Fig. 15.12. A more detailed view of a solar prominence is shown in Fig. 15.14. Spectacular motion pictures showing motion of gases in prominences have been made by combining a series of coronograph pictures taken over a period of hours or even days.

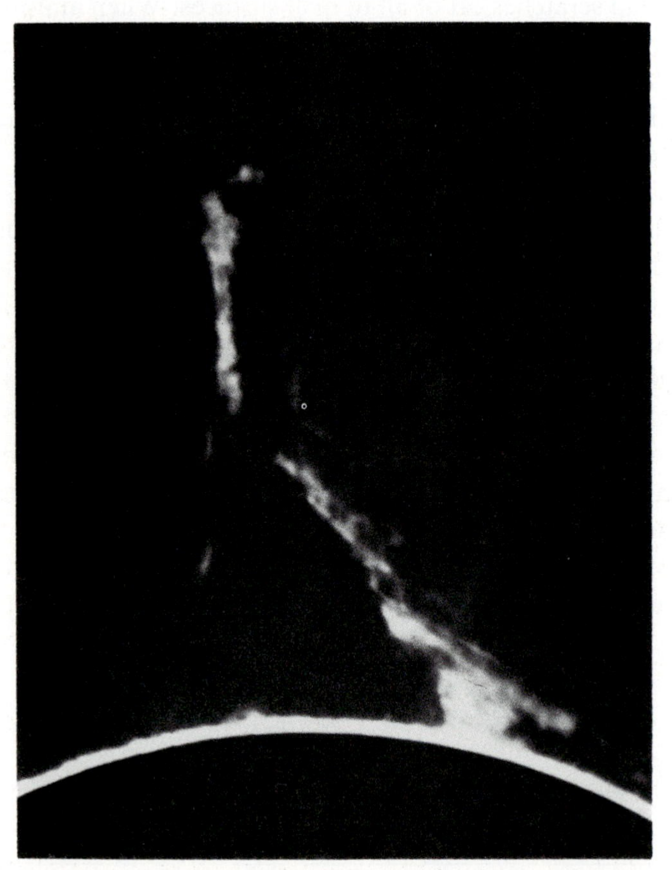

Fig. 15.14. Solar prominences photographed at the wavelength of the K line. This prominence extended 570 000 miles above the sun's surface. (The Observatories of the Carnegie Institution of Washington.)

High-altitude observations

Solar astronomers were fully aware of the effects that the earth's atmosphere had on their observations even as early as the years of Hale's work at Yerkes Observatory. Turbulence in our atmosphere blurs the details that should be resolvable in images of the sun, and absorption by atoms and molelcules places limits on our observable range of wavelengths. The solution was obvious: get the telescopes above the atmosphere. The advantages of such a possibility were outlined by the Indian physicist, M. Saha in 1937, and limited experiments using balloons had been performed even before that. Not surprisingly for that time, major problems prevented development of observations above even a large fraction of the atmosphere. These problems may be summarized as problems of technology and of commitment. The equipment for controlling a balloon-borne telescope and retrieving information from it would have been too bulky and too heavy until the age of micro-electronics began in the 1960s. And second, the amounts of money required would have been even beyond all expectation until governments in many countries saw that certain kinds of projects were so important that they should receive national support. Fortunately, projects involving the sun received adequate funding in the age of 'big science'.

The long sequence of steps leading to observations that were truly outside of the earth's atmosphere began in the late 1950s with the Stratoscope project organized and operated by scientists at Princeton University. A telescope with an aperture of 25 cm was carried to an altitude of 24 kilometers by a large balloon. At this altitude the telescope could hang quite steadily since it was above all winds and above ninety-six per cent of our atmosphere. The photograph in Fig. 15.15 was returned from Stratoscope and shows clearly the structure which causes the penumbral region of a sunspot to be so different from the umbra. These photographs also gave us clear pictures of the granulation. Even at the altitudes of Stratoscope molecules in the thin atmosphere above the telescope scatter so much light that the brightness of the sky near the limb of the sun is too great for the outer corona to be seen or photographed with a coronagraph.

During the 1960s and 1970s observations were made from considerably higher altitudes. NASA mounted instruments for solar studies in rockets which went to altitudes ten times higher than those achieved by the balloons. The rockets are only at their peak altitudes for brief periods of time, so there is not any flexibility in their programs of observations.

Nevertheless, the rockets showed that prolonged observations from even higher altitudes would be valuable.

A series of orbiting observatories were put into orbit by NASA beginning in the late 1960s and ending with Orbiting Solar Observatory 8 (OSO-8) in 1973. These were small satellites that were in orbits about 550 km above the earth's surface. They carried equipment for making observations in the ultraviolet and x-ray regions, and they produced spectacular pictures of the x-ray emission which originates in the corona.

A major advance in observations from above the atmosphere began in May, 1973 with the launch of Skylab, a large spacecraft equipped with the Apollo Telescope Mount. The first of three crews of astronauts entered the spacecraft eleven days later and repaired some serious

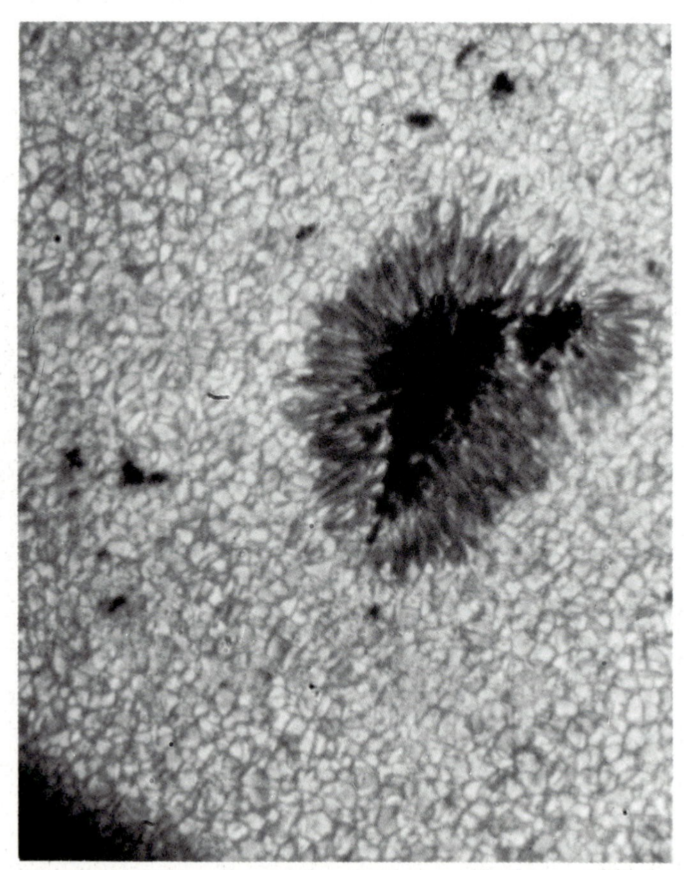

Fig. 15.15. The surface of the sun photographed at an altitude of 24 kilometers from the Stratoscope balloon-borne telescope. This famous photograph gave us a clear picture of granulation and of the structure in the penumbral region of a sunspot. (Project Stratoscope photograph, Princeton University. Supported by NASA, NSF and ONR.)

damage sustained during the launch. The three crews manned the station for a total of nine months, and the entire project was an unqualified success. Extensive records of the sun were obtained in the x-ray and ultraviolet regions, and for the first time observations of the corona were made outside of solar eclipses. A typical photograph of the sun at x-ray wavelengths is shown in Fig. 15.16. An excellent summary of the entire Skylab mission is found in John Eddy's book, *A New Sun*, which is listed among the references at the end of this chapter.

Tracking the sun

Observers who are planning to make regular observations of sunspots may wonder about the correct rate at which their telescope's drive should run when tracking the sun. It is not difficult to see what the correct solar rate should be. We recall that the solar day is about four minutes longer than the sidereal day. Therefore, the drive should run more slowly when tracking the sun than when tracking the stars. Since there are 1440 solar minutes in a solar day, there will be 1436 solar minutes in one sidereal day. The ratio of these two is 359/360. In order to

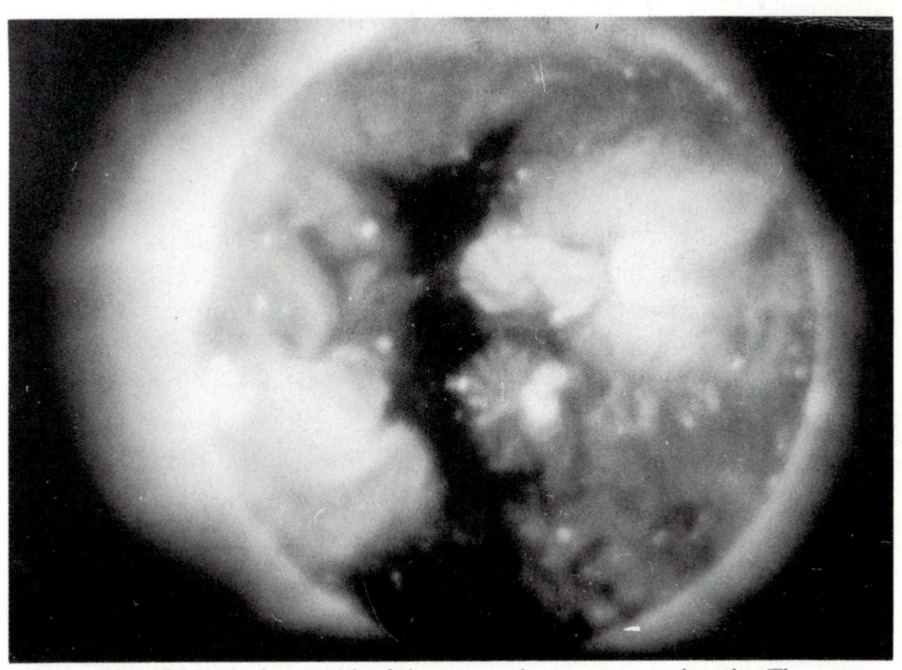

Fig. 15.16. A photograph of the sun made at x-ray wavelengths. The camera was mounted in a rocket which carried it above the earth's atmosphere. (NASA.)

know how the drive rate of a telescope is affected by the above ratio, we must digress briefly to describe the actual motors that are used on most modern telescopes.

The synchronous motor is one in which the rotational speed of the motor depends upon the frequency of the alternating voltage which is supplied to it. Many parts of the world use systems which operate at 60 hertz (60 cycles per second), and some use systems which operate at 50 hertz. Synchronous motors are simple and reliable, so it is not surprising that they are widely used in all kinds of applications. They are suitable for use in small telescopes if the user understands one limitation. Electric power companies try very hard to be sure that AC power at exactly 60 hertz is supplied to the distribution lines at all times. They succeed most of the time. When usage of electric power is especially heavy, however, the load may cause the generators to slow down very slightly. The frequency may drop below 60 hertz for short periods. The power companies make sure that the average frequency in a twenty-four hour period is exactly 60 hertz, so they speed up the generators when the demand has lightened. For this reason, clocks – and telescopes – driven by synchronous motors may run a little bit slow or fast from time to time.

To return now to the rate for tracking the sun, we can express the solar rate in terms of the 60 hertz AC power supplied to the telescope. If we let X be the frequency for tracking the sun, we may say

$$X/60 = 359/360$$
$$X = 59.833333 \text{ hertz}$$

This means that we may track the sun precisely by reducing the frequency by $\frac{1}{6}$ hertz. This is so small a difference that most observers need not bother with it. If the field of an eyepiece was one half degree in diameter, it would take the sun six hours to move halfway out of the field.

In large installations in which precise tracking at both the sidereal and solar rates is necessary, a precision AC power supply is usually provided. This is an electronic unit in which the output frequency is controlled by a crystal oscillator regardless of the input frequency. A variable-frequency power supply can be used to provide the precise solar rate.

Heliostat and coelostat

Applications of solar telescopes are such that it is often useful to have the telescope in a fixed position and reflect sunlight into it. Such arrangements were described above in the cases of the Mt. Wilson tower telescopes and the McMath telescope. Two types of mirror systems have been frequently used for this. In the first, the heliostat, two mirrors can be arranged as shown in Fig. 15.17. The north point of the horizon is to the right. Mirror A is equatorially mounted, and is driven westward at the solar rate. Mirror B is fixed in position, and in this case reflects sunlight into the objective of a vertical telescope. The tilt of Mirror A must be adjusted for the declination of the sun from day to day. This arrangement has the advantage of simplicity, but it has the disadvantage that the solar image rotates in the focal plane of the objective. Furthermore, the rotation is not at a uniform rate. For visual observation of sunspots this is not a problem.

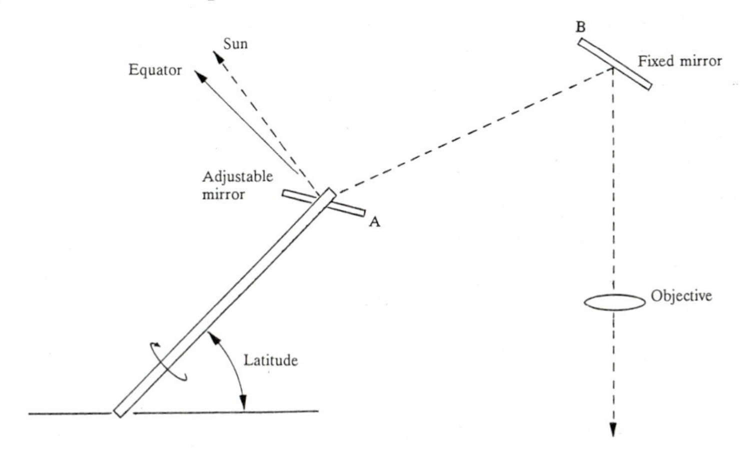

Fig. 15.17. The mirrors and light path in the heliostat.

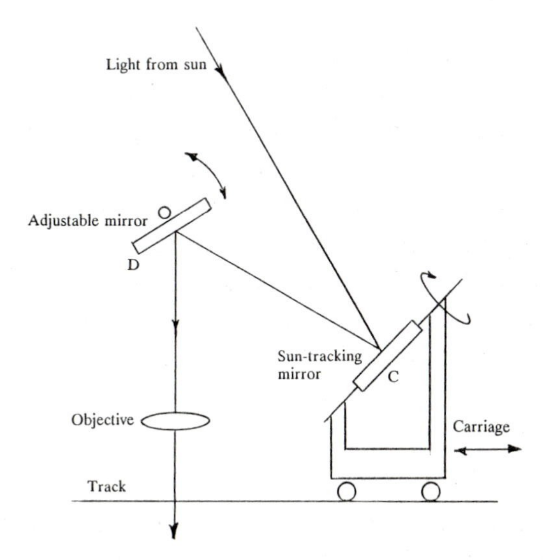

Fig. 15.18. The arrangement of the parts in the coelostat. In this sketch the carriage would have to be either in front of or behind the plane of the page. The adjustable mirror must move east and west as well as north and south.

In the second arrangement, the coelostat, the mirrors are arranged as in Fig. 15.18. Mirror C is fixed to a polar axis and rotates at one half of the solar rate to follow the sun. The tilt of Mirror D must be changed whenever the declination of the source is changed, and, in addition, Mirror C must be moved with respect to Mirror D. When the sun or other source is on or near the meridian, there is the chance that the shadow of Mirror D will fall on Mirror C. To avoid this, Mirror C must be moved to the east of Mirror D in the morning and to the west in the afternoon. The extra complexity of the coelostat over the heliostat is offset by the fact that the image in the focal plane of the objective does not rotate.

QUESTIONS FOR REVIEW

1. Why must the observer always exercise great care when making observations of the sun?
2. Imagine that you wish to equip a six-inch telescope with a screen for observing sunspots. The focal length of the telescope is seventy-five inches, and you want the diameter of the sun's image to be eight inches. Where should the screen be placed? Where should the eyepiece be placed?
3. What kinds of features might you try to look at on the white-light image of the sun?
4. What special problems are encountered when one attempts to photograph the surface of the sun?
5. Why is it so difficult to photograph the solar corona? Under what circumstances can the corona be seen and photographed?
6. What can be learned from photographs of the flash spectrum? How is the flash spectrum best photographed?
7. Imagine that a solar image projected on a screen is fourteen inches in diameter. A sunspot on the image has a diameter of 0.72 in. What is the diameter of the spot in miles or kilometers?
8. Calculate the orbital speeds of the earth and the moon. Then calculate the time that it will take for the moon's shadow to move across the earth. Finally, calculate the duration of a solar eclipse for an observer on the earth. Assume that the moon's orbit is circular and that the moon's shadow moves along the earth's equator.
9. Using some of the data from the previous questions, calculate the maximum duration of the umbral portion of a lunar eclipse.
10. Make a sketch of a coronagraph and describe the function of each of its parts.

Further reading

Eddy, J. (1979). *A New Sun, The Solar Results from Skylab*. National Aeronautics and Space Administration. The solar experiments aboard Skylab are described with many excellent illustrations.

Kitchin, C. R. (1984). *Astrophysical Techniques*, Adam Hilger Ltd. For other methods of observing the sun, see the material which begins on page 354.

Miczaika, G. R. and Sinton, W. M. (1961). *Tools of the Astronomer*. Harvard University Press. Observations of the sun are well treated in Chapter 7.

Muller, R. (1975). The Sun. In *Astronomy: A Handbook*, ed. G. D. Roth, p. 221. Sky Publishing Co. This is Chapter 9 in Roth's book. In addition to notes on observational methods, Muller includes methods of specifying coordinates on the sun.

Noyes, R. W. (1982). *The Sun, Our Star*. Harvard University Press. This comprehensive book contains more material on methods of observing the sun. See especially Chapter 2.

Pannekoek, A. (1961). *A History of Astronomy*. Interscience Publishers. In Chapter 37 Pannekoek recounts the progress of observations of the sun and of theories of the sun.

Petri, W. (1975). Observation of Total Solar Eclipses. In *Astronomy: A Handbook*, ed. G. D. Roth, p. 255. Sky Publishing Co. This material, Chapter 10, will tell you how to observe a solar eclipse and what to look for.

16

Radio astronomy

In the 1950s radio astronomy became the most exciting area of research in all of astronomy. Astronomers who for years had made their observations in conventional ways were suddenly face to face with some new aspects of physics and new types of telescopes that did not look at all like the optical ones. Radio radiation from the sky and especially from the Milky Way had been discovered in 1932 by Karl Jansky and confirmed in 1939 by Grote Reber, and the accomplishments of these two men will be described in more detail at the end of this chapter. Interest grew slowly at first, but spread rapidly after 1950. Two factors were at least partially responsible for the broadened interest. One was the refinement of electronic equipment during the 1940s, and the other was the discovery in 1951 of an emission line from atomic hydrogen at a radio wavelength.

The radio portion of the electromagnetic spectrum begins with wavelengths of a few millimeters and extends to wavelengths measured in meters. Among those who regularly work with radio it is conventional to define specific portions of the radio spectrum in terms of frequency rather than wavelength. As for all electromagnetic radiation, the relation between frequency, f, and wavelength, λ, is

$$\lambda = c/f$$

where c is the velocity of light. In Table 16.1 we list the ranges of

Table 16.1. *The radio bands*

Band	Frequency	Wavelength
AM radio	55 to 160 kHz	5454 m to 1875 m
Shortwave radio	3 to 30 MHz	100 m to 10 m
FM and TV (vhf)	30 to 300 MHz	10 m to 1 m
(uhf)	300 to 3000 MHz	1 m to 10 cm
	3000 to 30 000 MHz	10 cm to 1 cm

frequency and wavelength for the conventional radio bands. The frequencies which are of interest in ground-based radio astronomy are limited by our atmosphere and range from 30 MHz to 30 000 MHz (10 cm to 1 cm). Radar used by the police, aircraft and the military operates in the centimeter range.

Sources of radio waves

Today we understand very clearly several processes by which radio waves are produced in space and are, therefore, of interest in astronomy. First, there is thermal or black-body radiation which comes from hot or warm objects such as the sun, the planets or the cosmic background. This is continuous radiation, and the relation between intensity and wavelength is described by Planck's Law just as it is in the optical wavelengths. Second, there is line radiation which comes from transitions within atoms or molecules. This type of radiation must come from clouds of gas in space and can be detected both in emission and absorption. Third, there is synchroton radiation which is emitted from charged particles moving at high velocities in magnetic fields. One of the important tasks of the radio astronomer is to examine the radio radiation as it is received from the sky and try to deduce the conditions under which the radiation was emitted. The temperatures, compositions and distances (in some cases) can be determined. For extended sources the size and distribution of brightness can be mapped.

Radio telescopes

A large radio telescope is pictured in Fig. 16.1, and at first glance it bears no resemblance to a conventional telescope. When we examine Fig. 16.2, however, we see that, in fact, the radio telescope functions in much the same way as the optical reflector. Radiation is collected by a paraboloidal reflecting surface, or dish, reflected from a secondary surface, and brought to a Cassegrain focus[1] at some point near the primary reflecting surface. The receiver is placed at this focus in order to receive the incoming signal. In the earliest radio telescopes the receiver was placed at the prime focus.

[1] In the seventeenth century a French physician, N. Cassegrain, devised the folded optical system which bears his name. After entering the telescope, light is reflected from a concave primary mirror to a convex secondary. The light then passes through a hole in the primary to form an image at the Cassegrain focus behind the primary.

Fig. 16.1. The 140-foot diameter radio telescope of the National Radio Astronomy Observatory at Green Bank, West Virginia, USA. (The National Radio Astronomy Observatory, operated by Associated Universities, Inc. under contract with the National Science Foundation.)

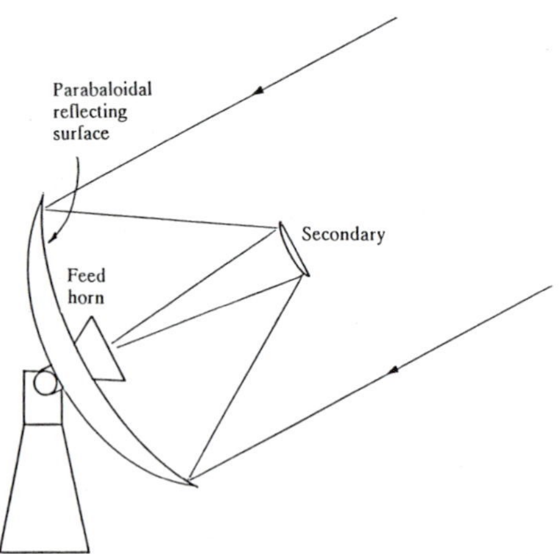

Fig. 16.2. Schematic diagram of a radio telescope.

The paraboloidal dish serves several important functions. First, it is a collector which reflects radiation from a beam equal to its own diameter to the receiver. Since we are always dealing with very weak signals, a reflector with a large diameter focuses large amounts of energy. Second, its focal length is short compared to its diameter, or in other words, it has a small focal ratio (about one). This means that the dish itself shields the detector from radiation that may be coming to it from the rear direction. Third, the dish provides a great improvement in angular resolution.

Just as in the optical case, the angular resolution depends upon the wavelength being used and the diameter of the reflecting surface. If we let a be the minimum angle in seconds of arc that can be resolved, we can show that

$$a = 2.1 \times 10^5 \times (\lambda/d)$$

where λ is the wavelength and d is the diameter. λ and d must be expressed in the same units. As an example we could calculate the angular resolution of a radio telescope with a diameter of forty meters when it is being used to detect the 21 cm line of neutral hydrogen. The result would be 17.2 minutes of arc. The resolution of the unaided eye is about 15 seconds of arc for comparison. As a means of improving angular resolution, astronomers have built a number of very large reflectors. The largest fully steerable one is operated by the Max Planck Institute for Radio Astronomy in the Effel mountains of Germany. Its diameter is one hundred meters. In the mountains of Puerto Rico there is a non-steerable reflector with a diameter of about three hundred meters. The receiver supported above it can be moved in the north–south direction to adjust to limited values of declination, and the earth's rotation sweeps it across the sky in right ascension.

The radio astronomer is also concerned with 'beam width', a term which describes the angular size of the portion of the sky from which the radio telescope can receive. This is very much like the field of view in an optical telescope. A major difference is that the radio telescope receives best from points directly in front of it, and its sensitivity falls off rapidly for other directions. This is illustrated in Fig. 16.3 by the solid line marked 'pattern'. As the angle between the axis and the direction to

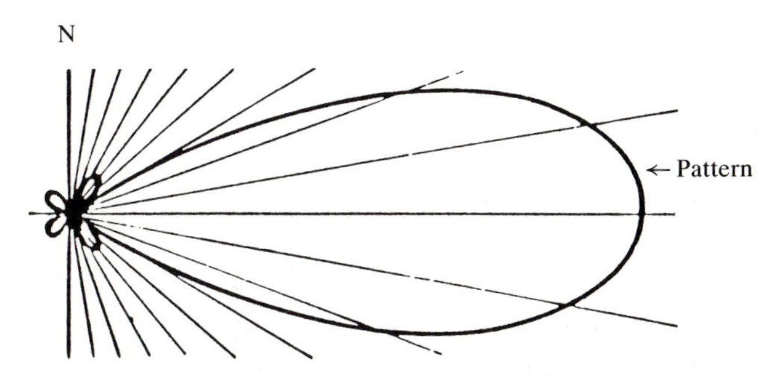

Fig. 16.3. The sensitivity pattern or beam of a radio telescope.

some point increases, the sensitivity decreases. For a point in the direction marked N the sensitivity would be zero. The small lobes pointing toward the side and toward the rear are called 'side lobes'. The main beam is usually symmetrical about the axis. It is customary to describe the angular size of an antenna beam or pattern in terms of its half-power beam width. This is the angle between the points at which the sensitivity is half of what it is along the axis.

Another point that has a counterpart in optical telescopes relates to the smoothness of the reflecting surface. It is impossible to generate an optical surface that is absolutely perfect, so there must be some criterion by which one can decide that a surface is good enough. That is, further improvement in the surface will not cause a noticeable improvement in the images. For optical surfaces this criterion is one-quarter wavelength, so the maker of a lens or a mirror tries to keep the finished surface within one-quarter wavelength of the ideal surface. At radio wavelengths a similar but more stringent standard applies. Those who build radio telescopes try to keep the reflecting surface to within one-tenth wavelength. This means that the reflecting surface of a radio telescope can have sizeable irregularities if it is being used at wavelengths of some tens of centimeters. It also means that the reflecting surface does not have to be solid. The surface may have holes in it or be made of a wire mesh. This reduces both the weight and the resistance of the antenna to winds.

Anyone who has stood near a large radio telescope is bound to be impressed by its size. Those who have not had this experience can try to imagine a structure as large as a football field that can be moved about to point at any part of the sky. The design of such a structure presents prodigious problems, and the construction is necessarily costly. As a result, the one-hundred-meter reflector mentioned earlier is likely to be the largest that will be built for many years.

Receivers and amplifiers

The receiver placed at the focus of a radio telescope bears only remote resemblance to the devices of the same name that most of us use every day. The signals that are being recorded by the radio astronomer are unbelievably weak, and so the receivers must be designed and built to the highest standards. The weak signal is fed to the receiver by means of an antenna which is usually in the form of a horn such as the one sketched in Fig. 16.4. At the small end of the horn the signals are reflected along a waveguide to a resonating cavity. Here the energy of the radio waves is

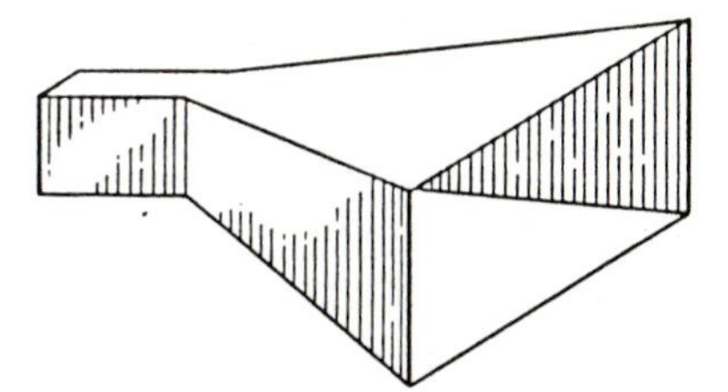

Fig. 16.4. A horn antenna. The small end is attached to a waveguide.

changed into electrical signals which are amplified in the electronics. At longer wavelengths the horn might be replaced by a simple dipole which is just a pair of metal rods connected to the receiver's circuitry. Today a maser is often used as the first stage of the receiver.

When a spectrum line such as the 21-cm line of hydrogen is to be studied, some different techniques must be used. Here we wish to measure the intensity of the radiation in a range of wavelengths that includes the line to be studied. We could imagine using a receiver that could be tuned through a range of wavelengths just as we would tune a common radio receiver up to and beyond a particular radio station. As we turned the dial and listened we would hear a station become louder and clearer and then fade away. In the case of the radio telescope we could record the receiver's output on a chart recorder and watch it change as we tuned the receiver through the wavelengths near an emission line. Or we might try to record the necessary range of wavelengths all at once as we do with an optical spectrograph. The latter is the method actually used in radio astronomy, and there are several ways of doing this. The simplest method in terms of explanation requires that the receiver be equipped with a series of electronic filters and a means for rapid switching from one to the next. Each filter transmits a specific band or range of wavelengths (frequencies) as shown in Fig. 16.5. If the bands are narrow, then the specific wavelength at which the maximum of an emission line occurs can be determined. If such an observed wavelength is different from the wavelength known for some emission line, the difference can be interpreted as a Doppler shift. This is the means by which the radial velocities are determined for interstellar clouds.

Interpreting the radio data

In Fig. 16.6 we have sketched the kind of signal that might be recorded when two radio sources are allowed to drift through the beam

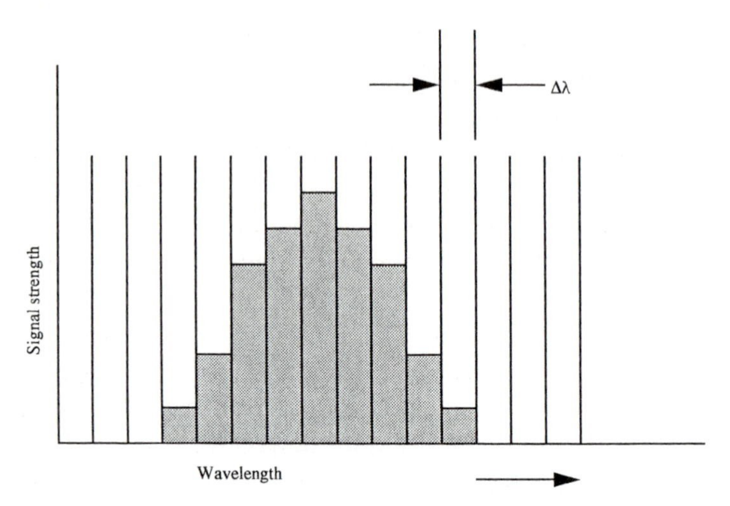

Fig. 16.5. A series of transmission bands defined by electronic filters in the receiver. The shaded sections represent an emission line.

of a fixed radio telescope. Time is increasing toward the right. At the left we see clear evidence that a strong source passed through the beam. A short time later there is the barest suggestion of a second source. Throughout the duration of this record there are rapid, small-scale fluctuations in intensity, and these are best described as 'noise'. Since the noise is present all of the time, we are very much concerned with knowing how it originates and how to keep it as small as possible. We often speak of the 'signal-to-noise ratio' as a measure of the ease with which a signal might be recognized. In Fig. 16.6 the signal-to-noise ratio is about five for the strong source and just slightly larger than one for the weak

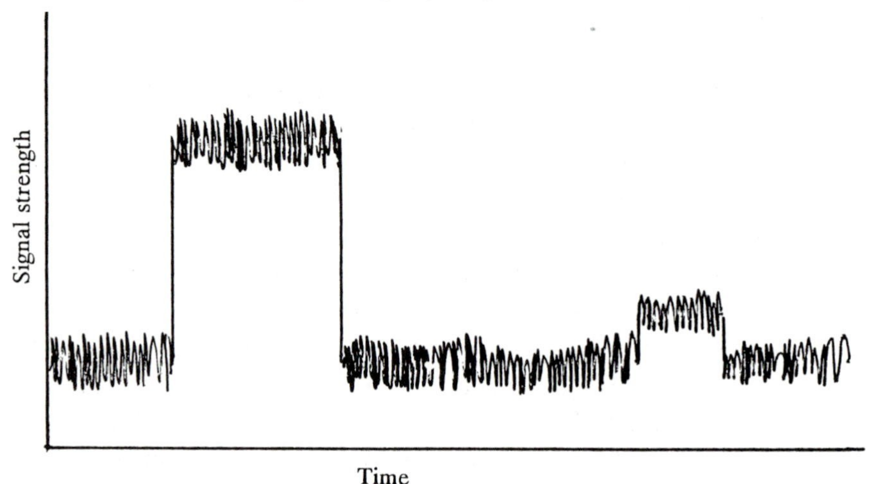

Fig. 16.6. A drift curve which includes two radio sources.

one. We want to see here how the signal can be interpreted once it has been separated from the noise.

Radio engineers have known for many years that resistors contribute noise in electronic circuits and that the level of the noise depends upon the resistor's temperature. The antenna has a resistance of its own, and because of this, the strength of a signal may be thought of in terms of the temperature of the antenna. Radio astronomers came to develop their terminology around the concept of temperature.

Let us begin with the assumption that we have an antenna that receives equally well from all directions. We place the antenna inside of a spherical cavity that radiates as a perfect radiator, that is, as a black body. Planck's Law governs the relationship between intensity of the radiation and the wavelength. The enclosed antenna will come to the same temperature as the cavity and will add an appropriate level of noise to the circuit. Now if we increase the temperature of the cavity, the radiation emitted and received will increase and a new equilibrium and a new level of noise will be reached. Thus, we can calibrate the output of the receiver directly in terms of temperature.

Next we replace the omni-directional antenna with one that has a pattern typical of those found in radio telescopes. Now we receive radiation from only a small part of the surface of the cavity, but we can still relate the strength of the signal to temperature. The sky could replace the cavity in this reasoning, and so the measured signal becomes a temperature of a portion of the sky. This kind of a temperature is referred to as the antenna temperature, T_A.

In an actual case at a radio observatory, the measured signal would include some unwanted signals which must be removed. First, there will be a contribution due to the noise of the receiver, and second, there will be some signal received from the background sky near the source. It is common to make a measurement with the telescope pointed slightly off the source in an area that is as free as possible of radio noise and then make a measurement with the telescope on the source. This is equivalent to observations of sky and sky-plus-star in optical photometry. The radio signal from the point away from the source is subtracted from the signal from the source-plus-sky.

There is some attenuation of the received signal because the signal must pass through the earth's atmosphere, so the antenna temperature must be corrected for this. It is similar to the extinction correction applied in photographic and photoelectric photometry. In addition, a correction can be introduced because of the unique characteristics of the particular radio telescope being used. These characteristics define the 'gain' of the telescope, that is, the amount by which the pattern increases the signal over an antenna which received equally well in all directions. When these corrections have been properly determined, the resulting antenna temperature for a given source should be independent of the telescope used. It is this fully corrected temperature that is referred to as the brightness temperature, T_B.

The interpretation of T_B depends upon the nature of the source, and here there are three possibilities. The source may have a very small angular diameter which would lead us to describe it as a point source or a 'radio star'. Second, it might be referred to as a discrete source which means that it has a measurable angular diameter but one that is smaller than the angular size of the telescope's beam. The third case is that in which the angular size of the source is larger than the beam of the telescope. Fortunately, the observations can tell us which kind of source is being observed.

In the case of the point source the brightness temperature gives us only the flux density of the radiation. In the second case the brightness temperature leads to the average temperature across the source. This would be the case for a planet. Finally, if the source is very large, temperature may be determined at many points, and contours of equal temperature may be plotted.

Radio interferometry

As the diameters of their reflectors continued to increase over the years radio astronomers saw continued reduction in the beams of their telescopes, but with a single antenna they were never able to approach the resolution which was possible in the optical wavelengths. They did, however, find an interesting way to combine the signals from two antennae to achieve remarkable improvement. The solution in this case was the radio interferometer.

We may understand the operation of the interferometer if we first examine Fig. 16.7 in which we indicate two identical antennae on an east–west line. Transmission lines from both go to a common receiver. Parallel lines indicate the direction to a radio source, and two of a series of wavefronts from the source are indicated. Clearly each wavefront will arrive at antenna B before it arrives at antenna A. If we know the distance D between the two antennae and the angle, θ, then we can calculate the length of the line CA to be equal to $D \sin \theta$. If CA is an even number of the wavelengths that are being received, then a wave will reach A at the same time that the next wave reaches B. Therefore, the signals from two successive waves will reach the receiver at the same time. This is the condition for constructive interference. If CA is a half wavelength longer or shorter than above, a crest of one wave will reach one antenna at the same time as a trough reaches the other antenna. This is the condition for destructive interference.

Now as the earth rotates, the source in the direction, S, will appear to move across the sky. As the source approaches the meridian, angle θ will become smaller and CA will become progressively shorter. The output of the receiver will show a series of fluctuations or interference fringes. When the source is actually crossing the meridian, there will be constructive interference. Therefore, we shall expect to record a peak at that time. If we know the sidereal time when the two antennae are pointed toward the meridian, then we also know the right ascension of the source.

The declination of the source can be found from the frequency of the interference fringes. Think about the celestial sphere and the way in which the altitude and azimuth of a star will change as the earth rotates. Only at one of the celestial poles would an object appear to be fixed.

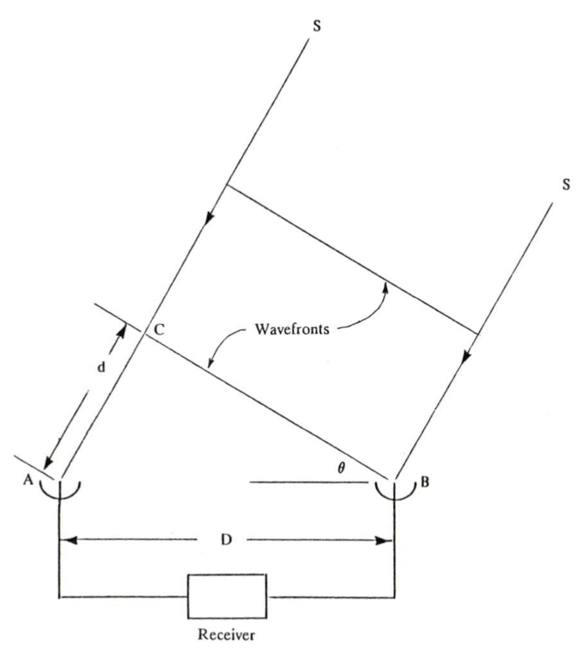

Fig. 16.7. The radio interferometer.

Azimuth will change more rapidly for a source near the equator than for a source near one of the poles, and so will the frequency at which the fringes appear. Thus the spacing of the fringes will be a clue to the declination of the radio source.

The interferometer effectively produces a beam with very small dimensions depending on the wavelength being observed and the length of the baseline, the distance between the antennae. As the distance is increased, the resolution becomes progressively better. To make use of the longest possible baseline, astronomers have learned to record signals simultaneously at widely separated telescopes. Antennae in the USSR, Massachusetts, Cambridge (UK), West Virginia and California, for example, have been used as ends of various baselines. By means of accurate time markers on the tapes, the signals can be combined at one location to give resolution comparable to that which can be achieved at optical telescopes.

Aperture synthesis

Not long after the first applications of interferometry in radio astronomy, astronomers developed an extension of interferometry which may be described as earth-rotation synthesis, supersynthesis or aperture synthesis. Using these techniques they were able to produce high-resolution maps of extended regions of radio emission. Fig. 16.8 is an example of such a map.

Again let us consider an interferometer in which the two telescopes can be pointed directly toward a source and are able to track it for periods of up to twelve hours. The conditions for interference will not have changed, and we can imagine a pattern of interference fringes projected onto the celestial sphere. As the earth rotates, the line between the two telescopes will rotate with respect to the sky, so the fringe pattern will also rotate on the sky. The result will be a received signal which changes in both frequency and amplitude with time. If four telescopes had been placed along the same east–west line, then there would have been several simultaneous signals received from different pairs of receivers. The signal from each pair would have shown its own fluctuations in frequency and amplitude with time. Over a period of hours the series of interference fringes would rotate around the point toward which the telescopes were pointed.

In a complicated process these simultaneous signals can be combined and analyzed to produce the high-resolution maps mentioned above. The

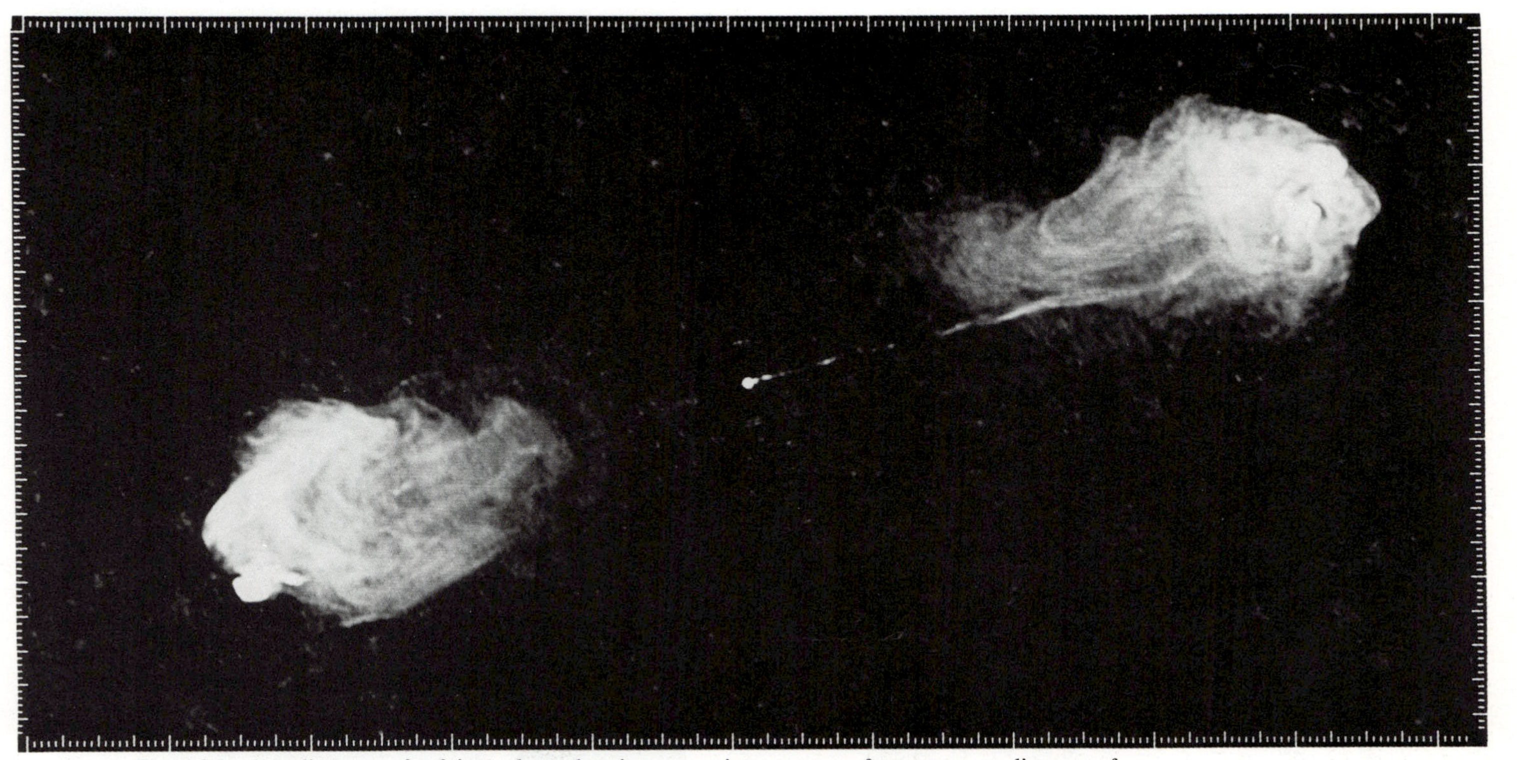

Fig. 16.8. A radio map of a faint galaxy showing extensive streams of gas at great distances from the visible galaxy. (Courtesy NRAO/AUI. Observers: R. A. Perley, J. W. Dreher and J. J. Cowan.)

resolution of the map is the resolution that could be expected from a single telescope with an aperture equal to the longest dimension of the array of telescopes. This is truly a remarkable achievement.

The same type of results can be obtained in less observing time if the telescopes are arranged in a two-dimensional pattern rather than just along the east–west line. This was done on a grand scale with the Very Large Array (VLA) which went into operation on the Plain of Saint Augustin near Socorro, New Mexico in the southwestern part of the United States. Here three tracks each extend twenty-one kilometers from a central location, and there are twenty-seven fully steerable radio telescopes which can be moved along the tracks. Each of the telescopes has an aperture of twenty-five meters. There are twenty-four stations on each arm of this immense 'Y', so there are a great many possible baselines. Five wavelengths were chosen for normal operations, and these are 1.3, 2, 6, 18 and 21 cm. Fig. 16.9 shows the VLA from the air, and Fig. 16.10 is a view of one of the twenty-seven individual telescopes.

Beginnings of radio astronomy

At the opening of this chapter we mentioned only briefly that Karl Jansky first recognized the celestial nature of some radio noise and

Fig. 16.9. The VLA seen from the air. (Courtesy NRAO/AUI.)

Fig. 16.10. One of the radio telescopes which make up the VLA. (Photograph by H. W. Lamb.)

that Grote Reber confirmed Jansky's discovery a few years later. Let us now say a bit more about these two men and discuss some of the other pioneers in radio astronomy as well.

Karl Jansky was an engineer who worked for the Bell Telephone Laboratories in the late 1920s. He was assigned to do some research on the sources of static or noise that interferes with radio reception. We often notice this when we try to listen to a distant (and therefore, weak) station or when there is a thunderstorm nearby. Jansky found that he could attribute some of the noise to nearby thunderstorms and some to very distant ones. However, he found that there was always a faint hiss that did not seem to be related to thunderstorms at all. He observed this for over a year and recorded the rises and falls in the intensity of the weak signal. His antenna was a large, directional one mounted on wheels so that it could be rotated through 360° in twenty minutes. Part of one of his records is reproduced in Fig. 16.11, and we readily see peaks in the recorded signal at roughly twenty-minute intervals. The clue to the extra-terrestrial origin of the signal lies in the slow movement of the peak from south at the beginning to southwest after two hours. The azimuth of a source located on the earth would not change in an interval such as this.

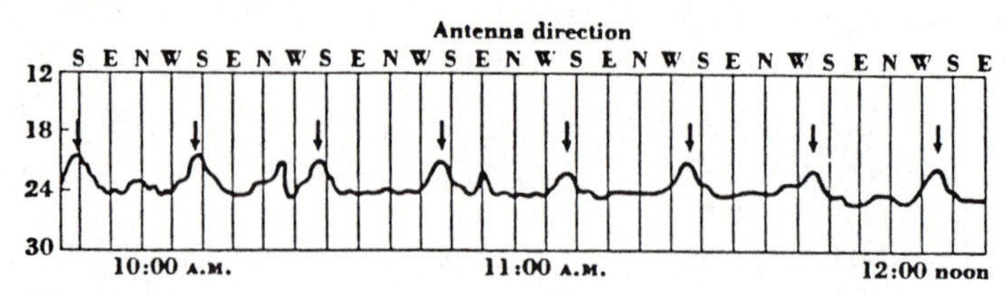

Fig. 16.11. A portion of one of Jansky's records from February, 1932. The changing direction of the peak indicates that the source cannot be on the earth.

Jansky first noticed this effect in January 1932, and he published his results in December of that year.

By the end of 1935 Jansky had published two more papers on his discovery of what he once called 'star static'. He was now able to show that the Milky Way was the source of the signal and that the signal was strongest in approximately the direction of the constellation Sagittarius. Jansky made some suggestions about the need for an instrument with a smaller beam, and described a radio telescope very much like those that actually were built in later years. Other research was assigned to him, however, and he was not able to pursue the new field.

It is perfectly understandable that the discovery of radio noise from the Milky Way did not attract much immediate attention. After all, the engineers who read his papers had more practical projects to work on, and it seems doubtful that many astronomers even heard about it. Fortunately, there was one person who did read Jansky's papers and found ways in which he could begin his own observations. This person was another radio engineer, Grote Reber, who lived in the town of Wheaton, Illinois, near Chicago.

In the backyard of his home Reber built his first radio telescope, a paraboloidal reflector with a diameter of 9.5 meters (about 31 feet). It was a meridian telescope, so it could only be moved in declination. Knowing that Jansky had worked at 14.6 m and at 10 m wavelengths, Reber decided to try shorter wavelengths. He assumed that he was looking for radiation that followed Planck's Law, so he could expect it to be stronger at shorter wavelengths. Early in 1939 Reber found radiation at 1.87 m and confirmed Jansky's discovery that the signal was strongest in the direction of the Milky Way. Some of his preliminary results were published in the *Astrophysical Journal* in 1940. The circumstances which led to the publication of this paper make an interesting story, and the paper itself opened the doors to important work which was to follow.

According to reminiscences of Otto Struve, Reber took a course in astrophysics at the University of Chicago because he felt that he needed to know some astronomy in order to interpret his observations. There he met three notable astronomers, Bengt Stromgren, Gerard Kuiper and Struve, and soon showed them his radio data. After studying Reber's charts and visiting his radio telescope, their skepticism gave way to the recognition that an important new field of research was before them. Struve was at that time the editor of the *Astrophysical Journal*, and it is a tribute to him that he made sure that Reber's unusual paper was published.

Grote Reber continued his observations, and in 1944 he published the map shown in Fig. 16.12. This map shows contours of equal intensity of radio noise on the celestial sphere. These contours fall right on top of the Milky Way and match very nicely the contours of the optical brightness. Reber was also the first person to understand antenna temperature as a measure of the equivalent temperature of the distant source.

In June of 1940 the United States had not yet entered World War II, so communication with European astronomers was still possible. In the Netherlands, astronomy was very much restricted by the German occupation, but regular discussions among astronomers took place at the observatory in Leiden. In 1944 Jan Oort, the director, brought Reber's paper to the attention of his colleagues and asked one of them, H. C. van

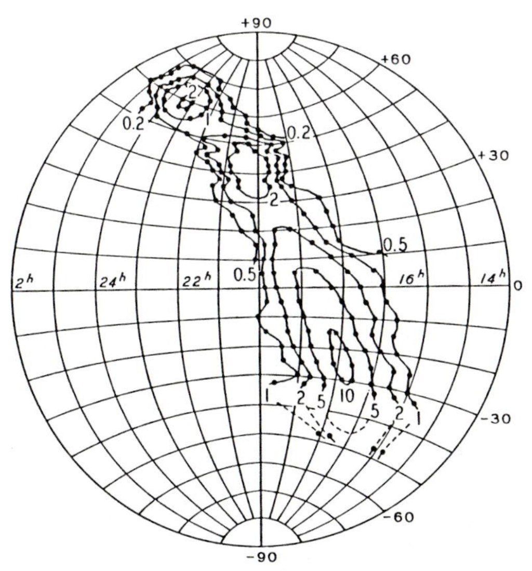

Fig. 16.12. Grote Reber's map of the radio emission from the Milky Way. This was published by Reber in 1944. (Reprinted with permission from *The Astrophysical Journal*.)

de Hulst, to study the paper and lead a discussion of it. Oort also suggested that if there could be an emission line in the radio part of the spectrum, that line could be useful in studies of the motions of the gas in which the line originated. In his report van de Hulst showed that there was, in fact, a possible transition that could occur in an atom of hydrogen when the spin of the electron spontaneously reversed itself. The likelihood of such a transition was very small, and so was the energy released. But the photon would be in the radio region with a wavelength of 21.2 cm. If the amount of hydrogen in interstellar space was sufficiently large, then a signal at 21 centimeters (1420 megacycles) might be detectable.

It was five or six years before the Leiden astronomers were able to start to assemble the equipment to search for the 21 cm emission. They were not alone, for groups in the United States and in Australia were also preparing to make the search. A fire in their receiver delayed the Dutch, and the initial discovery was made by H. I. Ewen and E. M. Purcell in the United States. Confirming observations by the Dutch and the Australians followed within weeks. Surprisingly, the strength of the 21 cm signal was such that unexpectedly large amounts of hydrogen had to be filling the central plane of the galaxy. This discovery was certainly one of the most significant in astronomy of the twentieth century, and it made observations at radio wavelengths a necessary part of future astronomy.

QUESTIONS FOR REVIEW

1. What is the range of wavelengths which is of most interest to ground-based radio astronomers?
2. What is the significance of the term, 'ground-based', in the previous question?
3. What are the advantages of a large diameter for the reflector in a radio telescope?
4. Explain how the antenna temperature is related to the actual temperature of some portion of the sky. What is it that the antenna temperature is actually indicating?
5. Using a single radio telescope, how could you know whether a radio source was a point source, a discrete source or an extended source?
6. Why is it that in a radio interferometer greater distance between the receivers permits better angular resolution?
7. How is the technqiue of aperture synthesis related to simple radio interferometry?
8. Outline the contributions of Karl Jansky and of Grote Reber to the begin-

nings of radio astronomy. Be sure to include dates to preserve an understanding of historical context.

Further reading

Kellerman, K. and Sheets, B. (1983). *Serendipitous Discoveries in Radio Astronomy.* National Radio Astronomy Observatory. This book is the proceedings of a workshop held to honor the fiftieth anniversary of Jansky's discovery of cosmic radio noise. Many of the papers are by astronomers who knew and worked with Jansky, and some are by members of his family. Once you have even a modest background in radio astronomy, you will enjoy going through this book.

Kitchin, C. R. (1984). *Astrophysical Techniques.* Adam Hilger Ltd. Look into Section 1.2 (page 75) for a brief discussion of radio detectors and fundamentals of radio astronomy.

Miczaika, G. R. and Sinton, W. M. (1961). *Tools of the Astronomer.* Harvard University Press. Chapter 8 is devoted to radio astronomy. The basics are good, but some of the methods described here have become obsolete.

Struve, O. and Zebergs, V. (1962). *Astronomy of the Twentieth Century.* Macmillan Company. Read Chapter VI for a review of the beginnings of radio astronomy. Struve's account is especially important, because he was actively involved with the publication of Reber's original paper.

Wellman, P. and Schmid, H. A. (1975). Radio Astronomy For Amateur Astronomers. *Astronomy: A Handbook*, ed. G. D. Roth, p. 126. Sky Publishing Co. Chapter 4 in Roth's book contains material on the equipment as well as on the sources of radiation. This is rather technical in places, and can best be utilized by the reader with a strong background in electronics.

Appendix 1

The photographic process

Ever since its development as a practical process in the middle of the nineteenth century, photography has played an important role in astronomy. The photographic film gave astronomers the ability to record by means of long exposures celestial scenes that were too faint to be seen visually at the telescope. Also, the film provided a permanent record of the sky at the time of the photograph. As a result of this, variable stars have been discovered through comparisons between photographs taken at two or more times, and stars of large proper motion have been found. Stellar spectra are routinely recorded on photographs as well. Astronomers and lay-persons have studied with wonder photographs of the moon, planets, interstellar clouds and galaxies. For more than one hundred years astronomers have employed photographic films and plates as routine tools of the trade. It is important for all students of astronomy to be familiar with photographic procedures and the theory behind them even though the day may come when electronic methods will partially replace photography in astronomy. Let us outline the important parts of the overall process.

The photographic emulsion

It had been discovered before 1800 that compounds of silver would darken when exposed to sunlight. By 1840 experiments in England and France eventually had led to the first practical methods of producing photographic images. In the years since then our understanding of the chemistry and physics of the process have increased steadily, and this has led, in turn, to vastly improved photographic materials. The modern photographic emulsion consists of a large number of very small crystals of one of the silver halides suspended in a layer of gelatin. Silver bromide (AgBr) is probably the one most commonly used today. The gelatin is spread evenly over a backing of either glass or film for support. The simplest emulsions are most sensitive to blue light, but by the addition of dyes, emulsions can be made sensitive to visual (yellow) light or to red light. Other additives increase sensitivity to x-rays or extend sensitivity to the infrared.

The latent image

Within a camera or a telescope a lens or a mirror is used to project an image onto the photographic emulsion. It is believed that when a photon strikes an individual crystal an electron is added to the silver ion to produce an atom of metallic silver. During the period in which light is allowed to fall on the emulsion atoms of silver are formed in many crystals. If the light is intense, silver is produced in many adjacent crystals. Where the light is faint, the affected crystals are relatively far apart. An examination of the emulsion under a microscope would not actually show that any change had occurred, but the patterns and intensities of the optical image formed by the lens have, in fact, been imprinted upon the emulsion. At this point the image is referred to as the 'latent image'.

Development

It is through the process which we know as development that the latent image is turned into a visible image. The exposed film is immersed in a bath of a reducing agent. There are many formulations for these baths, and they are all covered by the general term, 'developer'. The developer supplies extra electrons which can change the ions of silver into atoms, and the process affects crystals of the latent image before it affects crystals that have not been struck by a photon. Apparently the minute speck of silver which resulted from the action of light is necessary to start the action of the developer in a particular crystal. The chemical reactions of development continue until an affected crystal that was originally silver bromide has been converted entirely to silver. The silver appears black in the developed emulsion and thus produces a negative; that is, where there was light in the optical image, there will be dark, opaque silver in the developed emulsion. The bromine ions end up in water-soluble bromides.

Stop bath

It is now customary to transfer the film from the developer to a weak acid solution which neutralizes the developer and very quickly stops its further action. This is referred to as a stop bath, and it is a weak solution of acetic acid. An indicator is added to commercial stop baths so that the color changes when the pH of the solution has become greater than 6.8. At this level the color of the solution changes from yellow to purple, and the user is warned that it is time to discard the solution.

Fixer

After the developer has done its work on the crystals which made up the latent image, the emulsion still contains crystals of silver bromide which were never struck by light. These crystals are still sensitive to light and must be removed from the emulsion. This is accomplished by immersing the film in a third solution, the fixer. The fixer is sodium thiosulfate, and it reacts with the silver bromide to produce two water-soluble salts neither of which is sensitive to light. Thus the emulsion now contains small fibrous masses of metallic silver and several salts which have been left from the developing and the fixing reactions.

The term, hypo, is often used in reference to the fixer. This term comes from the early photographic work of John Herschel and Fox Talbot in England when sodium thiosulfate was erroneously called sodium hyposulfite.

Washing

The final stage in the processing of the emulsion is a thorough washing in water to remove the salts mentioned above. If the washing is not complete, the salts which remain will cause stains in the negative. The washed film or plate may then be allowed to drain and dry, and it is ready for permanent storage or use.

The brief outline given above should give the student an understanding of the photographic process in its simplest terms. In reality each step is much more complicated than we have implied, but from the point of view of practical use the details of the chemistry are not required. References to standard works on photography may be found at the end of this appendix.

Darkroom procedures

Most emulsions in use today are sensitive to a broad range of colors. This is desireable since it makes for the greatest efficiency in producing the latent image. On the other hand, it also means that processing must be done in total darkness. The beginner should look carefully for light leaks in the darkroom before opening any container of film or plates.

It is standard practice for photographers to store prepared mixtures of developer, stop bath and fixer in bottles in the darkroom so that they can be ready for use quickly. The chemicals for developers and fixer are usually sold as dry powders which are simply mixed with water. The stop bath is sold as a concentrate which is diluted with water as needed. There are many formulations for developers, and the choice of one over another depends upon the application. In most cases the manufacturer of a film includes a recommendation for type of developer and time of development in the package, and the user cannot go wrong in accepting the recommendations. Table A1.1 lists the times and other data for a number of emulsions which are often used by astronomers.

The manufacturer also specifies a temperature at which the processing should take place, and this is important. The reactions of the developer go much more quickly at higher temperatures. In addition, crystals which had not been part of the latent image will be affected and darkened if the developer is too warm. This causes the background to be darker than it should be, and such darkening is referred to as 'fog'. Fog may also be caused by improper storage over long periods of time.

During processing care should be taken not to contaminate one solution with another any more than necessary. When the film is removed from the developer, for example, it should be allowed to drip for a second or two before being put into the stop bath. And the stop bath should be allowed to drain off before the film is put into the fixer. Other contaminates should also be kept under control if best results are to be obtained, and photographers universally insist that their darkrooms are kept clean and free from dust. Dust particles which settle on a drying emulsion can cause confusion later when the emulsion is being studied under a microscope.

Density of exposed emulsion

In considering the emulsion on which an optical image has been projected we know that all areas of the film would be exposed for the same period of time and that the intensity of light would probably vary from place to place on the film. Let us now define 'exposure' as the product of the time and the intensity of the light. Then if the exposure varies from place to place on the film, we should expect the density of the developed image to vary in the same way. If we let I be the intensity of the light and t be the time of exposure, the product $I \times t$ implies that we should obtain the same density if we reach the same product with a longer time and a lower intensity. With the short exposure times that

Table A1.1. *Processing films, plates and papers*

Developer

Adjust the developing time according to the temperature.

		Time (mins)				
Emulsion	Developer	65°F	68°F	70°F	72°F	75°F
Sheet films	D19	4.5 min	4.0	3.75	3.50	3.0
Plates, 103a	D19	5.75	5.0	4.5	4.25	3.75
Plates, IIa	D76	17.25	15.0	13.75	12.5	11.0
Rolls, Pan X	D76	8.0	7.0	6.5	6.0	5.0
	D76 1:1	11.0	9.0	8.0	7.0	6.0
Rolls, Plus X	D76	6.5	5.5	5.0	4.5	3.75
	D76 1:1	8.0	7.0	6.5	6.0	5.0
Rolls, Tri X	D76	9.0	8.0	7.5	6.5	5.5
	D76 1:1	11.0	10.0	9.5	9.0	8.0
Papers	Dektol 1:2		1.5			

Stop bath

Immerse all materials for 30 seconds at any temperature. Temperature of the stop bath should be close to the temperature of the developer.

For 103a and IIa emulsions a half-strength stop bath should be used.

Fixer

Material	Solution	Time
Plates, 103a, IIa	Rapid fixer	3 to 5 min
	Fixer	10 to 15 min
Films	Rapid fixer	2 to 4 min
	Fixer	5 to 10 min
Papers	Rapid fixer 1:2	2 min

Wash

Temperature of the water should be between 70 and 75°F.

Material	Time
Plates and films	20 to 30 min
Paper (plain)	20 to 30 min
Paper, resin coated	4 min

Alternate process using Kodak hypoclearing agent (HCA): water 1 min; HCA 1 min; water 1 min.

Rinse (Optional)

To help prevent water spots, rinse all materials in Kodak Photo-Flo for 20 s.

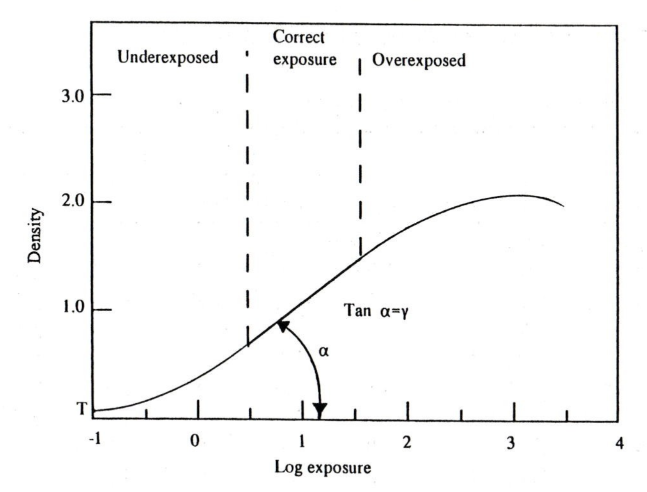

Fig. A1.1. The characteristic curve of a hypothetical photograph emulsion.

we use in personal cameras the expected densities are normally obtained. In astronomical photographs, however, where the exposure times might be from ten minutes to one hour the expected density is not obtained. We speak of this as 'failure of the reciprocity law'. The cause is not well understood, and the effect is usually not a problem in black and white photographs. We shall mention it again in connection with color photography.

It is very useful to establish the relation between exposure, *It*, and density in the developed image, and this can be done with simple equipment. We can devise an instrument which will permit us to project a series of spots onto an emulsion, and we can control the intensity of the light in each spot. By exposing an emulsion through such a device, we obtain in the developed film a series of spots of graded densities.[1] We can now measure these densities and examine the relationship between density and exposure. Fig. A1.1 is a plot of density versus the logarithm of the exposure. (See also Fig. 8.3.) This plot is referred to as the *characteristic curve*, and it holds some important information.

The characteristic curve

As we examine Fig. A1.1 we note that from log It=0.5 to about log It=1.5 the slope of the line is constant. This means that within this range of exposures we can expect greater density for greater exposure. This part of the curve has been marked 'correct exposure'. To the left the curve has been marked 'underexposed', and we note that we need to achieve some minimum exposure before we can expect useful information. We note also that a certain level of fog is indicated by the fact that the curve does not cross the density scale at zero. At the right we see that if the emulsion is overexposed, the rate of increase in density becomes less. In extreme cases of overexposure, the density can even begin to decrease. At this level the developed image is said to be saturated. Clearly, where we expect to make precise measurements of density and relate them to intensity, we should try to limit exposures to those which fall on the straight portion of the characteristic curve.

[1] Density is defined as the logarithm of the opacity, and opacity is defined as the reciprocal of the transmission. Transmission is defined as the ratio of the transmitted to the incident light, i.e., It/Io.

In Fig. A1.1 we also see that the slope of the straight portion of the curve has been marked with the Greek letter, α. The tangent of this angle is defined as the speed and is specified with the Greek letter, γ. If γ is large, then a small increase in exposure will result in a large increase in density. Photographic emulsions are made with a variety of speeds, and the choice of speed must depend upon the application. For high contrast, that is, distinct blacks and whites, a fast emulsion would be chosen. This would be the case in most astronomical applications. For a variety of subtle shadings an emulsion with a lower γ would be chosen. This would be the case in portrait photography, for example.

The student should keep in mind that in applications such as those in astronomy we are usually interested in the intensity of the incident light. On the photograph we can only measure density, so we must always go through the characteristic curve if we are to attempt to determine relative intensities from photographs. It is useful to note that in Fig. A1.1 the straight portion of the characteristic curve extends from about 0.5 to 2.0 on the log *It* scale. This represents a range in intensities of about thirty to one for the emulsion in this hypothetical example.

Graininess

When we look at a well-focused negative, boundaries between light and dark seem to be sharply defined. If we examine the same boundary under progressively higher magnification, however, we eventually see that the boundaries are irregular and splotchy. This variation in density is what photographers refer to as graininess, and it is always present at some degree of magnification or enlargement. When we go to even higher magnifications as with an electron microscope, we find that the grains are actually clusters of developed masses of silver that were once crystals of silver halide. Thus, the graininess that limits the resolution that may be attained is on a much different scale than the size of the individual crystals. So resolution is not limited by the size of the crystals of silver halide but the clumpiness of the developed crystals.

Graininess is the term used to describe the subjective impression of the inevitable irregularities described above. When some sort of measurements are made to describe the same irregularities in a quantitative way the term 'granularity' is used. Photographic engineers have learned to control the size of the crystals of silver halide in particular emulsions, and they have also found that the larger the size of the crystals, the faster will be the speed of the film. This presents something of a dilemma: an increase in speed is going to be accompanied by an increase in granularity.

The granularity can also be affected by exposure and by development. A minimal exposure will give minimal granularity. The same is true of development, so one should take care not to overdevelop the film and to avoid the most rapid developers.

Positive prints

Astronomers and x-ray technicians are among the few people who find it convenient to work directly with negative films or glass plates. Star images actually are best studied as black dots on a nearly clear background. For most people, however, a positive print is closer to the real scene and is therefore more easily understood. Prints are made in either of two ways. The negative may be put into an enlarger which projects an enlarged image onto a sheet of photosensitive paper, or the paper may be exposed to light while in contact with the negative. In both cases dark areas on the negative will produce light areas on the print.

All details of the process are essentially the same for papers as for films. The backing is just paper instead of a transparent medium. Processing follows the same steps as well. The exposed paper goes into the developer, then into the stop bath, and finally into the fixer. Again, one should use the developer recommended by the manufacturer of the paper. There are papers that give high contrast and those that give soft tones and gradations of density. The choice depends upon the application, and in prints of star fields high contrast is usually desired.

Photographic papers are not sensitive to red light, so it is possible to work with a red light on when making prints and enlargements. Special lamps and filters are made for use in darkrooms. One should not use a lamp just because it is red as it may have some yellow or blue to which the paper is sensitive.

Color films

Many people are intrigued by the idea of using color films for photographs of astronomical objects, and when properly handled these materials can give spectacular results. The color films are based on the same principles that have been described for black and white films, but are necessarily more complex. Colors are produced by combining three layers of emulsion on one backing. Each layer is sensitive to a different range of colors as indicated in Fig. A1.2. Notice also that a yellow filter is incorporated as another layer. Blue light exposes the top emulsion but is stopped by the yellow filter. Green light exposes the next layer, and red light exposes the final layer. Yellow light exposes both the second and third layers of the emulsion. If the film is now developed in a more or less normal way, silver will be produced in the affected areas in each layer. The film is not put into the fixer at this stage, however. Each of the layers of emulsion still contains crystals that are light-sensitive, and three more stages of exposure and development are required.

The film is next exposed to red light which now can affect only the undeveloped crystals in the lowest layer. It is then developed in a special developer which produces not only silver but also a cyan[2] dye in areas of the third layer which were not part of the original exposure. The next exposure is to blue light, so undeveloped crystals in the top layer are exposed. The next developer produces silver and a yellow dye in the top layer. The final exposure activates crystals in the center layer, and the final development produces a magenta dye.

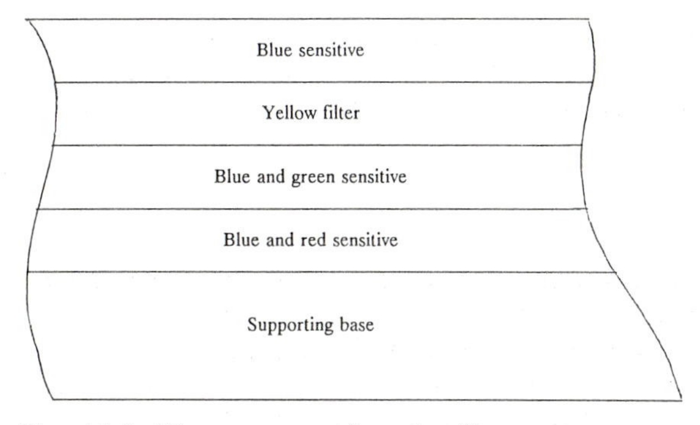

Fig. A1.2. The structure of a color film.

[2] Cyan is a color obtained by mixing blue and green light.

In the final stages of this complex process the silver and the yellow filter are bleached out. This leaves what amounts to three colored filters, cyan, yellow and magenta. When white light passes through this combination, the original colors are produced by the subtractive process. Films such as Kodachrome are made and processed as described here, and they result in color-positive slides.

The films for producing color negatives are also made with three color-sensitive layers and a yellow filter. Each layer now also contains chemicals which produce the appropriate dyes during development, and only one stage of development is needed. Again, the silver must be bleached out of the emulsion to give the color negative. Prints are made using paper on which the same layers have been coated, and the paper is processed in essentially the same way as the film to produce a positive print.

When color films are used in long exposures, as they are in astronomy, the final colors in the developed film do not match the colors in the original subject. This is due to the reciprocity effect which was mentioned on p. 300. Each of the three emulsions is affected differently by reciprocity failure during long exposures, so the relative exposure in the three colors is not the same. This problem is lessened if the emulsions are kept as cold as possible, and special refrigerated cameras are often used when accurate colors of astronomical subjects are desired. Correction of the colors can also be achieved by using a combination of filters when a print or an enlargement is being made.

Further reading

di Cicco, D. (1988). Astrophotography then and now. *Sky And Telescope*, **76,** 5, 463. This article by an acknowledged authority contains many interesting and valuable suggestions.

Dobbins, T. A., Parker, D. C. and Capen, C. F. (1989). *Introduction to Observing and Photographing the Solar System.* Willman–Bell. In addition to the information suggested in its title, this book contains much more including material on darkroom procedures.

Appendix 2

Time signals

Time standards were mentioned in Chapter 2 in connection with the computation of sidereal time and again in Chapter 14 in connection with the determination of heliocentric time. Very often the most convenient standards are those broadcast by radio from certain national time services. These can be received almost anywhere on the earth and are broadcast on many frequencies. Table A2.1 is a list of some of them.

Each of these stations uses its own scheme for identifying the hours, seconds and minutes. When tuned to one of these stations, the user hears a series of ticks which mark the seconds. These ticks are actually short, but precisely timed, pulses. The minutes are often marked by longer pulses, and there are occasional voice announcements on some stations.

With the exception of MSF these station all broadcast in the short-wave band, so they can be received on any receiver designed for this band. It is possible to buy an inexpensive

Table A2.1. *Broadcast time signals*

Station	Location	Nation	Frequency (kHz)	Wavelength (meters)	Schedule
WWV	F t Collins	USA	2 500	120	Continuous
			5 000	60	
			10 000	30	
			15 000	20	
			20 000	15	
			25 000	12	
WWV H	Hawaii	USA	2 500	120	Continuous
			5 000	60	
			10 000	30	
			15 000	20	
			20 000	15	
CHU	Ottawa	Can	3 330	90.1	Continuous
			7 335	40.9	
			14 670	20.45	
MSF	Rugby	UK	60	5 000	Continuous
DIZ	Nauen	GDR	4 525	66.3	Continuous
ZUD	Oliphants-fontein	RSA	5 000	60	Continuous
VNG	Lyndhurst	Aust	7 500	40	Continuous

receiver that will only pick up the time signals. With a proper degree of care the clock-/calendar in a computer may be synchronized to the broadcast time signal by ear-hand coordination. For greater precision, a receiver may actually be designed to start the computer's clock. When even greater precision is needed, the time signal and data from the telescope may be recorded at the same time on separate tracks of a magnetic tape.

In the United States the time signals may also be accessed by telephone from the U.S. Naval Observatory. The number to call is 1-202-653-1800. The same time is available through the telephone company at 1-900-410-TIME. There is a charge for this service.

Appendix 3

Some statistical principles and the method of least squares

One of the important actions that is in nearly all of the foregoing chapters is measurement. We measure time, coordinates, proper motions, parallaxes, magnitudes, the positions of lines in spectra and shifts in positions of spectral lines, for example. After a measurement has been made we want to know how good the measurement really is, and in order to evaluate our measurements, we must turn to statistics. We would also like to use our measured data to make predictions either within or beyond the range of the measurements. Here we shall describe some of the principles which permit us to achieve these two goals in practical situations.

As an example, let us assume that a student is asked to determine the position of a spectrum line by means of a screw-type measuring machine such as the one in Fig. A3.1. By turning the crank on the right, the user can move the stage to the left or to the right. A

Fig. A3.1. A very simple measuring machine. A precision screw moves the stage when the drum at the right is rotated.

cross-wire or reticle in the eyepiece can be aligned with the spectrum line, and the position of the microscope can be read from the graduated drum. The screw has a pitch of one half a millimeter, so it takes two turns of the crank to move the microscope a distance of one millimeter. There are five hundred divisions on the drum, so we should be able to read the setting to one one thousandth of a millimeter. We should also be able to estimate the nearest half division. The student matches the cross-wire to the spectrum and reads the drum. Just as a check she makes a second setting and reads a new value from the drum. She tries again and again until she has made fifty tries, and she never makes quite the same reading twice. Which of these many readings should be adopted as the correct one?

Mean, median and mode

Most of us when faced with the question asked above would answer that we should calculate the average value i.e. the mean. We would add up all of the measurements and divide by the number of tries, or we could write

$$\bar{x}=\Sigma x_i/n$$

Here $\bar{x}$ is the mean; x_i is the value of measurement number i; Σx_i is the sum of all values of x; and n is the number of tries. A little bit of experience can show us that while the mean may be a useful quantity by itself, we can easily see just how comfortable we should be with it. For example, we can look at the range of the values of the measurements. If the range is large, then all we really know is that the true value is somewhere in that range. If the range is small, we can be more confident that the true value is close to the mean.

Further insight may be gained if we plot the measurements in a bar graph such as those in Fig. A3.2. We assume some interval in the range of measurements and count the number of values in each increment of the interval. In the first plot the range is large and the envelope of the bar graph is relatively flat. In the second plot the range is small, and the bar graph shows a high central peak. Unless there are some special factors influencing the measurements, we should expect the bar graph to be symmetrical about the central interval as it is in these two examples. Furthermore, the mean value will fall in the central interval.

In Fig. A3.2(*c*) we show an unsymmetrical or skewed distribution of measured values. We note that the mean in this case does not coincide with the peak of the curve. On the figure we have indicated the positions of two other quantities, the median and the mode. The median is defined as the central value in the range, and the mode is defined as the most popular value. When the distribution is symmetrical, as in (*a*) and (*b*), the mean median and mode should all be the same. In what is to follow we shall not be concerned with the median and the mode, as we are assuming that our measurements fall in a symmetrical distribution.

If the number of measurements, the sample, is large, and the interval is small, then the bar graph begins to resemble a smooth curve. In statistics there are a number of theoretical discussions of the concept of the *normal* distribution of values, and the normal distribution is the smooth, symmetrical distribution which we have been describing. At times this curve is also referred to as a bell-shaped distribution from its resemblance to the cross section of a bell, or simply as a bell curve. A figure of this shape may be quite closely represented by the equation

$$y=Ce^{-h^2(x_i-\bar{x})^2}$$

C is a measure of the height of the curve, for when $x_i=\bar{x}$ the exponent of e is zero, and

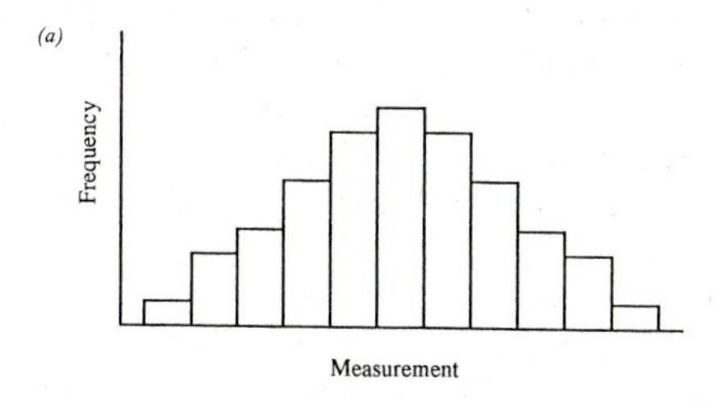

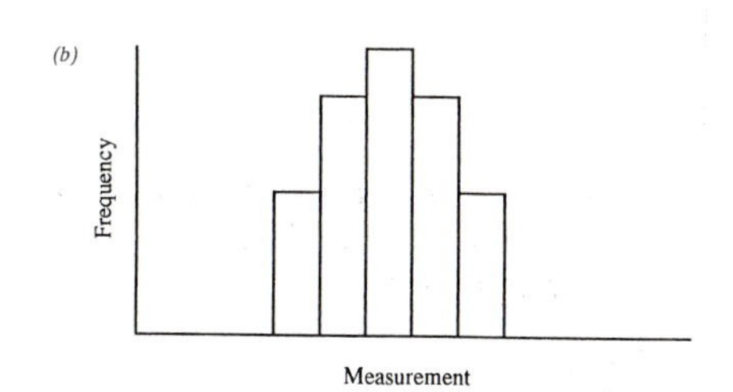

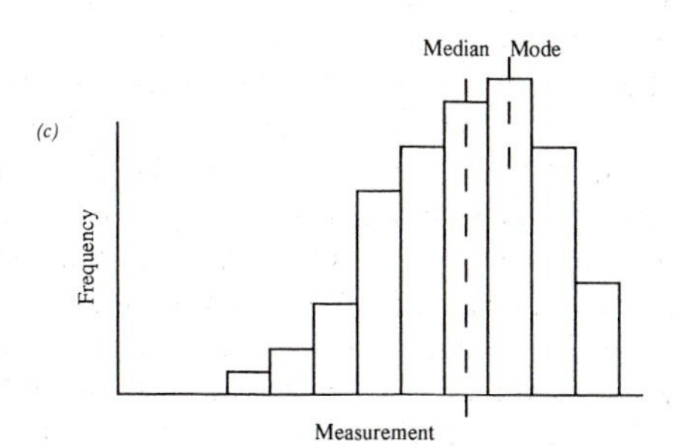

Fig. A3.2. (*a*) A bar graph showing the number of measurements in each small interval. The range of values is large. (*b*) The bar graph when the range is small. (*c*) A bar graph in which the data are unsymmetrical or skewed.

$e^o=1$. h is a measure of the shape of the curve, for when h is large, the curve will drop off sharply as x increases. In practical problems we do not have to deal with C and h. We have other means of computing some simple quantities which give all the necessary insight into the shape of the curve. Let us define five new terms.

1. *deviation*: the difference between any value and the mean

$$x_i - x$$

2. *mean deviation*: sum of all deviations divided by n

$$\Sigma|x_i - x| \, /n$$

3. *variance*: sum of the squares of the deviations divided by n

$$\Sigma[(x_i - \bar{x})^2\,]/n$$

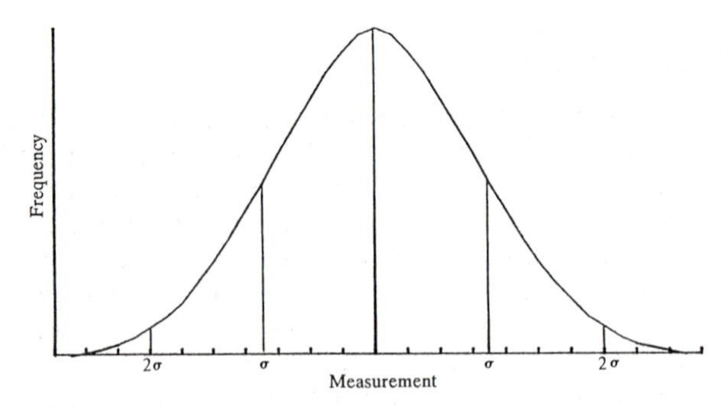

Fig. A3.3. Frequency distributions showing the meaning of standard deviation.

4. *standard deviation, σ:* square root of the variance

$$\sigma = \sqrt{\Sigma[(x_i - \bar{x})^2]/n}$$

68% of the values can be expected to lie in the range $\pm\sigma$, and 96% of the values will be in the range $\pm 2\sigma$. (See Fig. A3.3)

5. *probable error, p.e.*: a range of values such that half of the values lie inside the range and half lie outside of it. These ranges have been marked in Fig. A3.4. There is an equal likelihood of any value being in area I or in area II. Probable error is related to standard deviation by

$$\text{p.e.} = 0.67\sigma$$

Both the standard deviation and the probable error may be calculated from the measured data, and both are valid measures of the precision of the data. Of the two, the standard deviation is the more widely used. If we return now to the question of how much confidence we should have that the mean value represents the true value, we see that our answer lies in the value of the standard deviation. It is customary to include the standard deviation along with a numerically determined value to give the reader some idea of how good the computed value really is, for example, 178 mm±0.12 mm.

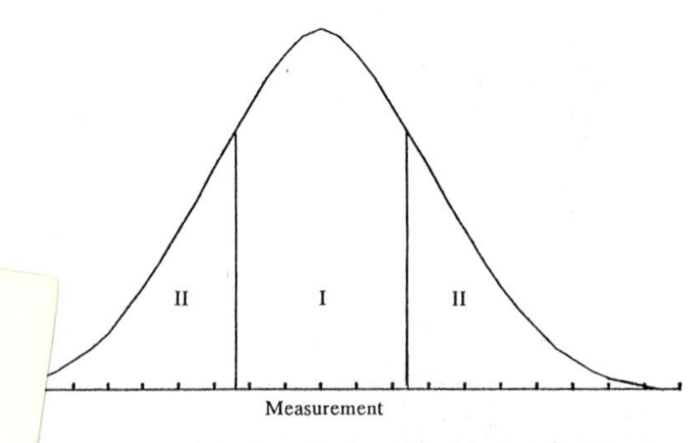

.4. Frequency distribution showing the meaning of probable error. obable error is one half of the width of Area I.

Method of least squares

In many areas of life, not just in the sciences, we often have a series of observations of two variables, and we are interested in looking for the relationship between them. We saw this in Chapter 6 when we looked for the relationship between iris reading and magnitude, and again in Chapter 13 when we plotted wavelength against measurements of the positions of spectrum lines. In both cases we were trying to determine the value of some quantity from a series of values of two variables. Fig. A3.5 is repeated from Chapter 8 and shows the relation between magnitude and iris reading. We would like to be able to find the equation of the line drawn on the figure, so that we could calculate values of magnitude from other iris readings. We would also like to have some idea of just how good that calculated value of magnitude might be. Both of these goals may be met through the method of least squares.

The basic premise of the method of least squares is that the best line through a series of points is that line for which the sum of the squares of the residuals is a minimum. The residual is the difference between an observed quantity and the value computed for that quantity from the equation of the best-fitting line. The residual for one point has been indicated in Fig. A3.5. We shall outline the method for the case in which a straight line best represents the data, i.e. the relationship between the two variables is linear. Then we shall show how a non-linear relationship can be recognized and determined.

First, we recall that the equation of a straight line is

$$y=a+bx$$

where b is the slope of the line and a is the intercept of the line with the x axis. The problem at hand, then, is to find values of a and b such that they define the line that meets the criterion that the sum of the squares of the residuals is a minimum. An expression for calculating the residual, δy_i, takes this form

$$\delta y_i=y_i-(a+bx_i)$$

y_i is an observed value of y, and $(a+bx_i)$ is the computed value. If we now square both sides of this, we have

$$(\delta y_i)^2=[y_i-(a+bx_i)]^2$$
$$=y_i^2-2ay_i-2bx_iy_i+a^2+2abx_i+b^2x_i^2$$

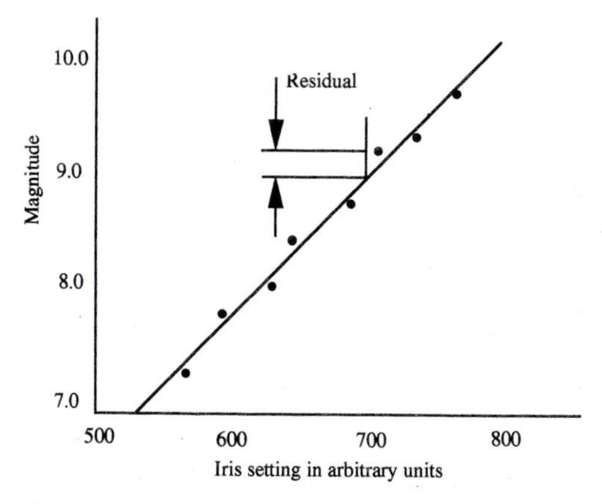

Fig. A3.5. Magnitude plotted against iris reading for data from Chapter 8. The line represents a least squares fit.

As before, we use the Greek letter sigma, Σ, to indicate a sum, so the sum of the squares of the residuals is given by

$$\Sigma(\delta y_i)^2 = \Sigma y_i^2 - 2a\Sigma y_i - 2b\Sigma x_i y_i + na^2 + 2ab\Sigma x_i + b^2\Sigma x_i^2$$

Now in order for the sum of the squares of the residuals to be a minimum, we find the partial derivatives of $(\delta y_i)^2$ with respect to a and with respect to b and set those derivatives equal to zero. When we do this and divide by 2, we have the two equations

$$na + \Sigma x_i b = \Sigma y_i$$

$$\Sigma x_i a + \Sigma x_i^2 b = \Sigma x_i y_i$$

These are known as the 'normal equations'. Since there are two equations and two unknowns, they may be solved simultaneously to find a and b.

$$a = \Sigma x_i^2 \Sigma y_i - \Sigma x_i \Sigma(x_i y_i)/(n\Sigma x_i^2 - (\Sigma x_i)^2)$$

$$b = n\Sigma(x_i y_i) - \Sigma x_i \Sigma y_i/(n\Sigma x_i^2 - (\Sigma x_i)^2)$$

From these two expressions we note that if we have a number of values of x and y we can easily find the sums and products in these two equations, and we have the equation of the best line through the points. In the magnitude problem, we could use the equation to find an unknown magnitude from a reading of the iris photometer.

The standard deviation in y can be found from

$$\sigma_y = \sqrt{\Sigma(\delta y_i)^2/(n-2)}$$

It should be noted that in some statistical nomenclature this analysis is referred to as linear regression, and b is called the regression coefficient.

Another useful quantity, the correlation coefficient, r, may be calculated from

$$r = \Sigma x_i y_i/(\sqrt{\Sigma x_i^2}\sqrt{\Sigma y_i^2})$$

This quantity gives a good measure of the scatter of the points about the line defined by $(y = a + bx)$. If all of the points lie exactly on the line, r will be equal to 1.0. As the scatter becomes greater, r becomes smaller.

Least squares fit to a curve

A set of data has been plotted in Fig. A3.6, and a straight line has been fitted to this data. When we examine the graph, however, we see right away that the straight line does not give the proper fit. At the top and bottom the data points are above the line, and in the central portion the data points are below the line. If we examined a table of residuals for this data, we would see the signs of the residuals change from positive to negative and back to positive. Clearly a curve would give a better fit to the data than does the straight line. The equation of the curve will be a second order equation and will have the form

$$y = a + bx + cx^2$$

There are now three unknowns, so we must have three normal equations.

The basic derivation of the normal equations in the case of the line can be extended to cover the case of the curve, but in practical terms the extension is very simple. We may write down the normal equations as

$$na + \Sigma xb + \Sigma x^2 c = \Sigma y$$

$$\Sigma xa + \Sigma x^2 b + \Sigma x^3 c = \Sigma xy$$

$$\Sigma x^2 a + \Sigma x^3 b + \Sigma x^4 c = \Sigma x^2 y$$

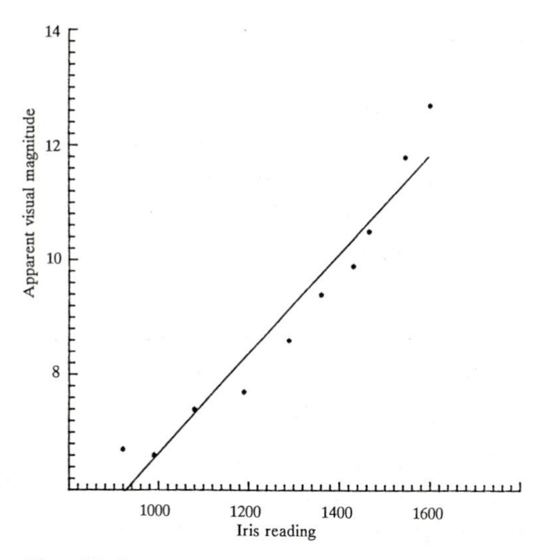

Fig. A3.6. A set of data in which a curve would clearly give a better fit than the straight line.

Comparing these three equations with those for the case of the line, we can notice the pattern in the coefficients of the three unknowns. This allows us to write directly the normal equations for a curve of third order (or any higher order). The equations may be solved by any of the standard methods such as matrix inversion or determinants.

Today there are many computer programs which can plot a set of values and fit a curve to the points. These are of great value in speeding up the analysis of data, and anyone who has much work of this sort to perform should acquire one.

Appendix 4

The charge-coupled device or CCD

A major new tool for astronomers came into use during the 1980s. This was the charge-coupled device, or CCD, a multi-element detector developed originally for use in light-weight television cameras. The large market for personal television led to rapid improvements in performance and reductions in cost, and astronomers quickly applied the new techniques to their own observations. Applications in spacecraft and at major observatories led to development of units intended solely for use in astronomy. Today, astronomers have found uses for the CCD in astrometry, photometry and spectroscopy. In addition, astronomers can sometimes record faint details in objects such as galaxies when only a very long exposure with a photograph could do the same thing. As one may notice in Fig. A4.1, however, the resolution in a CCD image is usually not as good as that in a fine-grain photographic emulsion.

In structure, the CCD is a two-dimensional array of photo-detectors, and each individual detector is described as a picture element or pixel. In one version made by Texas Instruments Corp. there are 800 rows and 800 columns of pixels in a detector only about twenty-five millimeters on a side. This makes 640 000 pixels in all. When some scene such as a telescopic image is projected onto the array, each pixel responds to the photons that fall on it. If the device was 100% efficient, each photon would produce an electron in a pixel. Unlike the photomultiplier or the photodiode, the pixels in the CCD can accumulate electrons over a period of many minutes. Thus, during the time of an exposure an electric charge builds up on each pixel, and the amount of charge is proportional to the intensity of the light and the length of the exposure. The pixels are so small that in a typical situation the image of a star may be spread out over several adjacent pixels. When an exposure has been completed, the charges on the pixels must be recorded in some manner so that the picture can be reconstructed. Under the control of a computer the charges are measured, digitized and read into a computer's memory one pixel at a time, row by row. The digitized image can later be reconstructed row by row on the computer's display or on a printer. The circuitry built into the CCD and the program which controls the read-out are such that the image can be displayed very quickly after the end of the exposure. In order to accomplish this, the computer must have a high processing speed and a large memory.

The advantages of the CCD over the photograph may be summarized as follows:

1. *high quantum efficiency*. About 60% of the photons which reach the CCD contribute to the recorded image compared to about 1% for photographic emulsions.

2. *recording of many stars at once*. The CCD, like the photograph, records all of the individual stellar images in the field simultaneously, while the photomultiplier usually measures the brightness of just one star at a time. The photomultiplier, of course, cannot discriminate when several stars are in the aperture at the same time.

3. *linear response*. There is a direct relationship between the exposure and the intensity of the recorded image. On photographic materials exposure and density are related through the characteristic curve.

4. *dynamic range*. The response of the CCD is linear over a broader range of exposures from threshhold to saturation than is the photographic emulsion. The dynamic range for photographic materials may be on the order of 100 to 1. The same number for the CCD can be as large as 100 000 to 1.

5. *broad color response*. The sensitivity of the CCD extends well beyond 10 000 Å, so it is considerably more sensitive in the red than photographic and photoelectric detectors. The early CCDs were not very good in the ultraviolet, but that situation is improving.

6. *image analysis by computer*. Since each pixel represents a specific location on the CCD, a computer can be programmed to perform an analysis of a star field automatically.

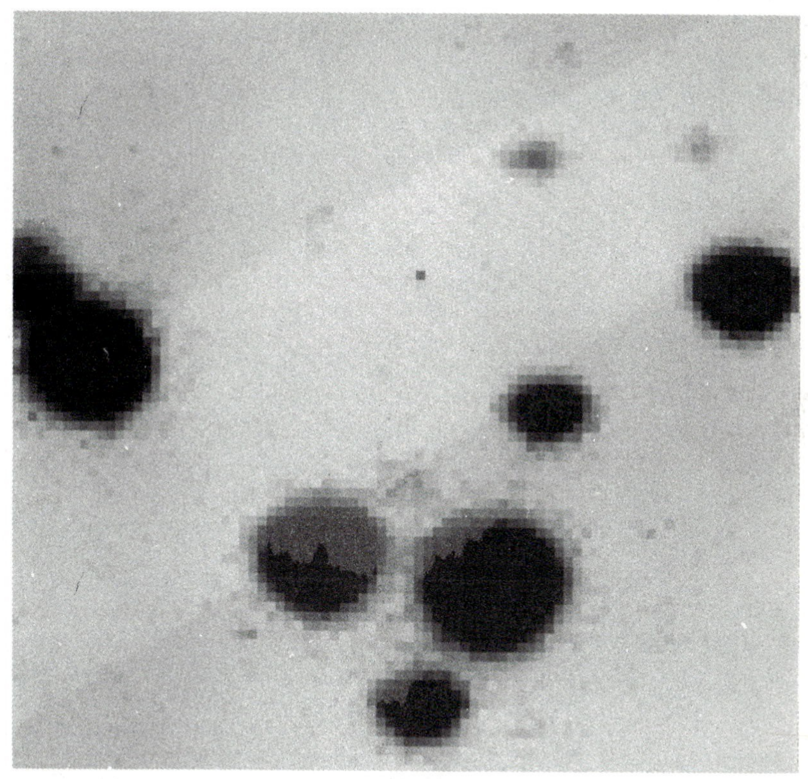

Fig. A4.1. Part of a CCD image of a star field. The mosaic appearance of each star is easily noticed in this photograph of the screen of a computer. These images are slightly elongated because the telescope was actually tracking an asteroid. The computer is able to resolve differences in intensity within the dark centers of the images even though the screen and the photograph cannot. (Courtesy of Linda French.)

On the other side, the CCD has some serious shortcomings which may be listed:

1. *small field*. A CCD which is only 25 mm on a side will cover an angular field of view of only a few minutes of arc when used with a telescope of several meters focal length. A photograph made with a similiar telescope might cover a degree or more. This gives the photograph an edge in variable star work, for example.

2. *large amounts of data*. Numbers representing location and charge for each of 640 000 pixels can be generated for each frame recorded by the CCD, and a frame may be recorded in a period of only a minute or so. As larger CCDs come into use, even larger amounts of data will have to be recorded and processed.

3. *calibration*. There are permanent variations in sensitivity across any CCD, and these must be thoroughly understood so that proper corrections may be applied.

4. *read-out noise*. A certain amount of noise or unwanted signal is introduced as the charges are moved out of the CCD, amplified, digitized and stored in the computer. This background places limits on the magnitudes of the faintest stars which can be recorded.

5. *cooling*. The readout noise can be reduced if the CCD is cooled to the temperature of −100 °C or lower. This complicates both the construction and operation of the CCD camera. The generation of CCDs which will be produced in the 1990s will give results at the temperature of melting ice, so this problem will be much more easily handled.

From these advantages and disadvantages, then, the astronomers must plan first in terms of their desired observation and then in terms of analysis, but these two parts of the sequence are not independent of each other. At the telescope the observer must adjust the duration of an exposure to record stars of some limiting magnitude without permitting the brighter stars to saturate the CCD. The center of the field must be chosen so that one or more reference stars will image at the same time as the star under study.

The computer software for the analysis must do at a minimum the following:

1. locate the images of all of the stars projected onto the CCD;
2. identify images of non-stellar objects such as cosmic ray tracks, flawed pixels, galaxies, asteroids and meteors;
3. measure and subtract sky brightness from stellar images;
4. perform differential photometry between stars of known and unknown magnitude;
5. permit the user to vary the contrast on the displayed image.

Several programs for accomplishing these complex tasks have been written and are available for users. Most of the programs have some common features, and it is these that we shall describe below.

The first task for the computer is to find the center of each stellar image which is of interest to the user. In some programs a cursor is moved to the approximate center of an image and at a command from the operator the program finds the center of the image. In other programs the centers of all of the stellar images in the field are found automatically. The program then determines the radius of a 'synthetic aperture' which in a way is the same as the aperture used in a photoelectric photometer. The size of the aperture is determined from the size of the brighter images in the field, and it is this aperture that is then used for all of the stars to be measured. The sky brightness is found from an area in the field that does not include any part of the image of a star.

When the average contribution of the sky to each pixel has been found, the brightness recorded by each pixel can be corrected. In other words, sky is subtracted from star-plus-sky just as in photometry with the photomultiplier. Then the stored charges for each pixel within the aperture are summed. It is this sum that is proportional to the brightness of the star.

In a crowded field the images may be so close to each other that some pixels record light from more than one star. In these cases a profile referred to as the 'point-spread function' (PSF) is found for an uncontaminated image. This profile should be symmetrical around the center of an image whereas the profile for a contaminated image will not be symmetrical. It is the PSF which is then used to correct the pixels affected by more than one star.

The sizes of the stellar images formed on the CCD appear to vary with the brightness of the stars, and this may seem to be at odds with the factors which affect the sizes of images. After all, each image represents a diffraction disc that has been smeared out or enlarged by the effects of the seeing, and each image should be affected in the same manner. We recognize, however, that there will be more light in the outer edges of images of bright stars than faint ones, so images of bright stars will appear to be larger.

With proper care given to all of many details, observers can expect magnitudes that are accurate to ±0.01 mag. This is slightly better than the precision that can be obtained with the best photographic photometry, but not quite as good as can be obtained with the best photoelectric photometry. It is to be expected that with the steady improvement of both the CCDs and the analysis programs the results obtainable with the CCDs will become even better.

Further reading

Eccles, M. J., Simm, M. E. and Tritton, K. P. (1983). *Low Light Level Detectors In Astronomy*. Cambridge: Cambridge University Press. See Chapter 8 for a brief but technical discussion of the CCD and other modern solid-state detectors.

Janesik, J. and Blouke, M. (1987). Sky on a chip: the fabulous CCD. *Sky and Telescope*, **74,** 238. This article will give the reader a good overview of the construction, function and application of the CCD.

Kristian, J. and Blouke, M. (1982). Charge-coupled Devices in Astronomy. *Scientific American*, **247,** 66. In addition to examples of images recorded by CCDs, this article contains interesting details on the structure and fabrication of these advanced detectors.

Appendix 5

Aligning the polar axis

When a small, personal telescope or a large professional one is in use, the user should have an expectation that over a period of some hours a star which has been centered in the field of view will remain centered. Several factors can affect the precision of the tracking, and should be understood by anyone using telescopes. Let us list several of these as follows:

1. the speed of the tracking motor;
2. the perpendicularity of the axes;
3. the perpendicularity of the optical axis to the declination axis;
4. the alignment of the polar axis with the celestial pole.

The first and second of these are not likely to present problems since they are part of the design and manufacture of the telescope's mounting. The motors and gears which drive the telescope in hour angle are chosen to provide the correct rate, and problems seldom arise here. Many small electric motors run at a speed dictated by the frequency of the power supplied by a local public utility. This frequency is sixty cycles per second in many parts of the world and is fifty cycles per second in a few. Under most conditions this frequency is carefully maintained by the power companies. At times of very heavy demand the frequency may decrease slightly, but when this happens, the power companies compensate in such a way that in any twenty-four hour period the total number of cycles is reliably constant. For small telescopes the so-called line frequency is good enough. A large observatory might have its own carefully regulated power supply.

The need for the polar and declination axes to be exactly at right angles to each other is easily recognized when we recall the details of the equatorial coordinate system described in Chapter 1. In today's world this seldom presents a problem since the principal manufacturers of telescopes take great pains to see that perpendicularity is maintained.

With telescopes purchased from the best known makers there is also an expectation that the optical axis of the telescope should be perpendicular to the declination axis. Anyone purchasing a used or a homemade telescope should check very carefully to see that the optical axis is, in fact, properly adjusted. In the references at the end of this appendix we have included several which describe methods for checking both the angle between the optical and declination axes and the angle between the polar and declination axes.

The fourth item on the above list, the alignment of the polar axis, can present an ongoing problem to users of portable telescopes since the polar axis must be realigned each time that the telescope is taken outdoors for use. Users of permanently mounted telescopes must know how to adjust the mounting when a telescope is first installed and whenever there is any reason to believe that the mounting might have been disturbed.

The user of the portable telescope needs to have a procedure which will be quick and simple, so that maximum use may be made of clear skies. We shall describe a quick method which can be used if the small telescope has setting circles and another method which can be used with telescopes which do not have setting circles. Then we shall describe an especially efficient method which can be used with large telescopes.

Aligning the small telescope

To begin with we need to know the right ascension of Polaris (2 h 19 m in 1988) and the local sidereal time. With these pieces of information one may calculate the hour angle of Polaris and then set the telescope to this position. The declination (89° 12' in 1988) may now be set, and the axis should be clamped. With the drive motor running the observer should adjust the entire mounting in both azimuth and altitude until Polaris is centered in the field of view of the eyepiece. It is assumed here that the mounting has been set for the latitude at which the telescope is being used and that the direction of north is known. The first-time user can expect to spend half an hour or more with these steps, but after even a little bit of experience, the observer should need only a few minutes to bring the polar axis into its proper position. If the user is not certain that the setting circles have been correctly adjusted, then the process will have to begin with the instructructions in the next paragraph.

When a small telescope is not equipped with setting circles, the procedure takes somewhat more time and effort. One should first make the tube parallel to the polar axis. To do this, first center Polaris in the field of a low-power eyepiece and clamp the declination axis. Then rotate the telescope around the polar axis. If Polaris moves away from the center of the field, the tube is not parallel to the polar axis, and the polar axis does not point to Polaris. Adjust the orientation of the mount and adjust the declination until Polaris remains as the polar axis is rotated. The tube is now parallel to the axis, and the axis points toward Polaris.

Now consult a map of the stars near Polaris and note the position of the north celestial pole. Without moving either axis readjust the mounting until the center of the field coincides with the position of the north celestial pole.

If an observer should find it necessary to check the adjustment of the setting circles on a telescope, then the above procedure should be followed. After the polar axis has been correctly aligned, the telescope can be moved to Polaris or any bright star. Then the coordinates of that star may be used in order to set the declination and hour-angle circles.

Aligning a permanent mount

A telescope large enough to warrant a permanent mount will probably have some provision for photography, and this makes possible an alignment method which is very positive and fast. Again, we begin with the assumption that the telescope has been designed for the latitude of its site and that it has been installed on a pier that is oriented toward the north. Only a few small adjustments in altitude and azimuth of the axis are to be expected. Telescopes are always provided with provision for these necessary adjustments. Bolts in the base can be turned as indicated in Fig. A5.1 to raise or lower the polar axis or to change its azimuth. Finally, the process to be described here is considerably speeded up if the photographic plateholder can be adapted for the use of Polaroid film or prints.

The telescope is now set to declination 90°, and the plateholder slipped in place. With

Fig. A5.1. The base of a small telescope showing bolts by which the mounting may be adjusted in altitude and azimuth. (Wellesley College photograph.)

the drive motor running the plateholder is opened to begin an exposure. After about ten minutes the drive motor is turned off, but the exposure is allowed to continue for another ten minutes. If the telescope's polar axis is exactly aligned with the pole, the developed film should show a pattern similar to that in Fig. A5.2. The images of stars should have their normal appearance, but each should be attached to a short trail formed after the drive motor was turned off. Let us now see what the images will look like if the polar axis is not properly aligned.

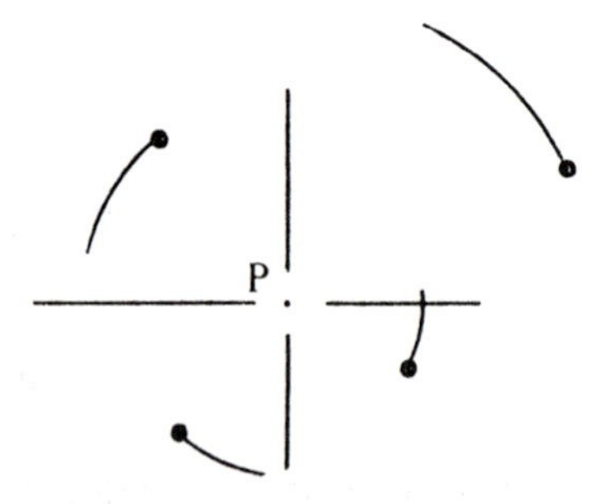

Fig. A5.2. Images and trails left by stars when the alignment of the polar axis is correct.

Fig. A5.3 represents a portion of the sky near the north celestial pole which is marked 'TP' for true pole. The polar axis of the telescope points toward the point 'IP', the instrumental pole. The sky rotates around the true pole at the rate of 15° per hour, and the telescope rotates around IP at the same rate. We now note what happens to the image of a star during a photographic exposure of one hour. The star should move through an angle of 15° around TP. The plate in the focal plane of the telescope moves through the same angle around IP. During the first part of the exposure the image of the star is at point A, and this point moves to A′ during the exposure. At the end of the exposure the star is at point B. On the plate the star's image has moved from A′ to B, so the developed plate shows a streak along this path. When the motor is turned off, the star's trail moves along the arc from B to C.

Every star in the field will leave a trail centered on the true pole, and every star will leave an exposed track similar to that from A′ to B. We now look at the star on the right hand side of the diagram. The exposed track now runs from D′ to E, and falls on the inside of the trail around TP. The overall appearance of this two-part track is noticeably different from the first one. With compass, protractor and pencil the reader could choose other points along the circle centered on TP and construct similar two-part tracks. In Fig. A5.4 we have drawn four tracks of this sort. As we study these tracks, we note that the shape of the track varies with the location of the star with respect to the line between IP and TP. The uniqueness of these tracks is a positive indication of the location of IP with respect to TP.

In the practical application of these principles one may photograph the area near the

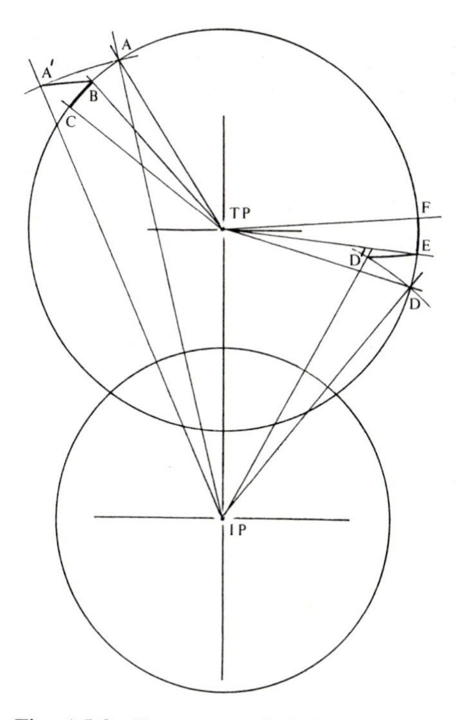

Fig. A5.3. Two-part trails left on a photograph when the polar axis is not properly aligned. The narrow trails which are centered on the true pole (TP) were left when the driving motor was turned off.

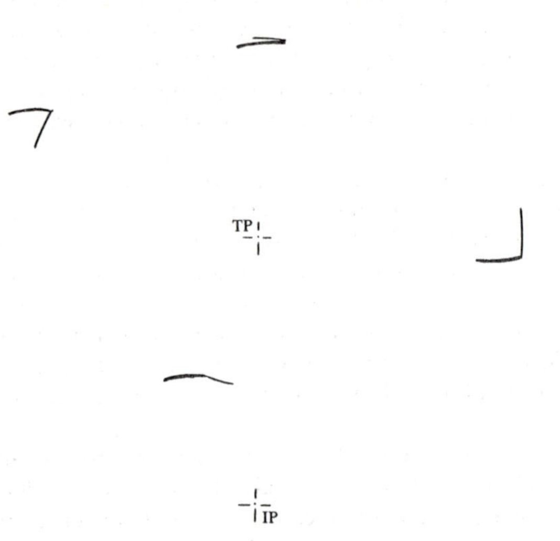

Fig. A5.4. The shapes of two-part trails in different positions with respect to the line between TP and IP.

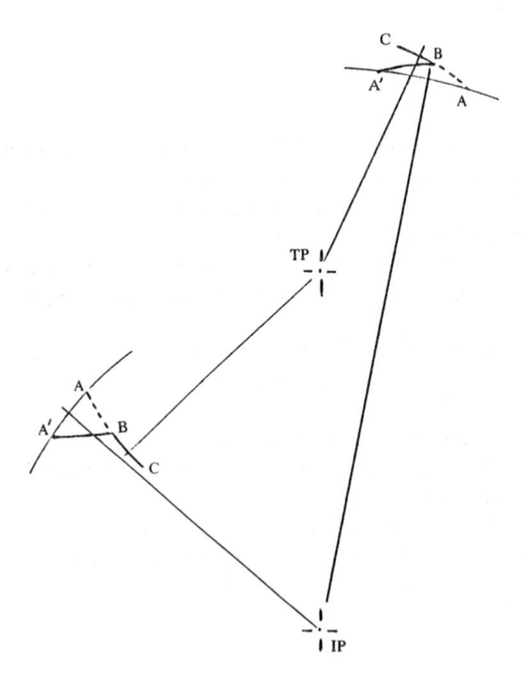

Fig. A5.5. The construction by which the exact positions of the true and instrumental poles may be located.

north celestial pole as described here and examine the resulting trails. The direction of the vertical should be noted at the time that the plate is removed from the telescope. Then a diagram such as that in Fig. A5.4 may be superimposed on the photograph (or compared to it). If the diagram is rotated with respect to the photograph until the patterns of the two-part tracks match, one immediately knows where the line between TP and IP lies.

A further simple construction can reveal the exact locations of the two poles on the photograph, and this construction is shown in Fig. A5.5. Let us assume that the two parts of the exposure were of the same duration. That is, the exposure with the drive turned on was the same as that with the drive turned off. The perpendicular bisector of the arc BC will point to the true pole, so the perpendicular bisectors of the circumpolar trails of two stars will locate the true pole. Point A may be located by drawing an arc backward from B and making it the same length as BC. Now the perpendicular bisector of a line from A to A′ will pass through the instrumental pole. Again, similar lines from the tracks of two or more stars will locate IP.

The plate scale should be known for the telescope being tested, so the measured distance between the poles may be converted into an angle. If the direction of the vertical was noted as the plate was removed from the telescope, then the errors in azimuth and altitude are known and appropriate adjustments to the polar axis may be made. One may also proceed just by trial and error after the first plate has been developed. An adjustment is made and a second plate is exposed and examined. If the track seems to be approaching the ideal appearance, the observer knows that the adjustment was made in the correct direction. It should not take very long to bring the polar axis into its correct position.

Further reading

Bell, L. (1922). *The Telescope*. New York: McGraw-Hill Book Company. This book may be hard to find, but it contains some very interesting material.

Liller, W. and Mayer, B. (1985). *The Cambridge Astronomy Guide*. Cambridge: Cambridge University Press. See page 170 in the appendix.

Newton, J. and Teece, P. (1988). *The Guide to Amateur Astronomy*. Cambridge: Cambridge University Press. These authors present some good suggestions in Section 4.3 beginning on page 217.

Sherrod, P. C. (1981). *A Complete Manual of Amateur Astronomy*. Englewood Cliffs, N J: Prentice-Hall. For additional ideas on the alignment of the polar axis see material which begins on page 26.

Sidgwick, J. B. (1980). *Amateur Astronomer's Handbook*. Hillside, N J: Enslow Publishers. This book contains chapters on testing optics, selecting a drive and choosing a site as well as material on aligning the mount.

Wallis, B. D. and Provin, R. W. (1988). *A Manual of Advanced Celestial Photography*. Cambridge: Cambridge University Press. The photographic method of aligning the polar axis is described in less detail than in the present book.

Index